Carsten Monka-Ewe

Mode Matching Solvers for Filters and Cavities

with 126 figures

Mode Matching Solvers for Filters and Cavities

Von der Fakultät für Elektrotechnik, Informationstechnik, Physik
der Technischen Universität Carolo-Wilhelmina zu Braunschweig

zur Erlangung des Grades eines Doktors
der Ingenieurswissenschaften (Dr.-Ing.)

genehmigte Dissertation

von Carsten Monka-Ewe
aus Uelzen

eingereicht am: 05.09.2018
mündliche Prüfung am: 18.12.2018

1. Referent: Prof. Dr.-Ing. Jörg Schöbel
2. Referent: Prof. Dr.-Ing. Achim Enders
3. Referent: PD Dr.-Ing. Thomas Kleine-Ostmann

Druckjahr: 2019

Dissertation an der Technischen Universität Braunschweig,
Fakultät für Elektrotechnik, Informationstechnik, Physik

Bibliografische Information der Deutschen Nationalbibliothek
Die Deutsche Nationalbibliothek verzeichnet diese Publikation in der Deutschen Nationalbibliografie; detaillierte bibliografische Daten sind im Internet über http://dnb.d-nb.de abrufbar.

1. Aufl. - Göttingen: Cuvillier, 2019
Zugl.: Braunschweig, Techn. Univ., Diss., 2018
ISBN: 978-3-7369-7086-1
eISBN: 978-3-7369-6086-2

Bildnachweis Coverhintergrund:
Oleksii / stock.adobe.com

Gedruckt auf umweltfreundlichem, säurefreiem Papier aus nachhaltiger Forstwirtschaft.

Revision main.tex 611 2019-10-21 19:16:16Z cmonka

To the memory of my grandma...

Acknowledgements

This thesis summarises my work on Mode Matching solvers and related subjects carried out at the Institut für Hochfrequenztechnik of the Technische Universität Braunschweig between the years 2013 to 2018. The completion of this work would not have been possible without the help and support of a large number of people:

I firstly would like to thank my supervisor Prof. Dr.-Ing. Jörg Schöbel, who not only sparked my interest in electromagnetics and microwave engineering in general, but also continuously encouraged my work on this thesis. I am really grateful for your support in both professional and personal matters, the freedom to develop my own research ideas and for countless discussions on the particularities of microwaves and (university) life in general. It is a pleasure to work with and for you.

Furthermore, I would like to thank Prof. Dr.-Ing. Achim Enders and PD Dr.-Ing. Thomas Kleine-Ostmann for being my second and third referees as well as Prof. Dr.-Ing. Thomas Kürner for charing the examination board.

Over the past years, I had the pleasure of working with several fine colleagues at our institute. In particular, my special thanks go to Fabian Schwartau and Dr.-Ing. Sebastian Brückner. I am really grateful for our discussions on radar and signal processing as well as for your honest criticism. Moreover, I want to thank Sebastian Paul for the joint work on material measurements and microwave plasmas as well as Markus Krückemeier for his advice on PCB layouting and electronics. Thank you for your precious support and the fun we had; I hope you guys stick around for a bit.

Some of the topics covered in this thesis were inspired by various industry cooperations. I would like to say thank you to Jan Fahlbusch, Dr.-Ing. Jan-Philipp Mai and Dr.-Ing. Jürgen Schmitz for our cooperation and the research questions you provided.

I am in debt to my mother Dagmar Monka for proofreading the manuscript for this thesis. Your corrections and suggestions on the text were an invaluable contribution to the finalisation of this thesis. Any errors, which may remain in the text, are due to my misinterpretation of your comments and shall not be attributed to you.

Ulrike Ewe gave insightful comments during the preparation of my thesis talk.

My deepest gratitude goes to my parents and family for their support throughout my studies. I also would like to say thank you to the Ewe family who truly made Braunschweig my new home over the last couple of years.

To our kids, Anna, Elia and Yunus, I say thank you for sparing a considerable amount of family time during the write-up of this thesis.

Finally, I would like to express my heartfelt appreciation to my wife Annika Ewe. Without your love and support this thesis most likely would not have been completed yet. Thank you for trying to share my enthusiasm on Eigenmodes and for your patience when I am getting absorbed too much in whatever interesting immediately-to-be-solved problem.

Carsten Monka-Ewe

Contents

Part I.

Introduction

1. Introduction

In 1864 James Clerk Maxwell published his "Dynamical Theory of the Electromagnetic Field" [1] where he postulates that light is an "electromagnetic disturbance" [1], whose velocity of propagation, i.e. the speed of light, can be deduced from purely electric measurements. Maxwell himself expressed this idea as follows:

"The agreement of the results seems to show that light and magnetism are affections of the same substance and that light is an electromagnetic disturbance propagated through the field according to electromagnetic laws." [1, p. 499]

Although Maxwell did not comment on the existence of "relatively long" [2] electromagnetic waves [3], today, Maxwell's theoretical work is nevertheless regarded as the foundation of modern electromagnetic theory in the context of microwave engineering. This is because Maxwell's paper introduced the radically new concept of displacement current [3], which is the key to all electromagnetic wave propagation phenomena.

Maxwell had developed his theory entirely from theoretical considerations and it took until 1888 that Heinrich Hertz proved the existence of electromagnetic waves. Regarding Maxwell's postulate on the nature of light, in his paper [4] Hertz concluded:

„Die schon durch viele Wahrscheinlichkeitsgründe gestützte Hypothese, dass die Transversalwellen des Lichtes electrodynamische Wellen seien, gewinnt feste Grundlage durch den Nachweis, dass es wirklich electrodynamische Transversalwellen im Luftraume gibt, und dass diese sich mit einer der Geschwindigkeit des Lichtes verwandten Geschwindigkeit ausbreiten.“[1] [4, p. 569]

Soon after Hertz' proof of the existence of electromagnetic waves the rapid development of communication systems began, which should revolutionise the way people live.

Since the earliest days of research on electromagnetic waves engineers have been confronted with the problem of guiding electromagnetic waves by means of transmission lines. While earliest transmission lines were two-wire lines [5], with the shift towards higher frequencies, hollow waveguides and coaxial cables have become standard [6].

Although today's widespread use of microwave systems would have been impossible without the transition to compact, low-cost planar transmission lines [6,7] such as microstrip lines [8] and coplanar waveguides [9], hollow waveguides are still used in a variety of applications [6]:

One typical application are satellite communication systems where waveguides are used both because of their low loss and high power handling capability [10] as well as due to their ability to handle two polarisations, thus enabling polarisation multiplexing [5]. Besides that, waveguides are often found both in millimeter wave test and measurement equipment [6] as well as in material characterisation setups (see chapter 15). Finally, waveguides are widespread in industrial microwave heating systems [11].

[1] "The hypothesis that transverse waves of light are electrodynamic waves, which has already been supported by several reasons of likelihood, gains solid foundation by the experimental proof that transverse electrodynamic waves indeed exist in free space as well as by the fact that they are propagating with a velocity related to the speed of light." [4, p. 569 - translated by the author of this thesis.]

The history of waveguides is a history of multiple rediscoveries [12]. Heaviside [13] considered the idea of electromagnetic waves propagating in a hollow tube, but came to the conclusion that two conductors are required in order to transfer electromagnetic energy.

The propagation of electromagnetic waves in hollow waveguides of rectangular and circular cross-section was predicted by Lord Rayleigh in his paper "On the passage of electric waves through tubes, or the vibrations of dielectric cylinders" [14] as early as 1897. Interestingly enough, Lord Rayleigh was already aware of the waveguides' infinite discrete spectrum of Eigenmodes. However, the theoretical discovery of waveguides was ahead of its time because high frequency sources were far away from common practice [12] and as a consequence, the idea got forgotten [12].

The "final" rediscovery of hollow waveguides occurred in 1936 when George C. Southworth [15] and Wilmer L. Barrow [16] independently discovered that the propagation of electromagnetic waves inside hollow tubes was indeed possible [12].

In the following, waveguides became common practice and considerable efforts were made to theoretically analyse various waveguide geometries as well as waveguide discontinuities. The possibly largest effort on analysing waveguide structures was carried out in the years 1942 to 1946 at the M.I.T. Radiation Laboratory [17] and based on these efforts, the possibly most complete work on the fine art of the analytical treatment of waveguides and discontinuities, Marcuvitz' Waveguide Handbook [17], was published.

With the advent of computers in the second half of the last century, the focus regarding the analysis of waveguides changed towards numerical techniques [7]. Today, the application of computational electromagnetics tools can be considered state of the art.

Computational electromagnetics methods can be grouped into four classes as is shown in figure 1.1 for some selected methods. In the following, we will discuss the general concepts as well as advantages and disadvantages of these methods. Because variational methods are of high importance for many computational electromagnetics methods, a concise overview of this subject is provided in appendix A.

Integral Equation methods

- Method of Moments

Partial Differential Equation methods

- Finite Difference Time Domain method
- Finite Integration technique
- Finite Element method

Modal Expansion techniques

- Mode Matching technique
- Spherical Wave Expansion

High Frequency methods

- Geometrical Optics method
- Physical Optics methods

Figure 1.1.: Some computational electromagnetics methods, with modifications from [18]

We will now discuss the first class of computational electromagnetics methods, **Integral Equation methods** such as the *Method of Moments* (MoM) [19].

In order to apply the Method of Moments, a suitable *integral equation* has to be found by careful analysis of the electromagnetic problem [20]. In the context of electromagnetic waves, such integral equations typically exhibit an unknown current distribution $\mathfrak{u}$ as well as a known forcing field distribution $\mathfrak{f}$ [21].

The "physical environment surrounding the radiator" [21, p. 132] or scatterer is included in the integral equation formulation by means of *Green's functions* [21], which provide the "essential link" [20, p. 290] between the partial differential equation and the integral formulation [20]. Green's functions can be interpreted as a spatial impulse response to a source or as field propagation function [21].

In order to determine the unknown current distribution, the field problem under form of the integral equation has to be inverted by converting it into a linear system of equations [20]. This is done by applying an *indirect variational method*, e.g. the *collocation* or *Galerkin methods*, which yields a matrix equation under form of (A.12) [20]. In order to populate the problem's matrix, the inner products in (A.12) have to be determined by integrating over the meshed surfaces of radiators or scatterers included in the problem [20].

The Method of Moments is well-suited for solving open-domain problems because it inherently implements radiation and only requires surfaces rather than the entire volume to be meshed [21]. However, for closed-domain problems, the method of moments is inferior to the partial differential equation methods discussed in the following because the method of moments yields a dense matrix [21], which is unfavourable both in terms of memory requirements and the time required for inverting the matrix.

In contrast, **Partial Differential Equation methods** such as the *Finite Element method* (FEM) [20], the *Finite Difference Time Domain* (FDTD) method [22, 23] and the *Finite Integration* (FI) technique [24] are the most common choice for solving closed-domain problems [21]. For these methods, a space-segmentation of the domain is required.

Similar to the method of moments, the finite element method is based on variational methods. While both direct and indirect methods can be used as an approach to the Finite Element method [21], here we will consider the *direct Rayleigh-Ritz method* only.

As discussed in appendix A, the Rayleigh-Ritz method strives to extremise the *functional* corresponding to the partial differential equation to be solved [20].

In order to obtain a solution to the partial differential equation for the domain of interest, the domain is firstly separated into subdomains [18], e.g. triangles or tetrahedra [20]. Inside these elements the fields are approximated by suitable polynomials [20, 25]. The corresponding coefficients of these polynomials can be expressed in terms of the values of the solution at the elements' vertices [20].

Careful analysis of the element governing equations [20] and subsequent combination of all elements' equations [20] yields an expression for the overall domain's value of the functional [20].

By forcing the expression's partial derivatives with respect to the values at the elements' vertices to be zero, a system of equations is obtained, which may be solved by standard means in order to determine the solution's values at the vertices of all elements, thus solving the field problem [20].

In contrast to the aforementioned techniques, the two other partial differential methods, namely the FDTD method and the FI technique both operate "directly" on Maxwell's equations rather than using a functional or an integral equation.

The FDTD technique uses finite difference equations to approximate the partial derivatives included in Maxwell's equations both in time and space [18]. The well-known *Yee grid* is used for the discretisation of the domain under consideration [22]. This *staggered grid* leads to finite difference equations for the spatial derivatives, which "are central-difference in nature and second-order accurate" [23, p. 60]. Moreover, the Yee grid is divergence free [23, 26] as discussed in greater detail e.g. in [26, p. 500f].

In contrast, the FI technique discretises an integral representation of Maxwell's equations [27]. When put under integral form, the curl and divergence operators become contour or surface integrals [27, 28], which may easily be evaluated for the grid cells, thus providing the so-called Maxwell grid equations [27, 29].

Since both methods operate on staggered grids [28], the field problem, which is either given under form of the FDTD method's finite difference equations or the FI technique's Maxwell grid equations, may be solved in a leapfrog manner [29, 30].

A backdraft of all three aforementioned techniques is the fact that if complicated or large geometries need to be segmented, the obtained meshes become very large and thus lead to both high memory requirements and long computation times [21].

The *Mode Matching technique*, which belongs to the third class of computational electromagnetics methods, **Modal Expansion techniques**, aims to avoid this issue for the price of reduced general applicability.

The idea of the Mode Matching technique is to decompose the structure to be analysed into sub-domains whose spectrum of Eigenmodes is known analytically. By performing orthogonal expansion of the yet unknown tangential field distributions at the interfaces between these sub-domains, a system of equations is obtained which can be solved for the Eigenmodes' amplitudes.

As the Eigenmodes of the geometries involved must be known, there is a certain limitation regarding the general applicability of the method. If, however, these limitations do not represent a problem for the structure under investigation, the Mode Matching technique outperforms partial differential equation methods by far as we will see later in this thesis.

Since Mode Matching generally treats individual discontinuities only, the Mode Matching technique is commonly used in combination with techniques such as the *generalised scattering matrix method* [31] in order to cascade multiple waveguide discontinuities.

However, in this thesis, a different approach is used, that is, all matrices obtained either by Mode Matching at the discontinuities or by analysing the Eigenmodes' propagation on waveguide sections of uniform cross-section are assembled into one large system matrix. This system matrix is a sparse matrix with a very narrow bandwidth, which only depends on the number of Eigenmodes included at the interfaces.

Another Modal Expansion technique, which can be used to apply the Mode Matching concept to certain open-domain problems, is *Spherical Wave Expansion*. [18].

For the sake of completeness, the fourth class, **High Frequency methods** shall be mentioned. This class includes ray-based methods such as *Geometrical Optics* (GO) and wave physical methods such as *Physical Optics* (PO), which are relevant for solving electrically large problems, for example calculating the radar cross-section of ships [18].

1.1. State of the Art of the Mode Matching Technique

The development of the Mode Matching technique was started in the 1960s by papers published by Wexler [32] and Clarricoats [33] and it was Wexler [32] who coined the term Modal analysis.

Apart from its limitation to a certain subset of geometries with known Eigenmodes, the most severe backdraft of the Mode Matching technique is the problem of *relative convergence* [34–39]:

As we will see later in this thesis, the series corresponding to the orthogonal expansions need to be truncated for the purpose of implementation. However, if this truncation is carried out in an improper manner, convergence occurs more slowly [38, 40, 41] and for some unfavourable geometries such as a bifurcation of a parallel plate waveguide containing an infinitely thin septum the solution may converge against an incorrect solution [34,39].

The problem of relative convergence has widely been studied [34–39] and it was found that this problem can be circumvented if the ratio of the numbers of Eigenmodes included on two waveguides adjacent to a discontinuity is chosen corresponding to the waveguides' dimensions.

The successful application of the Mode Matching technique has been reported for a variety of structures such as waveguide filters [42–44], waveguide transformers [43], antennas such as corrugated conical horns [45, 46] and stepped rectangular horns [47]. While we will investigate the analysis of rectangular and tubular filters in great detail, antennas are out of the scope of this thesis. Still, some remarks on the general concept appear imperative:

As discussed in [42, 43], one possible approach for solving antennas using the Mode Matching technique is to terminate the horn antenna in a very large waveguide, which approximates free space conditions. Because for $f \gg f_c$ the waveguide's wave impedance approaches the free-space wave impedance (see figure 10.3 on page 82), this is indeed a valid assumption. The far field is then calculated by means of equivalent Huygens sources for the aperture fields. After numerically solving the source integrals for the vector potentials describing the radiated fields, the electric and magnetic fields are obtained as derivatives of the vector potentials (see e.g. [48]).[2]

While the Mode Matching technique was originally developed for the analysis of discontinuities in hollow waveguides, the analysis of planar microstrip circuits has received attention as well. In order to derive a suitable set of orthogonal Eigenmodes, Wolff et. al. [49,50] modelled microstrip lines as dielectrically filled parallel-plate waveguides with perfectly magnetically conducting sidewalls [38]. Based on this model, for example in [51] planar filters were analysed.

Finally, it should be noted that the Mode Matching technique has also been applied for the investigation of accelerator structures (see [29] and references therein).

In order to alleviate the limitations regarding the geometries which can be treated using the Mode Matching technique, the method has been combined with various other computational electromagnetic techniques. Among those *Hybrid methods* are combinations of Mode Matching and Finite Element methods [52,53] the *Boundary Contour Mode Matching technique* extensively studied by Arndt and Reiter [54, 55] and other methods [10].

[2] In [42,43] it is said that the far field is obtained by integration of the aperture fields. This most likely refers to the aforementioned approach.

1.2. Organisation of this Thesis

The scope of the current thesis spans a wide range of topics from waveguide theory to the final applications, waveguide filters and electromagnetic cavities for dielectric material characterisation. The material presented in this thesis is grouped into three parts, which are organised as follows:

Part I: Introduction

After a brief introduction given in **chapter 1** with additional remarks provided in **appendix A**, the first part of this thesis covers the theory of cylindrical waveguides.

Cylindrical waveguides have a uniform cross-section in the wave's direction of propagation. We will investigate the boundary value problem imposed by such a waveguide and show that its solution is a discrete spectrum of orthogonal Eigenmodes in **chapter 2**.

In **chapter 3** we will then exploit the orthogonality property of the Eigenmodes in order to perform orthogonal expansion of transverse field components.

Part II: The Mode Matching Technique

In the second part of this thesis, the Mode Matching formalism is developed based on our insights gained in the two introductory chapters 2 and 3.

Firstly, in **chapter 4** we will study the treatment of an individual waveguide discontinuity. In this context, the edge condition and the problem of relative convergence will be discussed. Next, in **chapter 5** additional means for describing waveguide segments of uniform cross-section will be established.

Finally, in **chapter 6** we will discuss a method for cascading the building blocks of waveguide structures introduced in the two preceding chapters and investigate the structure's overall system of equations.

For the development of Mode Matching solvers a sound understanding of the coupling mechanisms at waveguide discontinuities is mandatory. This knowledge is not only important for the validation of the obtained results, but also for determining the smallest set of Eigenmodes which allows a complete treatment of the structure for a given excitation. This allows to drastically reduce the required computation time. In **chapter 7** we will exemplarily investigate coupling at a symmetric H-plane discontinuity. While the outcome of this investigation is well known, the obtained results are far from easy to establish. This exemplary discussion of a waveguide discontinuity should enable the reader to carry out appropriate analyses for other types of discontinuities.

In **appendix B** the Eigenmodes of all waveguides considered in this thesis are given.

Chapters 8 and 10 focus on the post-processing of the Mode Matching solution:

In **chapter 8** we will study the calculation of scattering parameters. As a matter of fact, the correct calculation of scattering parameters is far from easy because problems treated by the Mode Matching technique often include ports with different characteristic impedances. In order to correctly calculate scattering parameters, an in-depth understanding of the definition of voltage and current waves and of the waveguide's characteristic impedance is required. Supplementary information on this subject will be given in **appendix C**. The development of equivalent circuits for waveguide discontinuities will be covered in **chapter 10**. These equivalent circuits will prove highly valuable in the context of filter design later in this thesis.

Numerical aspects of the Mode Matching technique will be covered in chapters 9 and 11. In **chapter 9** means to assess the convergence of the Mode Matching solution are addressed.

Then, in **chapter 11** we will see that the Mode Matching representation of a structure leads to a band-limited sparse matrix, which can be exploited to reduce both the methods' memory requirements as well as the computation time for solving the structure's system of equations. One possible approach to minimising the computation time required to solve the structure's inhomogeneous system of equations is to apply an LU decomposition algorithm which is optimised for working on banded matrices. This will be investigated in **appendix E** in greater detail.

Although solving the structure's system of equations may become time-consuming for larger structures, the computational effort required for populating the structure's system matrix is considerably larger. Thus, in chapter 11 we will also investigate various means to speed up matrix population. We will then conclude chapter 11 by pointing out some possible approaches for accelerating Mode Matching solvers using parallelisation.

The final chapter of part II, **chapter 12**, will cover the extension of the Mode Matching technique for electromagnetic cavities, which allows to accurately determine the resonant frequency and unloaded quality factor of perturbed cavities.

Part III: Applications and results

The third part of this thesis will cover applications for the newly developed Mode Matching solvers and discuss the obtained results. Two main fields of application are considered, waveguide filters and electromagnetic cavities used for material characterisation.

In **chapter 13** we will compare the performance of the Mode Matching solver developed during the preparation of this thesis with the performance of a commercial FEM code, namely Ansys HFSS. We will consider three filter structures, a tubular stepped-impedance filter, a tubular direct-coupled bandpass filter with foreshortened transmission line resonators and a direct-coupled rectangular waveguide shunt-iris filter.

For all three filter structures we will find that the Mode Matching solver developed in the scope of this thesis outperforms the commercial FEM solver by far.

Although the above-mentioned filter structures were either kindly provided by SF Microwave GmbH or have been obtained from literature, it nevertheless appears appropriate to embed the obtained results in a concise discussion of waveguide filter theory. In doing so, the close interplay between the electromagnetic analysis of filter structures and the fine art of microwave filter design is nicely illustrated.

In tubular filter designs it might be desirable to insert dielectric tubes in order to both center the inner conductor as well as to increase the filter's power handling capability. In order to treat such a filter using the Mode Matching technique the Eigenmodes of such a partially filled coaxial waveguide must be known. **Chapter 14** provides a complete derivation and discussion of the Eigenmodes relevant in tubular filter designs.

Finally, in **chapter 15** we will investigate the impact of "secondary effects" on cavity perturbation material measurements, which are induced by approximations made during the method's derivation. In order to do so, approximated quantities provided by the analysis equations of the cavity perturbation material measurement technique are compared to exact ones obtained by performing full-wave solution of the perturbed cavity using a Mode Matching solver for resonant problems.

Chapter 16 concludes this thesis and provides an outlook on future work.

1.3. Contributions

This thesis offers the following contributions to the current state of knowledge:

- **A complete treatise of the Mode Matching Technique**

  Starting from graduate-level electromagnetics as discussed in [48], this thesis provides a detailed thorough development of a Mode Matching formulation. Important theoretical background topics such as the orthogonality relation are reviewed and critical pitfalls such as the problem of relative convergence are outlined.

  In later chapters this thesis focuses on topics related to the efficient implementation of the Mode Matching technique, e.g. the efficient population of a structure's system matrix. While implementation aspects are usually left out in standard literature on the subject [38], this thesis summarises the insights gained by the author in recent years in a comprehensible fashion. An outlook on the future parallelisation of the method concludes this topic.

- **A performance comparison between the Mode Matching Technique and a commercial FEM tool**

  In the scope of this thesis, three filter topologies, a stepped-impedance lowpass filter, a tubular direct-coupled bandpass filter with foreshortened transmission line resonators and a direct-coupled rectangular wave shunt-iris filter will be analysed using the Mode Matching solver developed in the scope of this thesis and a performance comparison with Ansys HFSS is carried out.

  While the Mode Matching solver discussed here typically solves the aforementioned structures in a few minutes at maximum, Ansys HFSS may easily require between half an hour and more than two hours of computation time depending on the complexity of the structure and on whether parallelisation is used.

  It will also be shown that at least for the given structures, the Mode Matching Technique is more robust in terms of convergence while the FEM code requires additional settings regarding the generation of the mesh in order to converge in a satisfying manner.

- **A demonstration of the close interplay between filter design and full-wave analysis of waveguide filters**

  In order to demonstrate the close interrelation between the electromagnetic analysis of waveguide filters and the art of filter design, the results on solving such filters using the Mode Matching technique are embedded in a concise review of filter theory.

  The interplay between these fields cannot be stressed enough because in the process of designing filters, aspects from both fields strongly influence the design process.

  One aspect of this close interplay is that computationally efficient full-wave solutions of filter structures are crucial for the filter design process because filter structures derived from prototype filters usually require manual optimisation in order to mitigate the effect of parasitics.

  A second aspect is the fact that filter polynomials often lead to symmetries in the filters' geometries, which may be exploited to accelerate the solution process of such filters, e.g. during the population of the filter's system matrix.

- **A complete rigorous discussion of the Eigenmodes of partially filled coaxial waveguides**

  In tubular filter designs, dielectric tubes might be desirable to increase the filter's power handling capability as well as to center the inner conductor. While such waveguides can be treated using the generalised theory of multi-layered cylindrical waveguides [48, 56, 57], a concise presentation of the waveguide's Eigenmodes relevant in tubular filter designs is not available in the body of literature.

  Moreover, applying the Mode Matching technique to such waveguides will be demonstrated.

- **A rigorous analysis of "secondary effects" in cavity perturbation material measurements**

  During the derivation of the equations required for cavity perturbation material measurements several approximations have to be made. These approximations are shown to introduce "secondary effects", which may skew measurements of the complex permittivity.

  We will highlight that for materials with a low loss tangent of $2 \cdot 10^{-4}$ the error[3] of the imaginary part may easily be in the region of 10 % even for dielectrics with a low permittivity if a cavity with a moderate unloaded quality factor is used.

  Moreover, it will be shown that at least the deviation of the real and imaginary part of the complex permittivity caused by the field approximation can be corrected by employing the concept of permittivity-dependent geometry factors introduced in this thesis.

  In order to do so, a Mode Matching solver for electromagnetic cavities, which was developed in the scope of this thesis, is used which is capable of calculating the resonant frequency, the unloaded quality factor and the field distribution inside a perturbed cavity in a quasi-analytical fashion.

Terminology, Nomenclature and List of Symbols

The current thesis has a strong theoretical focus and as a consequence, a rigorous nomenclature is mandatory. A detailed overview of the terminology and nomenclature used in this thesis is given on pages 235ff.

[3] See remark on terminology on p. 235.

2. The Cylindrical Waveguide

For the development of the Mode Matching formalism used in the scope of this thesis, a sound understanding of the Eigenmodes of the waveguides included is crucial. Fortunately, many waveguides such as rectangular, circular and coaxial waveguides can be treated in a generalised fashion:

Consider the PEC-bounded waveguide shown in figure 2.1. As the cross-section of this waveguide does not change along the z axis, such waveguides are called *cylindrical waveguides* [48, 58]. In order to maintain generality, this waveguide shall be treated using orthogonal curvilinear coordinates [58].

In the following sections, we will see that determining such a waveguide's Eigenmodes requires solving a scalar wave equation

$$\nabla_t^2 \psi_t + k_c^2\, \psi_t = 0 \tag{2.1}$$

for the transverse dependence ψ_t of the scalar vector potential ψ. Together with the boundary conditions imposed by the waveguide, (2.1) represents a *boundary value problem* defined over the waveguide's cross-section.

The solution to such a problem is difficult to establish unless the partial differential equation (2.1) can be separated, giving a solution under form of $\psi_t = \psi_{x_1}\psi_{x_2}$ [58]. Even if (2.1) is separable, the boundary value problem still is difficult to solve unless the waveguide's boundary "coincides with the constant coordinate curves" [58].

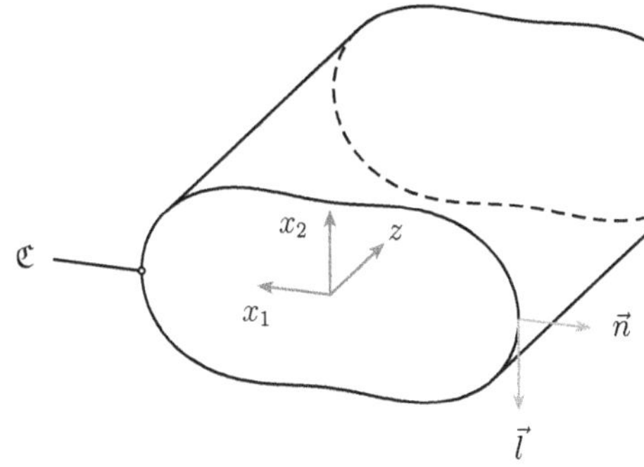

Figure 2.1.: The cylindrical waveguide

The remainder of this chapter is structured as follows:

Firstly, we derive the waveguide's boundary value problem using Lorenz[1]-gauged vector potentials [48].

We will then exemplarily show for rectangular and cylindrical coordinates that separation of (2.1) yields boundary value problems under Sturm-Liouville form. We will discuss Eigenfunctions and their corresponding Eigenvalues and see that solutions to Sturm-Liouville problems form a complete set of orthogonal functions.

Finally, starting from Lorentz'[2] reciprocity theorem, we will examine the orthogonality of the Eigenmodes' transverse field components. For homogeneously filled waveguides, we will see that the orthogonality of the transverse field components is in fact attributable to the orthogonality of the scalar vector potentials, which follows from Sturm-Liouville theory.

A brief discussion of orthogonality in partially filled waveguides as well as some remarks on the conservation of complex power or self-reaction during the Mode Matching process conclude this chapter.

2.1. The Boundary Value Problem

The Eigenmodes of the cylindrical waveguide are governed by Maxwell's equations, which denote[3]

$$\nabla \times \vec{H} = j\omega\varepsilon\vec{E} \tag{2.2}$$

$$\nabla \times \vec{E} = -j\omega\mu\vec{H} \tag{2.3}$$

if we assume the waveguide to be a source-free volume [48]. Maxwell's equations represent a coupled system of first-order partial differential equations (PDE).

2.1.1. The Scalar Wave Equation

It is well known that by introducing vector potentials, solutions to Maxwell's equations for source-free regions may be obtained by solving *Helmholtz wave equations* [48]

$$\nabla^2\vec{\mathfrak{A}} + k^2\vec{\mathfrak{A}} = 0 \qquad \nabla^2\vec{\mathfrak{F}} + k^2\vec{\mathfrak{F}} = 0, \tag{2.4}$$

which are (uncoupled) second-order PDEs. Note that (2.4) assumes Lorenz-gauged vector potentials [48][4].

Fortunately, in order to obtain the waveguide's Eigenmodes, it is sufficient to solve the scalar wave equations for one field component, e.g. [48]

$$\nabla^2\mathfrak{A}_z + k^2\mathfrak{A}_z = 0 \qquad \nabla^2\mathfrak{F}_z + k^2\mathfrak{F}_z = 0. \tag{2.5}$$

[1]Ludvig Lorenz, Danish physicists. See [59], p. 180, fn. 2

[2]Hendrik Antoon Lorentz, Dutch physicist. See footnote 6 on p. 24

[3]This thesis uses effective phasor notation similar to [48].

[4] [48] discusses the Lorenz Gauge without explicitly using this terminology. For an extended discussion, see e.g. [60].

Let us now rewrite (2.5) under a more general form [48]

$$\nabla^2 \psi(x_1, x_2, z) + k^2 \psi(x_1, x_2, z) = 0 \tag{2.6}$$

where ψ represents either $\mathfrak{A}_z$ or $\mathfrak{F}_z$. Using *separation of variables*, that is, by employing a product approach $\psi(x_1, x_2, z) = \psi_t(x_1, x_2)\,\psi_z(z)$, we can rewrite (2.6) as

$$\nabla_t^2 \psi_t(x_1, x_2) + k^2 \psi_t(x_1, x_2) + \left[\frac{1}{\psi_z(z)} \frac{\partial^2}{\partial z^2} \psi_z(z)\right] \psi_t(x_1, x_2) = 0 \tag{2.7}$$

where ∇_t^2 is the *transverse Laplacian* [48], whose formulation depends on the coordinate system.

By defining the wavenumber k_z, we obtain the ODE governing the Eigenmodes' z dependence [48],

$$k_z^2 = -\frac{1}{\psi_z(z)} \cdot \frac{d^2}{dz^2} \psi_z(z) \quad \rightarrow \quad 0 = \frac{d^2}{dz^2} \psi_z(z) + k_z^2 \psi_z(z), \tag{2.8}$$

which is under form of the differential equation of the harmonic oscillator.

Next, we rewrite (2.7) using (2.8) as

$$\nabla_t^2 \psi_t(x_1, x_2) + \left[k^2 - k_z^2\right] \psi_t(x_1, x_2) = 0,$$

which again can be rewritten using a *cut-off wavenumber* $k_c^2 = k^2 - k_z^2$ as [48, 58]

$$\nabla_t^2 \psi_t(x_1, x_2) + k_c^2\, \psi_t(x_1, x_2) = 0. \tag{2.9}$$

This is the boundary value problem's partial differential equation governing the transverse behaviour of the vector potentials and thus the derived fields. It should again be noted that $\psi_t(x_1, x_2)$ is dependent on x_1, x_2 only. For brevity's sake, we will omit the arguments of ψ_t from here on.

2.1.2. Deriving the Fields

Transverse Magnetic (TM) Eigenmodes

First let us derive the electric and magnetic fields from a magnetic vector potential

$$\vec{\mathfrak{A}} = \psi_t \, e^{\mp j k_z z} \, \vec{e}_z \tag{2.10}$$

where we have assumed (2.8) to yield a wave propagation term $e^{\mp j k_z z}$. It is well known that the magnetic field resulting from (2.10) may be obtained using

$$\vec{H} = \nabla \times \vec{\mathfrak{A}} \tag{2.11}$$

while the electric field is obtained from

$$\vec{E} = \frac{1}{j\omega\varepsilon} \nabla \times \left(\nabla \times \vec{\mathfrak{A}}\right), \tag{2.12}$$

which directly follows from (2.2) [48].

By analysing the curl operations imposed by (2.11) and (2.12) in orthogonal curvilinear coordinates, it can be shown that the resulting field components denote as [48, 58, 61]

$$\vec{E}_t = \frac{1}{j\omega\varepsilon}\;(\nabla_t\psi_t)\;\frac{\partial}{\partial z}e^{\mp jk_z z} = \mp\frac{jk_z}{j\omega\varepsilon}\;(\nabla_t\psi_t)\;e^{\mp jk_z z} \tag{2.13}$$

$$E_z = \frac{k_c^2}{j\omega\varepsilon}\;\psi_t\;e^{\mp jk_z z} \tag{2.14}$$

$$\vec{H}_t = -\left(\vec{e}_z \times \nabla_t\psi_t\right)e^{\mp jk_z z} \tag{2.15}$$

$$H_z = 0 \tag{2.16}$$

where $\vec{E}_t$, $\vec{H}_t$ denote the field components transverse to the direction of propagation while E_z, H_z denotes the electric field in the direction of propagation.

Since there is no magnetic field in the direction of propagation, that is, the magnetic field components are transverse only, these Eigenmodes are called *Transverse Magnetic* (TM).

Transverse Electric (TE) Eigenmodes

Similar to the previous section, from an electric vector potential

$$\vec{\mathfrak{F}} = \psi_t\;e^{\mp jk_z z}\;\vec{e}_z \tag{2.17}$$

using

$$\vec{E} = -\nabla \times \vec{\mathfrak{F}} \tag{2.18}$$

and (2.3), that is,

$$\vec{H} = \frac{1}{j\omega\mu}\nabla \times \left(\nabla \times \vec{\mathfrak{F}}\right) \tag{2.19}$$

we find [48, 58, 61]

$$\vec{E}_t = \left(\vec{e}_z \times \nabla_t\psi_t\right)\;e^{\mp jk_z z} \tag{2.20}$$

$$E_z = 0 \tag{2.21}$$

$$\vec{H}_t = \frac{1}{j\omega\mu}\;(\nabla_t\psi_t)\;\frac{\partial}{\partial z}e^{\mp jk_z z} = \mp\frac{jk_z}{j\omega\mu}\;(\nabla_t\psi_t)\;e^{\mp jk_z z} \tag{2.22}$$

$$H_z = \frac{k_c^2}{j\omega\mu}\;\psi_t\;e^{\mp jk_z z}. \tag{2.23}$$

As there is no electric field in the direction of propagation, i.e. the electric field components are transverse only, this type of Eigenmodes are referred to as *Transverse Electric* (TE).

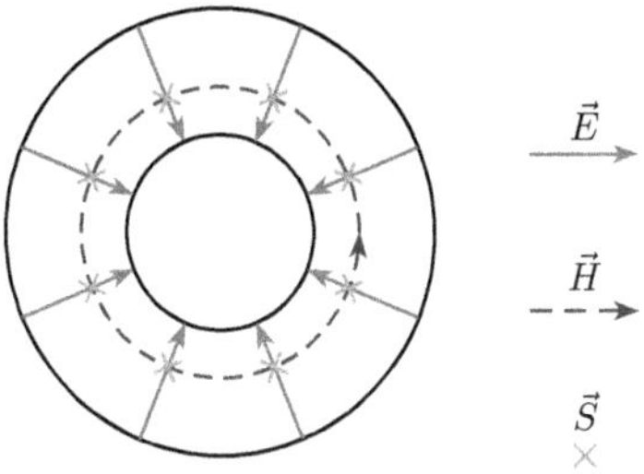

Figure 2.2.: The TEM Eigenmode of a coaxial waveguide

Transverse Electromagnetic (TEM) Eigenmodes

For completeness, it should be mentioned that a third type of Eigenmodes, *Transverse Electromagnetic* Eigenmodes (TEM), exist. This type of Eigenmodes neither exhibits an electric nor a magnetic field component in the direction of propagation and is most well-known as the fundamental Eigenmode of coaxial transmission lines (cf. figure 2.2).

In appendix B we will see that TEM Eigenmodes are in fact a special case of the TM Eigenmodes for $k_\rho = k_c = 0$ [48].

Alternative Mode-Sets

Obviously, the TM and TE Eigenmodes may be superposed to form so-called *hybrid Eigenmodes* [48]. If in addition the TM and TE Eigenmode share the same Eigenvalue, that is, the same cut-off wavenumber k_c, they are called *degenerated Eigenmodes* [48]. Since $k_z^2 = k^2 - k_c^2$ applies, degenerated Eigenmodes have the same propagation constant [48].

When dealing with rectangular waveguides, degenerated Eigenmodes are of high importance because *alternative mode-sets* being transverse to x_i may be constructed by combining degenerate Eigenmodes from the transverse-to-z mode-set [48]. Since the transverse-to-x_i mode-set contains linear combinations of Eigenmodes [48] from the transverse-to-z mode-set, the properties of the latter mode-set, which we will derive from Sturm-Liouville theory in section 2.3.1, apply equivalently.

The use of such alternative mode-sets greatly simplifies the implementation of Mode Matching algorithms if a suitable mode-set is chosen so that the number of overlap integrals to be calculated between TE and TM Eigenmodes is minimised.

Rather than constructing alternative mode-sets as linear combinations, it is more convenient to derive these mode-sets from suitable vector potentials $\vec{\mathfrak{A}} = \psi\, \vec{e}_{x_i}$ and $\vec{\mathfrak{F}} = \psi\, \vec{e}_{x_i}$ using the usual curl operations [48]. However, for these mode-sets, mixed boundary conditions are required in order to determine the scalar vector potential ψ.[5]

In the remainder of this thesis, the notation TM^{x_i}, TE^{x_i} is used in order to indicate the coordinate to which the fields of an Eigenmode are transverse to if specific clarification is required.

[5]This point becomes clear when comparing (4-21) and (4-34) in [48].

2.1.3. Boundary Conditions

In order to formulate the boundary value problem imposed by the waveguide, it is necessary to derive suitable boundary conditions for the scalar potential ψ_t.

Transverse Magnetic (TM) Eigenmodes

As TM Eigenmodes have an electric field component in the direction of propagation, which is obviously tangential to the waveguide's walls, we can directly conclude from $E_{tan}\Big|_{\mathfrak{C}} = 0$ that (cf. (2.14))

$$E_z\Big|_{\mathfrak{C}} = \frac{k_c^2}{j\omega\varepsilon}\,\psi_t\, e^{\mp jk_z z} = 0, \tag{2.24}$$

which leads to the desired *Dirichlet boundary condition* [61, 62]

$$\psi_t\Big|_{\mathfrak{C}} = 0 \tag{2.25}$$

for TM Eigenmodes [48].

Transverse Electric (TE) Eigenmodes

As TE Eigenmodes do not have an electric field component in the direction of propagation, the boundary condition $E_{tan}\Big|_{\mathfrak{C}} = 0$ must be expressed w.r.t. the Eigenmode's transverse field components.

As given in (2.20), the transverse electric field components of TE Eigenmodes may be denoted as

$$\vec{E}_t = (\vec{e}_z \times \nabla_t\psi_t)\; e^{\mp jk_z z}. \tag{2.26}$$

Using the unit vectors $\vec{l}$ and $\vec{n}$ depicted in figure 2.1, we may write the field components tangential to the waveguide's wall as [48]

$$E_{tan}|_{\mathfrak{C}} = \vec{l}(\vec{e}_z \times \nabla_t\psi_t)\; e^{\mp jk_z z} = \left(\vec{l}\times\vec{e}_z\right)\nabla_t\,\psi_t\, e^{\mp jk_z z} = \vec{n}\,\nabla_t\psi_t\, e^{\mp jk_z z} = 0, \tag{2.27}$$

which also can be denoted as

$$\frac{\partial\psi_t}{\partial\vec{n}}\Big|_{\mathfrak{C}} = 0, \tag{2.28}$$

which is the *Neumann boundary condition* [61, 62] for TE Eigenmodes [48].

2.1.4. Eigenfunctions and Eigenvalues

In section 2.1.1 we have derived the boundary value problem's PDE

$$\nabla_t^2 \psi_t + k_c^2 \, \psi_t = 0, \tag{2.29}$$

which together with the Dirichlet boundary condition for TM Eigenmodes

$$\psi_t \Big|_{\mathfrak{C}} = 0 \tag{2.30}$$

or the Neumann boundary condition for TE Eigenmodes

$$\frac{\partial \psi_t}{\partial \vec{n}} \Big|_{\mathfrak{C}} = 0 \tag{2.31}$$

forms the cylindrical waveguide's boundary value problem defined over the waveguide's cross-section.

The solution to this boundary value problem is a discrete spectrum of *Eigenvalues* [48], namely squared cut-off wavenumbers k_c^2, imposed by the boundary conditions as well as the corresponding *Eigenfunctions*, i.e. the waveguide's Eigenmodes [63].

At this point, it becomes obvious that the cylindrical waveguide can be interpreted as a two-dimensional resonator, for which the squared cut-off wavenumbers k_c^2 represent the resonator's Eigenvalues, similar to a cavity's Eigenvalues representing its squared resonant frequencies.

2.2. Solving the Boundary Value Problem

In the previous sections, we have derived the PDE

$$\nabla_t^2 \psi_t (x_1, x_2) + k_c^2 \, \psi_t (x_1, x_2) = 0 \tag{2.32}$$

of the cylindrical waveguide's boundary value problem. Note that ∇_t^2 denotes the transverse Laplacian and $k_c^2 = k^2 - k_z^2$ the Eigenmodes' Eigenvalue and its squared cut-off wavenumber, respectively.

In this chapter, we will exemplarily show for rectangular and cylindrical coordinates how to split this PDE into two ODEs of well-known types.

In the following section, we will then see that the obtained ODEs in conjunction with the boundary conditions introduced in the previous sections form a *Sturm-Liouville problem*. This class of problems is well understood and the solutions to this class of problems exhibit properties of fundamental importance for the Mode Matching technique.

2.2.1. Rectangular Coordinates

For rectangular coordinates ($x_1 = x$, $x_2 = y$, z), the transverse Laplacian denotes

$$\nabla_t^2 = \frac{\partial^2}{\partial x^2} + \frac{\partial^2}{\partial y^2}. \tag{2.33}$$

Using a product approach $\psi_t (x, y) = \psi_x (x) \psi_y (y)$, after dividing by $\psi_t (x, y)$, (2.32) denotes

$$\frac{1}{\psi_x (x)} \frac{\partial^2}{\partial x^2} \psi_x (x) + \frac{1}{\psi_y (y)} \frac{\partial^2}{\partial y^2} \psi_y (y) + k_c^2 = 0. \tag{2.34}$$

The two first terms in (2.34) depend on either x or y only and we thus define transverse wavenumbers k_x, k_y as

$$k_x^2 = -\frac{1}{\psi_x(x)} \frac{d^2}{dx^2} \psi_x(x) \quad \rightarrow \quad 0 = \frac{d^2}{dx^2} \psi_x(x) + k_x^2 \psi_x(x) \tag{2.35}$$

$$k_y^2 = -\frac{1}{\psi_y(y)} \frac{d^2}{dy^2} \psi_y(y) \quad \rightarrow \quad 0 = \frac{d^2}{dy^2} \psi_y(y) + k_y^2 \psi_y(y) \tag{2.36}$$

for the two ODEs (2.35), (2.36) [48] contained in (2.34). This immediately leads to the *separation condition* [48]

$$k_c^2 = k_x^2 + k_y^2, \tag{2.37}$$

which provides the Eigenmode's Eigenvalue k_c^2 after k_x and k_y were determined by solving (2.35) and (2.36) subject to the boundary conditions imposed by waveguide. Note that (2.35) and (2.36) are under form of the *differential equation of the harmonic oscillator*.

The Eigenmode's propagation constant then may be obtained by solving $k_c^2 = k^2 - k_z^2$ for k_z. Note that from k_z, we may determine the guided wavelength as $\lambda_g = 2\pi/k_z$.

2.2.2. Cylindrical Coordinates

For cylindrical coordinates ($x_1 = \rho$, $x_2 = \phi$, z), using the appropriate definition of the transverse Laplacian, (2.32) denotes as [48]

$$\frac{1}{\rho}\frac{\partial}{\partial\rho}\left(\rho\frac{\partial}{\partial\rho}\psi_t(\rho,\phi)\right)+\frac{1}{\rho^2}\frac{\partial^2}{\partial\phi^2}\psi_t(\rho,\phi)+k_c^2\psi_t(\rho,\phi)=0, \tag{2.38}$$

which after some manipulations may be rewritten as

$$\frac{\rho}{\psi_\rho(\rho)}\frac{\partial}{\partial\rho}\left(\rho\frac{\partial}{\partial\rho}\psi_\rho(\rho)\right)+\frac{1}{\psi_\phi(\phi)}\frac{\partial^2}{\partial\phi^2}\psi_\phi(\phi)+\rho^2k_c^2=0 \tag{2.39}$$

if a product approach $\psi_t(\rho,\phi) = \psi_\rho(\rho)\psi_\phi(\phi)$ is chosen.

By defining [48]

$$n^2=-\frac{1}{\psi_\phi(\phi)}\frac{d^2}{d\phi^2}\psi_\phi(\phi) \quad\rightarrow\quad 0=\frac{d^2}{d\phi^2}\psi_\phi(\phi)+n^2\psi_\phi(\phi) \tag{2.40}$$

we obtain the ODE governing the scalar vector potential's angular dependence. Note that (2.40) again is under form of the differential equation of the harmonic oscillator.

Using (2.40), after some manipulations, (2.39) may be put under a form of an ODE

$$\frac{d}{d\rho}\left(\rho\frac{d}{d\rho}\psi_\rho(\rho)\right)+\left[-\frac{n^2}{\rho}+\underbrace{\left(k^2-k_z^2\right)}_{k_c^2=k_\rho^2}\rho\right]\psi_\rho(\rho)=0 \tag{2.41}$$

governing the scalar vector potentials radial dependence. This ODE is *Bessel's differential equation* [48] where $k_c^2 = k_\rho^2$ denotes the Eigenmodes' Eigenvalues.

Solutions to Bessel's differential equations are *Bessel functions of first kind* $J_n(\rho)$, *Bessel functions of second kind* $N_n(\rho)$ (also: *Neumann functions*) or *Hankel functions* $H_n(\rho)$, which are linear combinations of the aforementioned two solutions [48].

The *separation condition* denotes [48]

$$k_c^2 = k_\rho^2 = k^2 - k_z^2 \tag{2.42}$$

and after solving (2.41) subject to the waveguide's boundary conditions, the Eigenmodes propagation constant may be obtained by rearranging (2.42) for k_z.

2.3. The Sturm-Liouville Problem

The ODEs (2.35) and (2.36) as well as (2.40) and (2.41) derived in the previous section are in fact Sturm-Liouville problems, which is a class of problems often encountered in the context of guided electromagnetic waves [58].

2.3.1. Definition and Solution

Most generally, the *Sturm-Liouville problem* may be denoted [58]

$$\frac{d}{dx_i}\left(p\left(x_i\right)\frac{d}{dx_i}\psi_{x_i}\left(x_i\right)\right)+\left[q\left(x_i\right)+\Lambda\sigma\left(x_i\right)\right]\psi_{x_i}\left(x_i\right)=0 \tag{2.43}$$

for a given domain of definition (e.g. $\{x_i|0\leq x_i\leq a\}$) at whose boundaries either a Dirichlet boundary condition

$$\psi_{x_i}=0 \tag{2.44}$$

or a Neumann boundary condition

$$\frac{d\psi_{x_i}}{dx_i}=0 \tag{2.45}$$

applies [58]. In (2.43) Λ denotes the solutions' Eigenvalues to be discussed in the following section. $\sigma\left(x_i\right)$ is a weighting function while $p\left(x_i\right)$ and $q\left(x_i\right)$ represent coefficients [58,64].

2.3.2. Properties of the Solution

The solution to such Sturm-Liouville problems is a *complete set* of Eigenfunctions $\{\psi_{x_i|m}\}$ with Eigenvalues Λ_m, which span a function space over the domain of definition [58].

Under the assumption that the Eigenfunctions are normalised, it can be shown that the Eigenfunctions are *orthogonal* w.r.t. the weighting function $\sigma\left(x_i\right)$ [58], that is,

$$\int_0^a\sigma\psi_{x_i|m}\psi_{x_i|n}\,dx_i=\begin{cases}1, & m=n\\0, & m\neq n.\end{cases} \tag{2.46}$$

We will exploit this property in the following section.

As a consequence of orthogonality, a function $f(x_i)$ lying in that function space may be expanded as *generalised Fourier series* [58]

$$f(x_i)=\sum_m a_m\psi_{x_i|m} \tag{2.47}$$

with coefficients [58]

$$a_m=\int_0^a\sigma\;f\;\psi_{x_i|m}\,dx_i. \tag{2.48}$$

2.3.3. The Cylindrical Waveguide's Boundary Value Problem as Sturm-Liouville Problem

Having briefly reviewed the Sturm-Liouville problem and the properties of its solution, we will now show that the cylindrical waveguide's boundary value problem is a Sturm-Liouville problem.

Rectangular Coordinates

Consider (2.43) again. If we let

$$\Lambda = k_{x_i}^2 \qquad p(x_i) = 1 \qquad q(x_i) = 0 \qquad \sigma(x_i) = 1$$

we find

$$\frac{d^2}{dx_i^2}\psi_{x_i}(x_i) + k_{x_i}^2\psi_{x_i}(x_i) = 0,$$

which is an ODE of similar form as (2.35) and (2.36) derived for the cylindrical waveguide in rectangular coordinates. Since in addition the Dirichlet boundary condition (2.30) or the Neumann boundary condition (2.31) applies, the boundary value problem imposed by the cylindrical waveguide in rectangular coordinates is under form of two Sturm-Liouville problems defined over the waveguide's cross-section.

Cylindrical Coordinates

Similarly, if we let $x_i = \rho$ and

$$\Lambda = k_\rho^2 = k_c^2 \qquad p(x_i) = x_i = \rho \qquad q(x_i) = -\frac{n^2}{x_i} = -\frac{n^2}{\rho} \qquad \sigma(x_i) = x_i = \rho$$

we find

$$\frac{d}{d\rho}\left(\rho\frac{d}{d\rho}\psi_\rho(\rho)\right) + \left[-\frac{n^2}{\rho} + k_\rho^2\rho\right]\psi_\rho(\rho) = 0,$$

which is Bessel's differential equation (2.41) derived for the cylindrical waveguide using cylindrical coordinates. The ODE (2.40) also can be shown to be of Sturm-Liouville form.

Again noting that the boundary conditions (2.30) and (2.31) apply, we see that the boundary value problem imposed by the cylindrical waveguide in cylindrical coordinates is composed of two Sturm-Liouville problems defined over the waveguide's cross-section.

2.4. Transverse Orthogonality of the Scalar Vector Potential

In section 2.2, we have seen that the transverse dependence of the scalar vector potential can be composed using a product approach $\psi_t(x_i, x_j) = \psi_{x_i}(x_i)\psi_{x_j}(x_j)$ where $\psi_{x_i}(x_i)$ and $\psi_{x_j}(x_j)$ are solutions to Sturm-Liouville problems.

Because $\psi_{x_i}(x_i)$ and $\psi_{x_j}(x_j)$ are solutions to Sturm-Liouville problems and thus are orthogonal, the product approach similarly is orthogonal because integration over the waveguide's cross-section $\mathfrak{S}$ yields

$$\iint_{\mathfrak{S}} \psi_{t|m}\psi_{t|n}dA = \int_{x_i} \psi_{x_i|m}\psi_{x_i|n}\,\sigma(x_i)\,dx_i \cdot \int_{x_j} \psi_{x_j|m}\psi_{x_j|n}\,\sigma(x_j)\,dx_j = 0 \qquad m \neq n. \tag{2.49}$$

2.5. Orthogonality of Transverse Fields

In the previous section, we have seen that the scalar vector potentials from which the waveguide's Eigenmodes are derived are orthogonal. In this section, we will investigate the orthogonality of the transverse field components of the cylindrical waveguide.

The orthogonality of the transverse field components is the key enabler of the Mode Matching technique as it allows expansion of the transverse field distributions in a Fourier-like sense.

2.5.1. Derivation of the Orthogonality Condition

Let $\vec{E}_n, \vec{H}_n$ and $\vec{E}_m, \vec{H}_m$ be the total fields of two linear independent solutions to Maxwell's equations. Using Maxwell's equations and some simple manipulations, it can be shown that [58]

$$\nabla\left(\vec{E}_n \times \vec{H}_m - \vec{E}_m \times \vec{H}_n\right) = 0, \tag{2.50}$$

which is *Lorentz reciprocity theorem*[6] for source-free regions [48].

By defining a transverse divergence operator ∇_t we can rewrite (2.50) as [58]

$$\begin{aligned}\nabla\left(\vec{E}_n \times \vec{H}_m - \vec{E}_m \times \vec{H}_n\right) &= \nabla_t\left(\vec{E}_n \times \vec{H}_m - \vec{E}_m \times \vec{H}_n\right) \\ &+ \vec{e}_z\frac{\partial}{\partial z}\left(\vec{E}_n \times \vec{H}_m - \vec{E}_m \times \vec{H}_n\right)\end{aligned} \tag{2.51}$$

Because of (2.50), the left side of (2.51) is zero. Let us now denote the waveguide's cross-section as $\mathfrak{S}$, its bounding contour as $\mathfrak{C}$ and apply the two-dimensional divergence theorem to (2.51). For the first term on the right side of (2.51) we find [58]

$$\iint_{\mathfrak{S}} \nabla_t\left(\vec{E}_n \times \vec{H}_m - \vec{E}_m \times \vec{H}_n\right) dA = \oint_{\mathfrak{C}} \vec{n}\left(\vec{E}_n \times \vec{H}_m - \vec{E}_m \times \vec{H}_n\right) dl = 0 \tag{2.52}$$

as can be seen by performing circular shift[7] of the scalar triple products on the right side because electric fields tangential to the waveguide's boundary must vanish [58], i.e.

$$\left(\vec{n} \times \vec{E}_n\right) dl = 0 \qquad \left(\vec{n} \times \vec{E}_m\right) dl = 0. \tag{2.53}$$

Consequently, if we assume the fields to be Eigenmodes with a wave propagation term, i.e. $e^{\mp jk_z z}$, we can rewrite[8] (2.51) as [58]

$$0 = \left(k_{z|n} + k_{z|m}\right) \iint_{\mathfrak{S}} \vec{e}_z\left(\vec{E}_{t|n} \times \vec{H}_{t|m} - \vec{E}_{t|m} \times \vec{H}_{t|n}\right) dA \tag{2.54}$$

where $\vec{E}_t$, $\vec{H}_t$ denote transverse field components.

[6] Hendrik Antoon Lorentz, Dutch physicist. Papas [65] cites Lorentz' original publication [66] where the reciprocity theorem $\iiint E_a J_b dV = \iiint E_b J_a dV$ can be found on p. 181.

[7] $\vec{A}\left(\vec{B} \times \vec{C}\right) = \vec{B}\left(\vec{C} \times \vec{A}\right) = \vec{C}\left(\vec{A} \times \vec{B}\right)$

[8] Here, we use the product rule for deriving the products of the wave propagation terms (which result from the cross-products) and anticipate the fact that only the cross-product of tangential fields, i.e. $\vec{E}_t \times \vec{H}_t$ is parallel to $\vec{e}_z$ (see also [58]).

Let us now rewrite the transverse field components as[9]

$$\vec{E}_t \quad = \quad \pm \quad \vec{\mathcal{E}}_t(x_1, x_2)\, e^{\mp j k_z z} \tag{2.55}$$

and

$$\vec{H}_t \quad = \quad \pm \quad \vec{\mathcal{H}}_t(x_1, x_2)\, e^{\mp j k_z z}. \tag{2.56}$$

$\vec{\mathcal{E}}_t$ and $\vec{\mathcal{H}}_t$ represent the transverse dependencies of the transverse field components and shall be considered unsigned, that is, their sign does not change if the direction of propagation is reversed.[10]

Using (2.55) and (2.56), we can rewrite[11] (2.54) as [58]

$$0 = \left(k_{z|n} + k_{z|m}\right) \iint_{\mathfrak{S}} \vec{e}_z \left(\vec{\mathcal{E}}_{t|n} \times \vec{\mathcal{H}}_{t|m} - \vec{\mathcal{E}}_{t|m} \times \vec{\mathcal{H}}_{t|n}\right) dA. \tag{2.57}$$

If we now assume the direction of propagation of the mth Eigenmode to be reversed, depending on the mode type, either the sign of the electric or magnetic field changes[12] and we obtain a second equation[13] [58], e.g.

$$0 = \left(k_{z|n} - k_{z|m}\right) \iint_{\mathfrak{S}} \vec{e}_z \left(-\left(\vec{\mathcal{E}}_{t|n} \times \vec{\mathcal{H}}_{t|m}\right) - \vec{\mathcal{E}}_{t|m} \times \vec{\mathcal{H}}_{t|n}\right) dA \tag{2.58}$$

where the sign of the magnetic field $\vec{\mathcal{H}}_{t,m}$ has been reversed.[14]

Finally, addition of (2.57) and (2.58) yields the *orthogonality relation*[15] [58]

$$\iint_{\mathfrak{S}} \vec{e}_z \left(\vec{\mathcal{E}}_{t|n} \times \vec{\mathcal{H}}_{t|m}\right) dA = 0 \qquad n \neq m \ . \tag{2.59}$$

For the loss-free case, we may similarly derive another orthogonality relation [58]

$$\iint_{\mathfrak{S}} \vec{e}_z \left(\vec{\mathcal{E}}_{t|n} \times \vec{\mathcal{H}}^*_{t|m}\right) dA = 0 \qquad n \neq m \ . \tag{2.60}$$

It is interesting to note that (2.59) provides a Mode Matching formulation where the self-reaction $\iint_{\mathfrak{S}} \vec{e}_z(\vec{E}_t \times \vec{H}_t)\, dA$ of the field is conserved over the discontinuity while using (2.60), a formulation is obtained which preserves complex power $\iint_{\mathfrak{S}} \vec{e}_z(\vec{E}_t \times \vec{H}^*_t)\, dA$ [67]. Both formulations are in fact equivalent [67].

[9] In order to comply with the notation used in the rest of this thesis, here we use a slightly different notation than [58].

[10] The $\pm$ in (2.55) and (2.56) accounts for this change of sign.

[11] Note that the wave propagation terms in (2.54) are eliminated by dividing by $e^{-jk_{z|n}z} \cdot e^{-jk_{z|m}z}$.

[12] This immediately becomes obvious when considering the Poynting vector.

[13] Again the wave propagation terms are eliminated by dividing by $e^{-jk_{z|n}z} \cdot e^{jk_{z|m}z}$.

[14] Note that the change of the field component's sign is due to deriving the vector potential while the change of sign of $k_{z|m}$ in the first bracket is due to the derivative of the fields as per (2.51).

[15] This result and its derivation requires some additional discussion for the case of degenerate Eigenmodes. The interested reader is referred to [58].

2.5.2. Proof of Orthogonality for a PEC-bounded Empty Waveguide

Let us now use the fact that $\vec{\mathcal{H}}_t = Z_{wave}^{-1}\left(\vec{e}_z \times \vec{\mathcal{E}}_t\right)$ [58] where Z_{wave} denotes the wave impedance to rewrite[16,17] the orthogonality relation (2.59) as [17, 48]

$$\iint_{\mathfrak{S}} \vec{\mathcal{E}}_{t|n} \cdot \vec{\mathcal{E}}_{t|m} \, dA = 0 \qquad n \neq m \tag{2.61}$$

and

$$\iint_{\mathfrak{S}} \vec{\mathcal{H}}_{t|n} \cdot \vec{\mathcal{H}}_{t|m} \, dA = 0 \qquad n \neq m \tag{2.62}$$

We shall now prove orthogonality in terms of the electric field. For the magnetic field, similar derivations are possible.

Two TM Eigenmodes

Using (2.13), we may rewrite (2.61) as [58]

$$\iint_{\mathfrak{S}} \nabla_t \psi_{t|n} \cdot \nabla_t \psi_{t|m} \, dA = 0 \qquad n \neq m \tag{2.63}$$

if all prefactors are cancelled out.

Two TE Eigenmodes

Similarly, using (2.20), we may rewrite (2.61) as [58]

$$\iint_{\mathfrak{S}} \left(\vec{e}_z \times \nabla_t \psi_{t|n}\right) \cdot \left(\vec{e}_z \times \nabla_t \psi_{t|m}\right) \, dA = \iint_{\mathfrak{S}} \nabla_t \psi_{t|n} \cdot \nabla_t \psi_{t|m} \, dA = 0 \qquad n \neq m \tag{2.64}$$

as can be shown by performing a circular shift of the scalar triple product and subsequent triple product expansion as given in footnote 16 [58]:

$$\begin{aligned} \left(\vec{e}_z \times \nabla_t \psi_{t|n}\right) \cdot \left(\vec{e}_z \times \nabla_t \psi_{t|m}\right) &= \vec{e}_z \left[\nabla_t \psi_{t|n} \times \left(\vec{e}_z \times \nabla_t \psi_{t|m}\right)\right] \\ &= \vec{e}_z \left[\left(\nabla_t \psi_{t|n} \cdot \nabla_t \psi_{t|m}\right) \vec{e}_z - \left(\nabla_t \psi_{t|n} \cdot \vec{e}_z\right) \nabla_t \psi_{t|m}\right] \\ &= \nabla_t \psi_{t|n} \cdot \nabla_t \psi_{t|m}. \end{aligned} \tag{2.65}$$

[16] The required manipulations can easily be carried out using
$\vec{a} \times \left(\vec{b} \times \vec{c}\right) = \vec{b} \cdot (\vec{a} \cdot \vec{c}) - \vec{c} \cdot \left(\vec{a} \cdot \vec{b}\right)$.

[17] While this step is possible for "well-behaved" waveguides, great care has to be taken when dealing e.g. with lossy or partially filled waveguides. This is because their Eigenmodes may be hybrid. [58]

In order to show that (2.63) and (2.64) are indeed true, we use *Green's first identity*[18] [48,58] to find [58]

$$\iint_{\mathfrak{S}} \nabla_t \psi_{t|n} \cdot \nabla_t \psi_{t|m}\, dA = -\iint_{\mathfrak{S}} \psi_{t|n} \cdot \nabla_t^2 \psi_{t|m}\, dA + \underbrace{\oint_C \psi_{t|n} \frac{\partial \psi_{t|m}}{\partial \vec{n}}\, dl}_{0} \tag{2.66}$$

where the second term on the right side vanishes due to the boundary conditions introduced in section 2.1.3 [58].

Using the wave equation (2.9), the first term on the right side of (2.66) can be denoted as [58]

$$-\iint_{\mathfrak{S}} \psi_{t|n} \cdot \nabla_t^2 \psi_{t|m}\, dA = k_{c|m}^2 \iint_{\mathfrak{S}} \psi_{t|n} \cdot \psi_{t|m}\, dA. \tag{2.67}$$

Consequently, we can rewrite (2.63) and (2.64) as [58]

$$\iint_{\mathfrak{S}} \psi_{t|n} \cdot \psi_{t|m}\, dA = 0 \qquad n \neq m, \tag{2.68}$$

which is true because the scalar potentials are solutions to Sturm-Liouville problems as discussed in section 2.3.1.[19]

TM and TE Eigenmode

While (2.63) and (2.64) consider the individual orthogonality of TM and TE Eigenmodes, the orthogonality of TM Eigenmodes to TE Eigenmodes and vice versa has not yet been discussed. The corresponding proof is possible in a conceptual similar fashion and left out for brevity. The interested reader is referred to [58].

2.6. Orthogonality in Partially Filled Waveguides

The classification of Eigenmodes of partially filled waveguides in TM and TE Eigenmodes is not always possible because such waveguides may have Eigenmodes representing linear combinations of the two aforementioned types [58], that is, the Eigenmodes are hybrid [48]. However, it is possible to group the Eigenmodes of such waveguides w.r.t. the air-dielectric interface plane, namely into *Longitudinal Section Electric* (LSE) and *Longitudinal Section Magnetic* (LSM) Eigenmodes [58]:

LSE Eigenmodes do not exhibit an electric field component normal to the interface plane; the electric field lies within the interface plane [58]. Similarly, LSM Eigenmodes do not have a magnetic field component normal to the interface plane and thus, the magnetic field lies within the interface plane [58].

An extensive discussion of the orthogonality of Eigenmodes in partially filled rectangular waveguides can be found in [58] where it is shown that LSE Eigenmodes are orthogonal while LSM Eigenmodes are orthogonal w.r.t. an additional weighting function $1/\varepsilon_r(x_i)$ [58].

[18] $\oiint \psi \frac{\partial \phi}{\partial \vec{n}}\, dA = \iiint \left(\psi \nabla^2 \phi + \nabla \psi \cdot \nabla \phi\right)\, dV$, see [48].

[19] The orthogonality relations, e.g. (2.46), contain a weighting function. Here, the weighting function is contained in the surface element dS as can be seen in section 2.4.

In this thesis, we will consider two types of partially filled waveguides:

In chapter 15.4.1, we will use the $TE^x_{m,0}$ Eigenmodes of a rectangular waveguide partially filled with a centered dielectric slab. The superscript x indicates that these Eigenmodes are transverse to x. These Eigenmodes are given in appendix B.2 for reference. Because similar to the corresponding Eigenmodes of the empty rectangular waveguide these Eigenmodes exhibit an E_y component only, they are of LSE type because E_y coincides with the air-dielectric interface.

In chapter 14, we will discuss the Eigenmodes of a partially filled coaxial waveguide [68]. We will see that the rotationally symmetric TM Eigenmodes of this waveguide are of LSM type because the Eigenmodes exhibit an H_ϕ component only, which coincides with the air-dielectric interface. During the implementation of the Mode Matching formalism for partially filled coaxial waveguides we have found that similar to the rectangular waveguide's LSM Eigenmodes, these Eigenmodes require an additional weighting function $1/\varepsilon_r(x_i)$ in order to maintain orthogonality.[20]

[20] While this is a result obtained in a practical fashion only, a proof should be possible by modifying the proof given in [58] for rectangular coordinates to apply for cylindrical coordinates. This is however considered future work.

3. Eigenwave Expansion of Transverse Field Components

The previous chapter was entirely devoted to studying the fundamental properties of the Eigenmodes of the cylindrical waveguide. We will now profit from the findings of the previous chapters, namely completeness and orthogonality, by introducing Eigenwave Expansion of transverse field components.

Firstly, orthogonal expansion will be introduced from a purely theoretical point of view. We will deploy a two-dimensional formulation for the orthogonal expansion of functions in vector spaces. Note that summaries covering the one-dimensional case can be found e.g. in [69,70]. We will see that orthogonal expansion in vector spaces of functions is conceptually similar to expanding "ordinary" vectors in coordinate spaces.

After that, we will introduce an electromagnetic formulation and derive equations for the orthogonal expansion of transverse field components in waveguides. During the development of the Mode Matching formalism used in this thesis, these equations will prove to be essential for the method.

3.1. Theoretical Formulation

3.1.1. Orthogonal Expansion in Vector Spaces of Functions

Consider a set of *Eigenfunctions* $\{\varphi_{m,n}(x_1,x_2)\} \in \mathbb{V}$ spanning a *vector space* of functions $\mathbb{V}$ over the *domain of definition* $\{x_1, x_2 \,|\, 0 \le x_1 \le a, 0 \le x_2 \le b\}$. As $\{\varphi_{m,n}(x_1,x_2)\}$ spans $\mathbb{V}$, the Eigenfunctions are also referred to as *basis functions*.

The *inner product* on $\mathbb{V}$ shall be defined as

$$\langle \mathfrak{f} | \mathfrak{g} \rangle = \int_0^b \int_0^a \mathfrak{g}^*(x_1,x_2)\, \mathfrak{f}(x_1,x_2)\, \sigma(x_1)\, \sigma(x_2)\, dx_1\, dx_2 \tag{3.1}$$

where $\sigma(x_i)$ denotes *weighting functions*. The inner product induces a *norm*

$$||\mathfrak{f}|| = \langle \mathfrak{f} | \mathfrak{f} \rangle^{1/2} \tag{3.2}$$

for $\mathbb{V}$.

The Eigenfunctions shall be orthogonal and normalised, that is, they maintain the orthogonality relation

$$\langle \varphi_{m,n} | \varphi_{p,q} \rangle = \delta_{m,p}\delta_{n,q} \tag{3.3}$$

over the domain of definition.

As a consequence, an arbitrary function $\mathfrak{f}(x_1, x_2) \in \mathbb{V}$ can be expanded as *generalised Fourier series* [71]

$$\mathfrak{f} = \sum_m^\infty \sum_n^\infty c_{m,n} \cdot \varphi_{m,n} \tag{3.4}$$

by means of a set of basis functions $\{\varphi_{m,n}\}$ with coefficients $c_{m,n} = \langle \mathfrak{f} | \varphi_{m,n} \rangle$.

3.1.2. Orthogonal Expansion in Coordinate Spaces

The concept of orthogonal expansion in vector spaces of functions is similar to representing an "ordinary" vector $\vec{w} \in \mathcal{W}$ in a coordinate space $\mathbb{W} = \mathbb{K}^N$ spanned by a set of orthonormal basis vectors $\{\vec{e}_{x_i}\} \in \mathbb{W}$ under form of a linear combination [62]

$$\vec{w} = \sum_i^N a_i \, \vec{e}_{x_i} \tag{3.5}$$

as is shown in figure 3.1.

The coefficients a_i are again calculated using the inner product [62]

$$a_i = \langle \vec{w} | \vec{e}_{x_i} \rangle \tag{3.6}$$

which is defined [62]

$$\langle \vec{v} | \vec{w} \rangle = \sum_i^N w_i \, v_i. \tag{3.7}$$

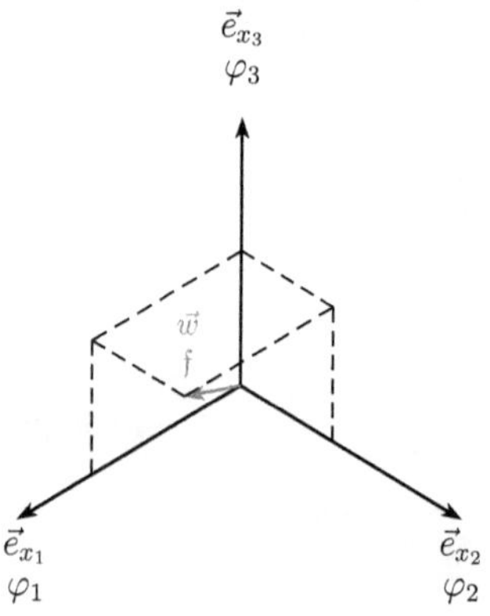

Figure 3.1.: Conceptual similarity between the expansion of a function $\mathfrak{f}$ and a vector $\vec{w}$

3.2. Electromagnetic Formulation

3.2.1. Unsigned Transverse Field Components

Similar to (2.55) and (2.56) an arbitrary transverse field component $F_{x_i|m,n}$ of a waveguide's m, nth Eigenmode may be denoted as

$$F_{x_i|m,n} = \pm \begin{Bmatrix} A_{m,n} \\ B_{m,n} \end{Bmatrix} \mathcal{F}_{x_i|m,n}\left(x_1, x_2\right) e^{\mp jk_z z}. \tag{3.8}$$

The unsigned transverse field component $\mathcal{F}_{x_i|m,n}(x_1, x_2)$ denotes the transverse dependency of $F_{x_i|m,n}$. Since calculating transverse field components from the scalar vector potential ψ may require deriving ψ for z, the field component's sign may change with the Eigenmode's direction of propagation (cf. (2.13) and (2.22)). This change in sign is accounted for by the $\pm$ notation in (3.8) while $\mathcal{F}_{x_i|m,n}$ shall be considered unsigned, that is, its sign does not change if the direction of propagation is reversed. This convention will prove useful in the following chapters.

$A_{m,n}$ represents the amplitude of the forward travelling Eigenmode while $B_{m,n}$ denotes the amplitude of the backward travelling Eigenmode.

3.2.2. Expansion of Transverse Field Components

Let us consider the Eigenmodes, more specifically $\mathcal{F}_{x_i|m,n}$, as a set of basis functions $\{\mathcal{F}_{x_i|m,n}\} \in \mathbb{V}$ spanning $\mathbb{V}$ over the domain of definition $\{x_1, x_2 \mid 0 \leq x_1 \leq a, 0 \leq x_2 \leq b\}$, namely the waveguide's cross-section.

As we have seen in the previous section, the Eigenmodes of the cylindrical waveguide have orthogonal field components and form a complete set. Thus, after normalising the Eigenmodes derived from ψ using the norm (3.2) so that they form an orthonormal set, we may expand a function $\mathfrak{f}_{F_{x_i}} \in \mathbb{V}$ as

$$\mathfrak{f}_{F_{x_i}} = \sum_m^\infty \sum_n^\infty \left\langle \mathfrak{f}_{F_{x_i}} \middle| \frac{\mathcal{F}_{x_i|m,n}}{||\mathcal{F}_{x_i|m,n}||} \right\rangle \frac{\mathcal{F}_{x_i|m,n}}{||\mathcal{F}_{x_i|m,n}||}. \tag{3.9}$$

By exploiting the linearity of the inner product, we rewrite (3.9) as

$$\mathfrak{f}_{F_{x_i}} = \sum_m^\infty \sum_n^\infty \frac{\left\langle \mathfrak{f}_{F_{x_i}} \middle| \mathcal{F}_{x_i|m,n} \right\rangle}{||\mathcal{F}_{x_i|m,n}||} \frac{\mathcal{F}_{x_i|m,n}}{||\mathcal{F}_{x_i|m,n}||}$$

and

$$\mathfrak{f}_{F_{x_i}} = \sum_m^\infty \sum_n^\infty \frac{\left\langle \mathfrak{f}_{F_{x_i}} \middle| \mathcal{F}_{x_i|m,n} \right\rangle}{||\mathcal{F}_{x_i|m,n}||^2} \mathcal{F}_{x_i|m,n},$$

which can be denoted using (3.2) as

$$\mathfrak{f}_{F_{x_i}} = \sum_m^\infty \sum_n^\infty \frac{\left\langle \mathfrak{f}_{F_{x_i}} | \mathcal{F}_{x_i|m,n} \right\rangle}{\left\langle \mathcal{F}_{x_i|m,n} | \mathcal{F}_{x_i|m,n} \right\rangle} \mathcal{F}_{x_i|m,n}.$$

As we have seen in the previous section, the inner product $\langle \mathfrak{f}_{F_{x_i}} | \mathcal{F}_{x_i|m,n} \rangle$ represents an integral under form of (3.1). As a matter of fact, this integral is of similar form as the orthogonality relations derived in the previous chapter. Because (3.1) contains the complex conjugate, the Mode Matching formalism used in this thesis assumes continuity of the complex power as discussed in section 2.5.1.

Since the integral (3.1) calculates the overlap of two field distributions, similar to quantum mechanics where the overlap of wave functions is calculated [72], these integrals shall be called *overlap integrals*.

Now consider an interface between two waveguides. If on one waveguide the field distribution $\mathfrak{f}_{F_{x_i}}$ is known, without further assumptions, $\mathfrak{f}_{F_{x_i}}$ can be expanded in terms of forward travelling Eigenmodes with amplitudes $A_{m,n}$, backward travelling Eigenmodes with amplitudes $B_{m,n}$ or a combination of both types.

In order to account for this additional degree of freedom, we interpret the quotient of inner products as a sum or difference amplitude[1]

$$\left[A_{m,n} \pm B_{m,n}\right] = \frac{\left\langle \mathfrak{f}_{F_{x_i}} | \mathcal{F}_{x_i|m,n} \right\rangle}{\left\langle \mathcal{F}_{x_i|m,n} | \mathcal{F}_{x_i|m,n} \right\rangle} \qquad \text{for all } m, n \tag{3.10}$$

and denote the field expansion as *generalised Fourier Series*

$$\mathfrak{f}_{F_{x_i}} = \sum_m^\infty \sum_n^\infty \left[A_{m,n} \pm B_{m,n}\right] \cdot \mathcal{F}_{x_i|m,n}. \tag{3.11}$$

In the following chapter, we will often encounter (3.10) and (3.11) and both equations will show themselves to be essential for developing the Mode Matching formalism used in this thesis.

[1] As we will see in the following chapter, the sign on the left side depends on the actual field component to be expanded.

Part II.

The Mode Matching Technique

4. The Single Interface - Fundamentals of the Mode Matching Technique

4.1. Introduction of the Electromagnetic Problem

Consider the interface between two waveguides bounded by a perfectly conducting surface shown in figure 4.1. The waveguides adjacent to the interface are infinitely short and displayed in an elongated fashion for illustrative purposes only. Let us call the volume of this structure $\mathcal{V}$ and refer to its surface as $\mathcal{S}$. Furthermore, we refer to the waveguides' cross-sections on $\mathcal{S}$ through which electromagnetic waves enter or leave the structure as *waveports*[1] I and II.

A common problem in electrical engineering is to calculate the fields inside such a structure for a given modal excitation as well as to determine the structure's scattering parameters for the set of Eigenmodes relevant in the appropriate practical application.

Uniqueness dictates that if on every part of $\mathcal{S}$ either the tangential electric or magnetic fields are known, the fields inside $\mathcal{V}$ are completely determined provided $\mathcal{V}$ is source-free[2] [48].

The tangential electric field on the perfectly conducting part of $\mathcal{S}$ must be zero and thus is known. In order to determine the fields inside $\mathcal{V}$ it is thus required to determine either the total electric or magnetic tangential field on waveport I and II only.

However, it is not sufficient to choose a field distribution for waveport I and II which satisfies the boundary conditions of the waveports, i.e. the waveguides boundary conditions, because the fields inside the structure also must maintain continuity at the interface between the two waveguides.

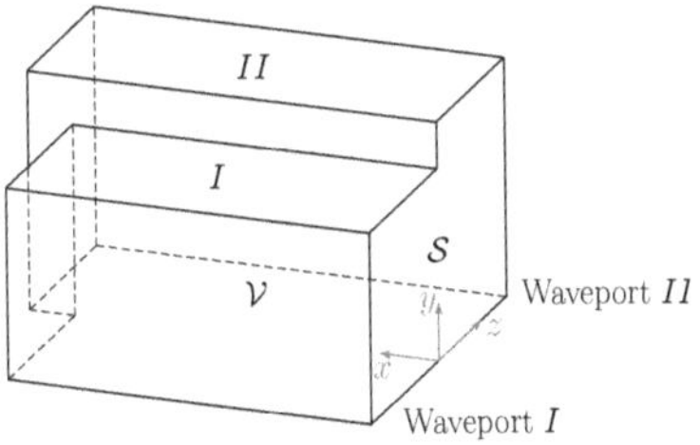

Figure 4.1.: Interface between two waveguides bounded by a perfectly conducting surface. The structure's volume shall be denoted as $\mathcal{V}$ and its outer surface as $\mathcal{S}$. Electromagnetic waves enter or leave the structure through waveports I and II.

[1] In this thesis, the term *waveport* is used in the context of interfaces and waveguides. The term *port* is used in a scattering parameter sense, that is, it refers to the terminals of a superordinate structure.

[2] If $\mathcal{V}$ is not source-free, the sources need to be known as well.

4.2. Concept of Mode Matching

In order to overcome the issue discussed in the previous section, a method has to be found which provides the complete field distributions on the waveports (and thus the total fields inside $\mathcal{V}$) from partly specified field distributions on the waveports (i.e. the Eigenmodes travelling towards the interface exciting the structure) so that continuity at the interface is maintained. This is the central concept of the Mode Matching technique.

By performing orthogonal expansion of the continuity conditions of the interface's tangential field components in terms of Eigenmodes, a system of equations can be set up which can be used to calculate the total field distributions on the waveports if the Eigenmodes travelling towards the interface are known. The coefficients of this system of equations are only dependent on the structure's geometry and the solution frequency. An obvious prerequisite for applying the Mode Matching technique is that the Eigenmodes of the waveguides must be known.

4.3. Deriving the Mode Matching Formalism

In the following, we will exemplarily derive the Mode Matching formalism for the structure shown in figure 4.2. We will use the $\{TE, TM\}^x$ mode-set in rectangular coordinates. This is an arbitrary selection, which will show favourable later as it reduces the number of tangential field components to be considered during Mode Matching. Similar derivations can be carried out for alternate mode-sets and different coordinate systems.

4.3.1. Continuity of Tangential Fields

At the interface at $z = 0$ tangential field components, namely those in the direction of x and y, must be continuous. Using a notation where A, B denote the Eigenmodes' amplitudes and $\mathcal{E}$, $\mathcal{H}$ denote the Eigenmodes' unsigned transverse field distributions, the four continuity conditions may be denoted using sums over the individual contributions of an infinite number of Eigenmodes as shown on the following page.

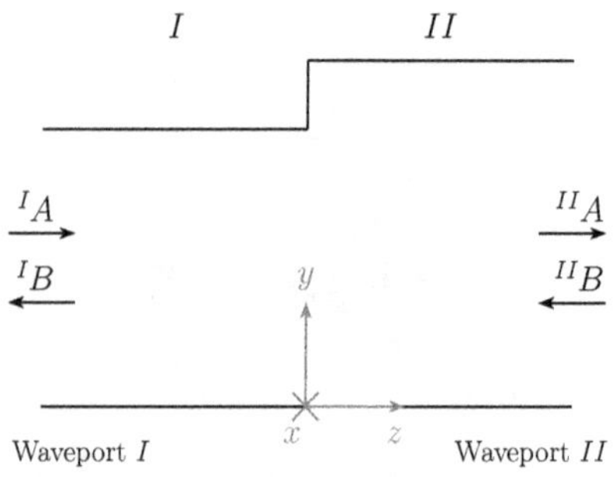

Figure 4.2.: Definition of the Eigenmodes' amplitudes at the interface

At $z = 0$, for E_x the continuity condition is

$$\sum_{v=0}^{\infty}\sum_{w=1}^{\infty}\left[{}^{I}A_{v,w}^{TM} + {}^{I}B_{v,w}^{TM}\right] \cdot {}^{I}\mathcal{E}_{x|v,w}^{TM} = \sum_{p=0}^{\infty}\sum_{q=1}^{\infty}\left[{}^{II}A_{p,q}^{TM} + {}^{II}B_{p,q}^{TM}\right] \cdot {}^{II}\mathcal{E}_{x|p,q}^{TM} \tag{4.1}$$

and similarly for E_y we find

$$\begin{aligned} &\sum_{m=1}^{\infty}\sum_{n=0}^{\infty}\left[{}^{I}A_{m,n}^{TE} - {}^{I}B_{m,n}^{TE}\right] \cdot {}^{I}\mathcal{E}_{y|m,n}^{TE} \\ + &\sum_{v=0}^{\infty}\sum_{w=1}^{\infty}\left[{}^{I}A_{v,w}^{TM} + {}^{I}B_{v,w}^{TM}\right] \cdot {}^{I}\mathcal{E}_{y|v,w}^{TM} = \\ &\sum_{s=1}^{\infty}\sum_{t=0}^{\infty}\left[{}^{II}A_{s,t}^{TE} - {}^{II}B_{s,t}^{TE}\right] \cdot {}^{II}\mathcal{E}_{y|s,t}^{TE} \\ + &\sum_{p=0}^{\infty}\sum_{q=1}^{\infty}\left[{}^{II}A_{p,q}^{TM} + {}^{II}B_{p,q}^{TM}\right] \cdot {}^{II}\mathcal{E}_{y|p,q}^{TM}. \end{aligned} \tag{4.2}$$

Again at $z = 0$, for H_x we find

$$\sum_{m=1}^{\infty}\sum_{n=0}^{\infty}\left[{}^{I}A_{m,n}^{TE} + {}^{I}B_{m,n}^{TE}\right] \cdot {}^{I}\mathcal{H}_{x|m,n}^{TE} = \sum_{s=1}^{\infty}\sum_{t=0}^{\infty}\left[{}^{II}A_{s,t}^{TE} + {}^{II}B_{s,t}^{TE}\right] \cdot {}^{II}\mathcal{H}_{x|s,t}^{TE} \tag{4.3}$$

and for H_y the continuity condition denotes

$$\begin{aligned} &\sum_{m=1}^{\infty}\sum_{n=0}^{\infty}\left[{}^{I}A_{m,n}^{TE} + {}^{I}B_{m,n}^{TE}\right] \cdot {}^{I}\mathcal{H}_{y|m,n}^{TE} \\ + &\sum_{v=0}^{\infty}\sum_{w=1}^{\infty}\left[{}^{I}A_{v,w}^{TM} - {}^{I}B_{v,w}^{TM}\right] \cdot {}^{I}\mathcal{H}_{y|v,w}^{TM} = \\ &\sum_{s=1}^{\infty}\sum_{t=0}^{\infty}\left[{}^{II}A_{s,t}^{TE} + {}^{II}B_{s,t}^{TE}\right] \cdot {}^{II}\mathcal{H}_{y|s,t}^{TE} \\ + &\sum_{p=0}^{\infty}\sum_{q=1}^{\infty}\left[{}^{II}A_{p,q}^{TM} - {}^{II}B_{p,q}^{TM}\right] \cdot {}^{II}\mathcal{H}_{y|p,q}^{TM}. \end{aligned} \tag{4.4}$$

The sums on either side of these equations represent the contribution of the waveguide's TE or TM Eigenmodes to the total field distribution at the interface.

Because the sums in (4.1) to (4.4) comprise an infinite number of Eigenmodes, these equations are exact ones (without any approximations). Later, for numerical purposes, the total number of Eigenmodes must obviously be limited. However, the electromagnetic behaviour of an interface often can be calculated with sufficient accuracy when a comparable small number of Eigenmodes per waveguide is considered.

Again note that where applicable, the change in sign of the field components due to the derivation of the wave propagation term w.r.t. z is directly included into (4.1) - (4.4) while the transverse field distributions $\mathcal{E}$, $\mathcal{H}$ are unsigned.

In the following, the required contribution of each set of TE or TM Eigenmodes of both waveguides to establish continuity shall be calculated. In doing so, it is assumed that the contribution of the appropriate remaining sets of Eigenmodes is known.

For the electric field components we find

$$\sum_{p=0}^{\infty}\sum_{q=1}^{\infty}\left[{}^{II}A_{p,q}^{TM}+{}^{II}B_{p,q}^{TM}\right]\cdot{}^{II}\mathcal{E}_{x|p,q}^{TM} = \sum_{v=0}^{\infty}\sum_{w=1}^{\infty}\left[{}^{I}A_{v,w}^{TM}+{}^{I}B_{v,w}^{TM}\right]\cdot{}^{I}\mathcal{E}_{x|v,w}^{TM} \tag{4.5}$$

$$\begin{aligned}\sum_{s=1}^{\infty}\sum_{t=0}^{\infty}\left[{}^{II}A_{s,t}^{TE}-{}^{II}B_{s,t}^{TE}\right]\cdot{}^{II}\mathcal{E}_{y|s,t}^{TE} = & \sum_{m=1}^{\infty}\sum_{n=0}^{\infty}\left[{}^{I}A_{m,n}^{TE}-{}^{I}B_{m,n}^{TE}\right]\cdot{}^{I}\mathcal{E}_{y|m,n}^{TE}\\ & +\sum_{v=0}^{\infty}\sum_{w=1}^{\infty}\left[{}^{I}A_{v,w}^{TM}+{}^{I}B_{v,w}^{TM}\right]\cdot{}^{I}\mathcal{E}_{y|v,w}^{TM}\\ & -\sum_{p=0}^{\infty}\sum_{q=1}^{\infty}\left[{}^{II}A_{p,q}^{TM}+{}^{II}B_{p,q}^{TM}\right]\cdot{}^{II}\mathcal{E}_{y|p,q}^{TM}\end{aligned} \tag{4.6}$$

and similarly for the magnetic field components

$$\sum_{m=1}^{\infty}\sum_{n=0}^{\infty}\left[{}^{I}A_{m,n}^{TE}+{}^{I}B_{m,n}^{TE}\right]\cdot{}^{I}\mathcal{H}_{x|m,n}^{TE} = \sum_{s=1}^{\infty}\sum_{t=0}^{\infty}\left[{}^{II}A_{s,t}^{TE}+{}^{II}B_{s,t}^{TE}\right]\cdot{}^{II}\mathcal{H}_{x|s,t}^{TE} \tag{4.7}$$

$$\begin{aligned}\sum_{v=0}^{\infty}\sum_{w=1}^{\infty}\left[{}^{I}A_{v,w}^{TM}-{}^{I}B_{v,w}^{TM}\right]\cdot{}^{I}\mathcal{H}_{y|v,w}^{TM} = & -\sum_{m=1}^{\infty}\sum_{n=0}^{\infty}\left[{}^{I}A_{m,n}^{TE}+{}^{I}B_{m,n}^{TE}\right]\cdot{}^{I}\mathcal{H}_{y|m,n}^{TE}\\ & +\sum_{s=1}^{\infty}\sum_{t=0}^{\infty}\left[{}^{II}A_{s,t}^{TE}+{}^{II}B_{s,t}^{TE}\right]\cdot{}^{II}\mathcal{H}_{y|s,t}^{TE}\\ & +\sum_{p=0}^{\infty}\sum_{q=1}^{\infty}\left[{}^{II}A_{p,q}^{TM}-{}^{II}B_{p,q}^{TM}\right]\cdot{}^{II}\mathcal{H}_{y|p,q}^{TM}.\end{aligned} \tag{4.8}$$

For the time being, the way (4.1) to (4.4) were rearranged appears to be a random choice. However, this choice was in fact made deliberately in order to enable the calculation of overlap integrals as will be discussed in section 4.3.2.

4.3.2. Expanding the Continuity Conditions

Eq. (4.5) to (4.8) provide the total contribution of each set of TE or TM Eigenmodes of both waveguides required to establish continuity. In order to determine the required contribution of an individual Eigenmode, modal expansion of (4.5) to (4.8) has to be performed.[3]

Let us rewrite (4.5) to (4.8) as

$$\begin{aligned}
\sum_{p=0}^{\infty}\sum_{q=1}^{\infty}\left[{}^{II}A_{p,q}^{TM} + {}^{II}B_{p,q}^{TM}\right]\cdot {}^{II}\mathcal{E}_{x|p,q}^{TM} &= \mathfrak{f}_{\mathcal{E}_x} \\
\sum_{s=1}^{\infty}\sum_{t=0}^{\infty}\left[{}^{II}A_{s,t}^{TE} - {}^{II}B_{s,t}^{TE}\right]\cdot {}^{II}\mathcal{E}_{y|s,t}^{TE} &= \mathfrak{f}_{\mathcal{E}_y} \\
\sum_{m=1}^{\infty}\sum_{n=0}^{\infty}\left[{}^{I}A_{m,n}^{TE} + {}^{I}B_{m,n}^{TE}\right]\cdot {}^{I}\mathcal{H}_{x|m,n}^{TE} &= \mathfrak{f}_{\mathcal{H}_x} \\
\sum_{v=0}^{\infty}\sum_{w=1}^{\infty}\left[{}^{I}A_{v,w}^{TM} - {}^{I}B_{v,w}^{TM}\right]\cdot {}^{I}\mathcal{H}_{y|v,w}^{TM} &= \mathfrak{f}_{\mathcal{H}_y}
\end{aligned} \qquad (4.9)$$

where the contributions $\mathfrak{f}_{\mathcal{E}_x}$ to $\mathfrak{f}_{\mathcal{H}_y}$ of the appropriate remaining sets of Eigenmodes, that is, the right sides of (4.5) to (4.8), are assumed to be known.[4]

The sums on the left side of these equations represent generalised Fourier series as introduced in section 3.2 (see (3.10) and (3.11)), for which the waveguide's Eigenmodes serve as orthogonal basis functions.

[3] In chapter 4-9 of Harrington's book [48] merely Modal Expansion rather than actual Mode Matching is discussed.
The expansion of the electric field discussed in [48] yields the amplitudes of the forward travelling Eigenmodes on the larger waveguide, which provide the longitudinal electric field components required to maintain the PEC boundary condition at the interface's surface. An alternative point of view is that these Eigenmodes provide the additional magnetic field components corresponding to the currents on the interface's surface. These parasitic field components dominate the reactive behaviour of the discontinuity.
However, the amplitudes of backward travelling Eigenmodes are not obtained by the discussed procedure. This is in contrast to the Mode Matching technique as can also be seen in section 7.1.2.

[4] Obviously, the contribution of the remaining sets is not yet known. However, this analysis will lead to a system of equations which allows to solve for the remaining Eigenmodes' amplitudes if the problem's excitation is properly predefined.

4.3.3. The Edge Condition for Truncating the Series

The sums in (4.9) are of infinite nature, that is, an infinite number of Eigenmodes is used to expand $\mathfrak{f}_{\mathcal{E}_x}$ to $\mathfrak{f}_{\mathcal{H}_y}$.[5] Obviously, for numerical purposes, the number of Eigenmodes used must be limited. These truncated sums are called best approximations as they indeed represent the best approximation of $\mathfrak{f}$ for a limited number of Eigenmodes [62].

Properly truncating the series (4.9) is a difficult task because two or more sums need to be truncated in an appropriate manner. As a matter of fact, for a number of unfortunate structures, it can be shown that the Mode Matching technique converges against entirely different solutions depending on how the individual series were truncated. This phenomenon is called *relative convergence* [34–39]. A classic example is the bifurcation of a parallel plate waveguide containing an infinitely thin septum [34, 39].

The underlying reason for this is that the Mode Matching formulation does not include the *edge condition* [73] (also see e.g. [29, 38, 39]), which states that field components perpendicular to the edge exhibit a singularity at the edge while the energy density in its vicinity must remain integrable[6], i.e. finite [73]. More precisely, if $\rho \ll \lambda$ denotes the distance to the edge as is illustrated in figure 4.3, the field strengths of the perpendicular field components may maximally increase with $\rho^{(\pi/\varphi - 1)}$ where $\pi \leq \varphi \leq 2\pi$ denotes the edge angle [59].

Mittra [34,39] has shown that the edge condition is maintained and thus proper convergence of the Mode Matching solution is guaranteed if the ratio of the number of Eigenmodes on the two waveguides is chosen similar to the ratio of the waveguides' dimensions.

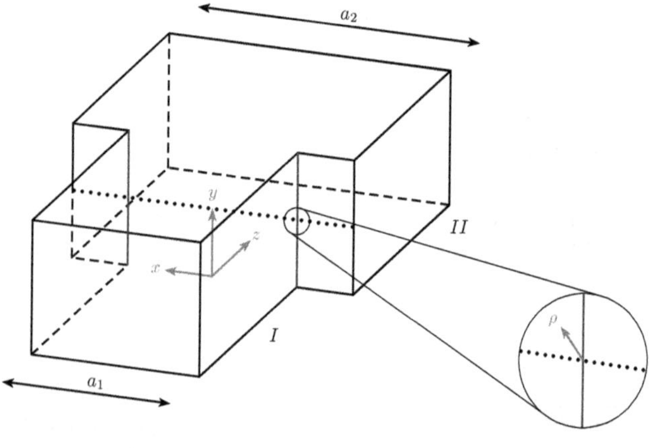

Figure 4.3.: H-plane step, $a_1/a_2 = 2/3$. At the blue edges, perpendicular fields are singular. The edge angle is $\varphi = 3\pi/2$ and thus the fields rise with $\rho^{-1/3}$. Also see fn. 7.

[5] The lower limits of the sums depend on the selection of the scalar vector potentials, c.f. e.g. [48, pp. 130, 153]

[6] This becomes obvious when considering the fact that the energy is calculated by integrating the square of the electric or magnetic field components with a differential volume element $dV = \rho\ d\rho\ d\phi$, i.e. $\int_{\mathcal{V}} \left(\rho^{((\pi/\varphi)-1)}\right)^2 \rho\ d\rho\ d\phi\ dz = \int_{\mathcal{V}} \rho^{((2\pi)/\varphi)-1}\ d\rho\ d\phi\ dz$. The integrand remains finite for $\rho \to 0$ provided $\varphi \leq 2\pi$. If the integrand remains finite, the energy integral remains finite as well.

Consider the H-plane step of a rectangular waveguide shown in figure 4.3.[7] If this structure is excited using the fundamental $TE_{1,0}$ Eigenmode, it may be treated taking only ${}^{I}TE_{m,0}$ and ${}^{II}TE_{s,0}$ Eigenmodes into account (also see chapter 7). Thus, the continuity condition at the interface can be denoted using two infinite series, which need to be truncated for the purpose of practical implementation. For this interface, according to [34, 39], the series have to be truncated according to

$$\frac{M}{S} \approx \frac{a_1}{a_2} \tag{4.10}$$

where M denotes the number of $TE_{m,0}$ Eigenmodes on waveguide I and S denotes the number of $TE_{s,0}$ Eigenmodes on waveguide II.

The fact that the number of Eigenmodes to be included in the analysis is dependent on the geometry of the interface is an intriguing outcome. However, this result becomes much more obvious when inspecting the numbers of maxima of the individual field distributions.

Consider for example the magnetic field component H_x, which denotes

$${}^{I}H_x \propto \sin\left(k_x\left(x + \tfrac{a_1}{2}\right)\right)\cos\left(k_y y\right) \qquad {}^{II}H_x \propto \sin\left(k_x\left(x + \tfrac{a_2}{2}\right)\right)\cos\left(k_y y\right).$$

Over the cross-section of waveguide I, the field distribution may maximally exhibit M maxima when depicted by the waveguide's first M Eigenmodes. In contrast, due to the larger geometry of waveguide II, the basis functions of waveguide II may only provide ${}^{a_1}/{}_{a_2}S$ maxima over the width of the interface's cross-section because the Eigenmodes' maxima are equally distributed [74][8] over the waveguide's width. It is undesirable that the Eigenmodes of waveguide I form a higher-order field distribution over the cross-section than the Eigenmodes of waveguide II may depict. We thus choose $M \approx {}^{a_1}/{}_{a_2}S$.

Figures 4.4 and 4.5 show the field distribution of H_x along the dotted line in figure 4.3. Blue lines indicate the edges at which the field distribution would become singular. The width ratio of the two waveguides is ${}^{a_1}/{}_{a_2} = {}^{2}/{}_{3}$. As can be seen from figure 4.4, if the mode ratio is chosen in accordance with (4.10), the field distributions over the cross-section are of similar order. In contrast, if the mode ratio is chosen ${}^{M}/{}_{S} = 1$, the cross-sectional field distribution on waveguide I is of higher order compared to the field distribution on waveguide II as depicted in figure 4.5. Note that in both figures only odd-order TE Eigenmodes are taken into account.

From figures 4.4 and 4.5 it is also interesting to note that the Mode Matching technique "approximates the solution of Maxwell's equations in the least square sense" [29] with respect to the inner products' integrals [29].

As the edge condition was not initially known to the author, some results presented in the scope of this thesis were obtained using codes which truncate all series equivalently. These codes provided good results for each and every structure discussed in this text. Other authors [38, 40, 41] have come to similar conclusions.

For example, an interesting investigation was carried out by Shih [40] regarding the susceptance of a H-plane step with the narrower waveguide being under cut-off. The authors compare the results obtained from a Mode Matching solver using various mode ratios to results from Marcuvitz' classic Waveguide Handbook [17][9] and show that all variations converge against the desired reference value if the number of modes is slightly increased.

[7] Non-zero length waveguides are depicted for illustrative purposes only.

[8] Piefke [74] considers zeros rather than maxima.

[9] Shih cites a different, earlier edition of the Waveguide Handbook:
N. Marcuvitz, Waveguide Handbook, M.I.T. Rad. Lab. Series, vol. 10. McGraw-Hill, 1951.

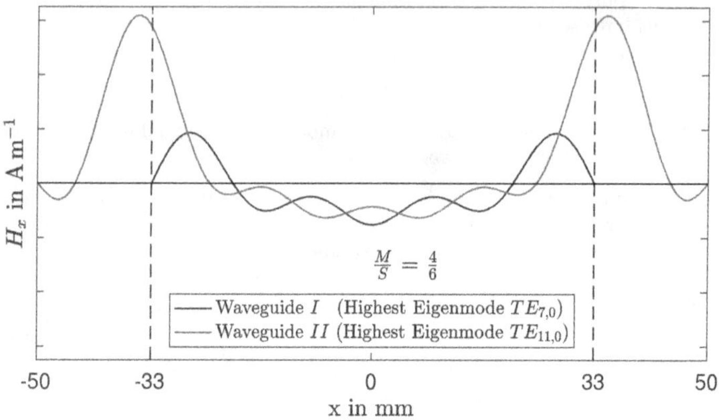

Figure 4.4.: Field component H_x along the dotted line in figure 4.3 for $M/S = 4/6$

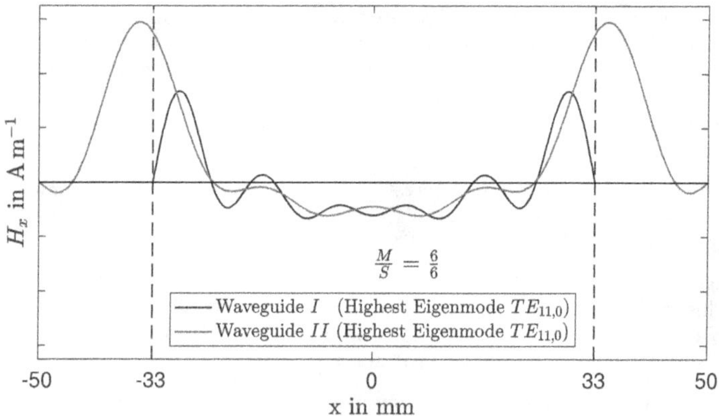

Figure 4.5.: Field component H_x along the dotted line in figure 4.3 for $M/S = 6/6$

4.3.4. Calculating the Expansion's Coefficients

We now truncate the infinite series (4.9) according to the edge condition introduced in the previous section and rewrite (4.9) as

$$\begin{aligned}
\sum_{p=0}^{P-1}\sum_{q=1}^{Q}\left[{}^{II}A^{TM}_{p,q}+{}^{II}B^{TM}_{p,q}\right]\cdot{}^{II}\mathcal{E}^{TM}_{x|p,q} &= \mathfrak{f}_{\mathcal{E}_x} && E_x : I\rightarrow II\\
\sum_{s=1}^{S}\sum_{t=0}^{T-1}\left[{}^{II}A^{TE}_{s,t}-{}^{II}B^{TE}_{s,t}\right]\cdot{}^{II}\mathcal{E}^{TE}_{y|s,t} &= \mathfrak{f}_{\mathcal{E}_y} && E_y : I\rightarrow II\\
\sum_{m=1}^{M}\sum_{n=0}^{N-1}\left[{}^{I}A^{TE}_{m,n}+{}^{I}B^{TE}_{m,n}\right]\cdot{}^{I}\mathcal{H}^{TE}_{x|m,n} &= \mathfrak{f}_{\mathcal{H}_x} && H_x : II\rightarrow I\\
\sum_{v=0}^{V-1}\sum_{w=1}^{W}\left[{}^{I}A^{TM}_{v,w}-{}^{I}B^{TM}_{v,w}\right]\cdot{}^{I}\mathcal{H}^{TM}_{y|v,w} &= \mathfrak{f}_{\mathcal{H}_y} && H_y : II\rightarrow I.
\end{aligned}\tag{4.11}$$

The total number of Eigenmodes per waveport (i.e. both mode types (TM or TE) and both directions of propagation) is denoted as

$${}^{I}\mathcal{M}=2\,(MN+VW)\qquad {}^{II}\mathcal{M}=2\,(PQ+ST)\,.\tag{4.12}$$

The Eigenmode's individual contributions, i.e. their amplitudes now can be calculated using (3.10) as[10]

$$\begin{aligned}
{}^{II}A^{TM}_{p,q}+{}^{II}B^{TM}_{p,q} &= \frac{\left\langle \mathfrak{f}_{\mathcal{E}_x}\middle|{}^{II}\mathcal{E}^{TM}_{x|p,q}\right\rangle_{II}}{\left\langle {}^{II}\mathcal{E}^{TM}_{x|p,q}\middle|{}^{II}\mathcal{E}^{TM}_{x|p,q}\right\rangle_{II}} && \text{for all } p,\ q\\
{}^{II}A^{TE}_{s,t}-{}^{II}B^{TE}_{s,t} &= \frac{\left\langle \mathfrak{f}_{\mathcal{E}_y}\middle|{}^{II}\mathcal{E}^{TE}_{y|s,t}\right\rangle_{II}}{\left\langle {}^{II}\mathcal{E}^{TE}_{y|s,t}\middle|{}^{II}\mathcal{E}^{TE}_{y|s,t}\right\rangle_{II}} && \text{for all } s,\ t\\
{}^{I}A^{TE}_{m,n}+{}^{I}B^{TE}_{m,n} &= \frac{\left\langle \mathfrak{f}_{\mathcal{H}_x}\middle|{}^{I}\mathcal{H}^{TE}_{x|m,n}\right\rangle_{I}}{\left\langle {}^{I}\mathcal{H}^{TE}_{x|m,n}\middle|{}^{I}\mathcal{H}^{TE}_{x|m,n}\right\rangle_{I}} && \text{for all } m,\ n\\
{}^{I}A^{TM}_{v,w}-{}^{I}B^{TM}_{v,w} &= \frac{\left\langle \mathfrak{f}_{\mathcal{H}_y}\middle|{}^{I}\mathcal{H}^{TM}_{y|v,w}\right\rangle_{I}}{\left\langle {}^{I}\mathcal{H}^{TM}_{y|v,w}\middle|{}^{I}\mathcal{H}^{TM}_{y|v,w}\right\rangle_{I}} && \text{for all } v,\ w.
\end{aligned}\tag{4.13}$$

The integrals required to calculate the inner products in (4.13) are overlap integrals (see section 3.2.2) of the field distribution to be expanded and the appropriate basis functions.

[10] The inner products' subscripts indicate the waveguide for whose dimensions the inner product has to be calculated.

Due to the way (4.1) to (4.4) were rearranged, the electric field of I is expanded into a series of Eigenmodes of II. Thus the overlap integrals in the first and second equation of (4.13) have to be calculated over the domain of definition of the Eigenmodes of II, i.e.

$$\{x, y \mid 0 \leq x \leq a_2, 0 \leq y \leq b_2\}. \tag{4.14}$$

For the cross-section of I, the electric field is assumed to be known. For the dashed surface in figure 4.6, the electric field must be zero since $E_{tan} = 0$. Thus, the total field distribution to be expanded is known.

Similarly, the magnetic field of II is expanded into Eigenmodes of I. The corresponding overlap integrals in the third and fourth equation of (4.13) have to be calculated over the domain of definition of the Eigenmodes of I, that is,

$$\{x, y \mid 0 \leq x \leq a_1, 0 \leq y \leq b_1\}. \tag{4.15}$$

Since the magnetic field of II is assumed to be known, the total field distribution to be expanded is known.

At this point, it becomes obvious why (4.1) to (4.4) were rearranged in the presented fashion: If (4.1) to (4.4) were rearranged so that the magnetic field of I would be expanded for the Eigenmodes of II, the overlap integrals could not be calculated over $\{x, y \mid 0 \leq x \leq a_2, 0 \leq y \leq b_2\}$ as the magnetic field distribution for the dashed surface is unknown because H_{tan} does not need to vanish at a PEC surface.[11]

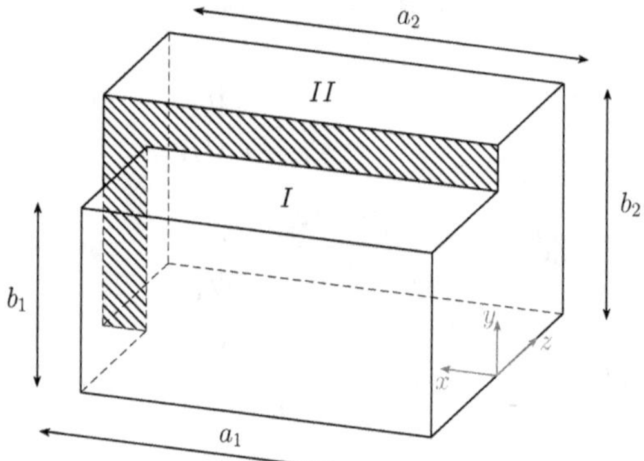

Figure 4.6.: On the direction of the field expansions at an interface between two waveguides bounded by a perfectly conducting surface. For the dashed PEC surface, E_{tan} must vanish. In contrast, H_{tan} has a non-zero value which cannot directly be determined.

[11]In [75, p. 99], Papas discusses a possible approximation for the magnetic field at an outer conductor step in a coaxial waveguide. While Papas considers letting $H_{tan} = \text{const}$ at the PEC surface a possible option, at the same time, he stresses that this is a somewhat arbitrary choice. This is especially true because the tangential magnetic field would need to be proportional to the surface current, which however cannot be constant due to current conservation over the cross-section of the PEC surface.

If we exploit the linearity of the inner product [70] $\langle \mathfrak{f}_1 + \mathfrak{f}_2 | \mathfrak{g} \rangle = \langle \mathfrak{f}_1 | \mathfrak{g} \rangle + \langle \mathfrak{f}_2 | \mathfrak{g} \rangle$ to individually calculate the inner product of the sums on the right side of (4.5) to (4.8) and introduce a new notation

$$\vartheta_{\mathfrak{g}}^{\mathfrak{f}} = \frac{\langle \mathfrak{f} | \mathfrak{g} \rangle}{\langle \mathfrak{g} | \mathfrak{g} \rangle}, \tag{4.16}$$

that is,

$$\begin{aligned}
\vartheta_{{}^{II}\mathcal{E}^{TM}_{x|p,q}}^{{}^{I}\mathcal{E}^{TM}_{x|v,w}} &= \frac{\left\langle {}^{I}\mathcal{E}^{TM}_{x|v,w} \middle| {}^{II}\mathcal{E}^{TM}_{x|p,q} \right\rangle_{II}}{\left\langle {}^{II}\mathcal{E}^{TM}_{x|p,q} \middle| {}^{II}\mathcal{E}^{TM}_{x|p,q} \right\rangle_{II}} && \text{for all } p,\ q \text{ and } v,\ w, \\
\vartheta_{{}^{II}\mathcal{E}^{TE}_{y|s,t}}^{{}^{I}\mathcal{E}^{TE}_{y|m,n}} &= \frac{\left\langle {}^{I}\mathcal{E}^{TE}_{y|m,n} \middle| {}^{II}\mathcal{E}^{TE}_{y|s,t} \right\rangle_{II}}{\left\langle {}^{II}\mathcal{E}^{TE}_{y|s,t} \middle| {}^{II}\mathcal{E}^{TE}_{y|s,t} \right\rangle_{II}} && \text{for all } s,\ t \text{ and } m,\ n, \\
\vartheta_{{}^{II}\mathcal{E}^{TE}_{y|s,t}}^{{}^{I}\mathcal{E}^{TM}_{y|v,w}} &= \frac{\left\langle {}^{I}\mathcal{E}^{TM}_{y|v,w} \middle| {}^{II}\mathcal{E}^{TE}_{y|s,t} \right\rangle_{II}}{\left\langle {}^{II}\mathcal{E}^{TE}_{y|s,t} \middle| {}^{II}\mathcal{E}^{TE}_{y|s,t} \right\rangle_{II}} && \text{for all } s,\ t \text{ and } v,\ w, \\
\vartheta_{{}^{I}\mathcal{H}^{TE}_{x|m,n}}^{{}^{II}\mathcal{H}^{TE}_{x|s,t}} &= \frac{\left\langle {}^{II}\mathcal{H}^{TE}_{x|s,t} \middle| {}^{I}\mathcal{H}^{TE}_{x|m,n} \right\rangle_{I}}{\left\langle {}^{I}\mathcal{H}^{TE}_{x|m,n} \middle| {}^{I}\mathcal{H}^{TE}_{x|m,n} \right\rangle_{I}} && \text{for all } m,\ n \text{ and } s,\ t, \\
\vartheta_{{}^{I}\mathcal{H}^{TM}_{y|v,w}}^{{}^{II}\mathcal{H}^{TE}_{y|s,t}} &= \frac{\left\langle {}^{II}\mathcal{H}^{TE}_{y|s,t} \middle| {}^{I}\mathcal{H}^{TM}_{y|v,w} \right\rangle_{I}}{\left\langle {}^{I}\mathcal{H}^{TM}_{y|v,w} \middle| {}^{I}\mathcal{H}^{TM}_{y|v,w} \right\rangle_{I}} && \text{for all } v,\ w \text{ and } s,\ t, \\
\vartheta_{{}^{I}\mathcal{H}^{TM}_{y|v,w}}^{{}^{II}\mathcal{H}^{TM}_{y|p,q}} &= \frac{\left\langle {}^{II}\mathcal{H}^{TE}_{y|p,q} \middle| {}^{I}\mathcal{H}^{TM}_{y|v,w} \right\rangle_{I}}{\left\langle {}^{I}\mathcal{H}^{TM}_{y|v,w} \middle| {}^{I}\mathcal{H}^{TM}_{y|v,w} \right\rangle_{I}} && \text{for all } v,\ w \text{ and } p,\ q
\end{aligned} \tag{4.17}$$

and

$$\begin{aligned}
\vartheta_{{}^{II}\mathcal{E}^{TE}_{y|s,t}}^{{}^{II}\mathcal{E}^{TM}_{y|p,q}} &= \frac{\left\langle {}^{II}\mathcal{E}^{TM}_{y|p,q} \middle| {}^{II}\mathcal{E}^{TE}_{y|s,t} \right\rangle_{II}}{\left\langle {}^{II}\mathcal{E}^{TE}_{y|s,t} \middle| {}^{II}\mathcal{E}^{TE}_{y|s,t} \right\rangle_{II}} && \text{for all } s,\ t \text{ and } p,\ q, \\
\vartheta_{{}^{I}\mathcal{H}^{TM}_{y|v,w}}^{{}^{I}\mathcal{H}^{TE}_{y|m,n}} &= \frac{\left\langle {}^{I}\mathcal{H}^{TE}_{y|m,n} \middle| {}^{I}\mathcal{H}^{TM}_{y|v,w} \right\rangle_{I}}{\left\langle {}^{I}\mathcal{H}^{TM}_{y|v,w} \middle| {}^{I}\mathcal{H}^{TM}_{y|v,w} \right\rangle_{I}} && \text{for all } v,\ w \text{ and } m,\ n,
\end{aligned} \tag{4.18}$$

we can rewrite the expressions obtained from E_x

$$
\begin{aligned}
{}^{II}A^{TM}_{p,q} + {}^{II}B^{TM}_{p,q} \quad = & \\
& \sum_{v=0}^{V-1} \sum_{w=1}^{W} \vartheta \frac{{}^{I}\mathcal{E}^{TM}_{x|v,w}}{{}^{II}\mathcal{E}^{TM}_{x|p,q}} \left[{}^{I}A^{TM}_{v,w} + {}^{I}B^{TM}_{v,w} \right]
\end{aligned}
\tag{4.19}
$$

for all p, q and from E_y

$$
\begin{aligned}
{}^{II}A^{TE}_{s,t} - {}^{II}B^{TE}_{s,t} \quad = & \\
& \sum_{m=1}^{M} \sum_{n=0}^{N-1} \vartheta \frac{{}^{I}\mathcal{E}^{TE}_{y|m,n}}{{}^{II}\mathcal{E}^{TE}_{y|s,t}} \cdot \left[{}^{I}A^{TE}_{m,n} - {}^{I}B^{TE}_{m,n} \right] \\
+ & \sum_{v=0}^{V-1} \sum_{w=1}^{W} \vartheta \frac{{}^{I}\mathcal{E}^{TM}_{y|v,w}}{{}^{II}\mathcal{E}^{TE}_{y|s,t}} \cdot \left[{}^{I}A^{TM}_{v,w} + {}^{I}B^{TM}_{v,w} \right] \\
- & \sum_{p=0}^{P-1} \sum_{q=1}^{Q} \vartheta \frac{{}^{II}\mathcal{E}^{TM}_{y|p,q}}{{}^{II}\mathcal{E}^{TE}_{y|s,t}} \cdot \left[{}^{II}A^{TM}_{p,q} + {}^{II}B^{TM}_{p,q} \right]
\end{aligned}
\tag{4.20}
$$

for all s, t as well as from H_x

$$
\begin{aligned}
{}^{I}A^{TE}_{m,n} + {}^{I}B^{TE}_{m,n} \quad = & \\
& \sum_{s=1}^{S} \sum_{t=0}^{T-1} \vartheta \frac{{}^{II}\mathcal{H}^{TE}_{x|s,t}}{{}^{I}\mathcal{H}^{TE}_{x|m,n}} \cdot \left[{}^{II}A^{TE}_{s,t} + {}^{II}B^{TE}_{s,t} \right]
\end{aligned}
\tag{4.21}
$$

for all m, n and similarly from H_y

$$
\begin{aligned}
{}^{I}A^{TM}_{v,w} - {}^{I}B^{TM}_{v,w} \quad = & \\
- & \sum_{m=1}^{M} \sum_{n=0}^{N-1} \vartheta \frac{{}^{I}\mathcal{H}^{TE}_{y|m,n}}{{}^{I}\mathcal{H}^{TM}_{y|v,w}} \cdot \left[{}^{I}A^{TE}_{m,n} + {}^{I}B^{TE}_{m,n} \right] \\
+ & \sum_{s=1}^{S} \sum_{t=0}^{T-1} \vartheta \frac{{}^{II}\mathcal{H}^{TE}_{y|s,t}}{{}^{I}\mathcal{H}^{TM}_{y|v,w}} \cdot \left[{}^{II}A^{TE}_{s,t} + {}^{II}B^{TE}_{s,t} \right] \\
+ & \sum_{p=0}^{P-1} \sum_{q=1}^{Q} \vartheta \frac{{}^{II}\mathcal{H}^{TM}_{y|p,q}}{{}^{I}\mathcal{H}^{TM}_{y|v,w}} \cdot \left[{}^{II}A^{TM}_{p,q} - {}^{II}B^{TM}_{p,q} \right]
\end{aligned}
\tag{4.22}
$$

for all v, w.

4.3.5. Assembling the Interface Matrix

To obtain the interface matrix $\boldsymbol{\vartheta}$, we rewrite (4.19) - (4.22) without using sums or differences of amplitudes. Applying vector matrix notation, we obtain

$$\vec{0} = \underbrace{\begin{bmatrix} 1 & 0 & 1 & 0 & -\boldsymbol{\vartheta}^{{}^{II}\mathcal{H}_x^{TE}}_{{}^{I}\mathcal{H}_x^{TE}} & 0 & -\boldsymbol{\vartheta}^{{}^{II}\mathcal{H}_x^{TE}}_{{}^{I}\mathcal{H}_x^{TE}} & 0 \\ \boldsymbol{\vartheta}^{{}^{I}\mathcal{E}_y^{TE}}_{{}^{II}\mathcal{E}_y^{TE}} & \boldsymbol{\vartheta}^{{}^{I}\mathcal{E}_y^{TM}}_{{}^{II}\mathcal{E}_y^{TE}} & -\boldsymbol{\vartheta}^{{}^{I}\mathcal{E}_y^{TE}}_{{}^{II}\mathcal{E}_y^{TE}} & \boldsymbol{\vartheta}^{{}^{I}\mathcal{E}_y^{TM}}_{{}^{II}\mathcal{E}_y^{TE}} & -1 & -\boldsymbol{\vartheta}^{{}^{II}\mathcal{E}_y^{TM}}_{{}^{II}\mathcal{E}_y^{TE}} & 1 & -\boldsymbol{\vartheta}^{{}^{II}\mathcal{E}_y^{TM}}_{{}^{II}\mathcal{E}_y^{TE}} \\ \boldsymbol{\vartheta}^{{}^{I}\mathcal{H}_y^{TE}}_{{}^{I}\mathcal{H}_y^{TM}} & 1 & \boldsymbol{\vartheta}^{{}^{I}\mathcal{H}_y^{TE}}_{{}^{I}\mathcal{H}_y^{TM}} & -1 & -\boldsymbol{\vartheta}^{{}^{II}\mathcal{H}_y^{TE}}_{{}^{I}\mathcal{H}_y^{TM}} & -\boldsymbol{\vartheta}^{{}^{II}\mathcal{H}_y^{TM}}_{{}^{I}\mathcal{H}_y^{TM}} & -\boldsymbol{\vartheta}^{{}^{II}\mathcal{H}_y^{TE}}_{{}^{I}\mathcal{H}_y^{TM}} & \boldsymbol{\vartheta}^{{}^{II}\mathcal{H}_x^{TM}}_{{}^{I}\mathcal{H}_y^{TM}} \\ 0 & \boldsymbol{\vartheta}^{{}^{I}\mathcal{E}_x^{TM}}_{{}^{II}\mathcal{E}_x^{TM}} & 0 & \boldsymbol{\vartheta}^{{}^{I}\mathcal{E}_x^{TM}}_{{}^{II}\mathcal{E}_x^{TM}} & 0 & -1 & 0 & -1 \end{bmatrix}}_{\boldsymbol{\vartheta}} \cdot \begin{bmatrix} {}^{I}\vec{A}^{TE} \\ {}^{I}\vec{A}^{TM} \\ {}^{I}\vec{B}^{TE} \\ {}^{I}\vec{B}^{TM} \\ {}^{II}\vec{A}^{TE} \\ {}^{II}\vec{A}^{TM} \\ {}^{II}\vec{B}^{TE} \\ {}^{II}\vec{B}^{TM} \end{bmatrix} \tag{4.23}$$

where the first set of rows results from the modal expansion of H_x, the second from E_y, the third from H_y and finally the fourth from E_x as presented earlier.

The coefficient matrix shall be defined as the *interface matrix* $\boldsymbol{\vartheta}$. If an additional amplitude vector $\vec{\alpha}$ is introduced, (4.23) can be written in short as

$$\vec{0} = \boldsymbol{\vartheta}\, \vec{\alpha}.$$

Interface matrices are of crucial importance for the Mode Matching technique because they allow to completely describe the field problem at an interface under form of a single matrix provided that the waveguides' Eigenmodes are known. $\boldsymbol{\vartheta}$ is geometry-dependent only and none of the Eigenmodes' amplitudes need to be known to calculate it.

Taking findings from section 4.2 into account, it becomes obvious that the field problem is solvable if the homogeneous system of equations $\vec{0} = \boldsymbol{\vartheta}\, \vec{\alpha}$ is solvable. This matter will be discussed in the following section.

4.4. Degrees of Freedom of the Field Solution

In the previous section, we have seen that the homogeneous system of equations (4.23) with an interface matrix $\boldsymbol{\vartheta} \in \mathbb{K}^{R \times C}$ completely describes the field problem.

Keeping in mind that ${}^{I}\mathcal{M}$, ${}^{II}\mathcal{M}$ denote the total number of amplitudes on waveport I and II, it becomes obvious that the interface matrix' number of columns is

$$C = {}^{I}\mathcal{M} + {}^{II}\mathcal{M}. \tag{4.24}$$

As (4.19) - (4.22) represent expansions in terms of sum or difference amplitudes, it is clear that the matrix' number of rows is

$$R = \frac{{}^{I}\mathcal{M} + {}^{II}\mathcal{M}}{2}. \tag{4.25}$$

Due to the orthogonality of the waveguides' Eigenmodes the matrix $\boldsymbol{\vartheta}$ has full rank and thus, the field solution has

$$\mathrm{DOF} = C - R = \frac{{}^{I}\mathcal{M} + {}^{II}\mathcal{M}}{2} \tag{4.26}$$

degrees of freedom, that is, half the number of amplitudes on each waveport must be predefined.[12]

Returning to the conceptual discussion of the Mode Matching technique presented in section 4.2, we remember that due to uniqueness, the fields inside a source-free volume are fully determined if either the tangential electric or magnetic fields are known on the entire surface of this volume. As a consequence, when considering an interface between PEC-bounded waveguides (where $E_{tan} = 0$) as shown in figure 4.1, the total tangential electric or magnetic fields on the structure's waveports must be known in order to obtain a unique field solution for $\mathcal{V}$.

In contrast, when the waveguides' Eigenmodes are known and an interface matrix $\boldsymbol{\vartheta}$ has been calculated by means of orthogonal expansion, it is sufficient to predefine half the number of amplitudes on each waveport rather than all amplitudes, as would be required to know the waveports' total fields.

For the predefinition, several choices are possible:

- In order to calculate scattering parameters, one would predefine ${}^{I}\vec{A}^{\,TE}$, ${}^{I}\vec{A}^{\,TM}$, ${}^{II}\vec{B}^{\,TE}$ and ${}^{II}\vec{B}^{\,TM}$, namely the Eigenmodes (i.e. the electric and magnetic field components) travelling towards the discontinuity.
- Alternatively, one could also completely define the electric and magnetic field distribution on one waveport by predefining the amplitudes of all Eigenmodes traveling through the waveport, for example ${}^{I}\vec{A}^{\,TE}$, ${}^{I}\vec{B}^{\,TE}$, ${}^{I}\vec{A}^{\,TM}$, ${}^{I}\vec{B}^{\,TM}$.
- Finally, there is the (theoretical) possibility to predefine all Eigenmodes travelling away from the discontinuity.

4.5. Solving the Interface Matrix

As already pointed out in the previous section, in order to solve the field problem presented in section 4.1, $4 \cdot \mathcal{M}$ amplitudes, i.e. degrees of freedom need to be predefined in order to obtain a unique solution to the field problem.

Since in most practical applications scattering parameters are to be determined, we choose to predefine ${}^{I}\vec{A}^{\,TE}$, ${}^{I}\vec{A}^{\,TM}$, ${}^{II}\vec{B}^{\,TE}$ and ${}^{II}\vec{B}^{\,TM}$, which are the amplitudes of the Eigenwaves travelling towards the discontinuity.

[12] This is a direct consequence of the rank-nullity theorem

$$\mathrm{DOF} = \dim\left(L_{\vartheta,\vec{0}}\right) = \dim\left(\ker(\mathfrak{f})\right) = C - \mathrm{rank}\left(\boldsymbol{\vartheta}\right)$$

where $L_{\vartheta,\vec{0}}$ denotes the system's solution space and $\ker(\mathfrak{f})$ represents the kernel of the mapping $\mathfrak{f} : \mathbb{K}^C \to \mathbb{K}^R,\ \vec{a} \to \boldsymbol{\vartheta}\,\vec{a}$ [76].

In order to solve the system of equations

$$
\vec{0} = \begin{bmatrix}
1 & 0 & 1 & 0 & -\boldsymbol{\vartheta}\frac{{}^{II}\mathcal{H}_x^{TE}}{{}^{I}\mathcal{H}_x^{TE}} & 0 & -\boldsymbol{\vartheta}\frac{{}^{II}\mathcal{H}_x^{TE}}{{}^{I}\mathcal{H}_x^{TE}} & 0 \\
\boldsymbol{\vartheta}\frac{{}^{I}\mathcal{E}_y^{TE}}{{}^{II}\mathcal{E}_y^{TE}} & \boldsymbol{\vartheta}\frac{{}^{I}\mathcal{E}_y^{TM}}{{}^{II}\mathcal{E}_y^{TE}} & -\boldsymbol{\vartheta}\frac{{}^{I}\mathcal{E}_y^{TE}}{{}^{II}\mathcal{E}_y^{TE}} & \boldsymbol{\vartheta}\frac{{}^{I}\mathcal{E}_y^{TM}}{{}^{II}\mathcal{E}_y^{TE}} & -1 & -\boldsymbol{\vartheta}\frac{{}^{II}\mathcal{E}_y^{TM}}{{}^{II}\mathcal{E}_y^{TE}} & 1 & -\boldsymbol{\vartheta}\frac{{}^{II}\mathcal{E}_y^{TM}}{{}^{II}\mathcal{E}_y^{TE}} \\
\boldsymbol{\vartheta}\frac{{}^{I}\mathcal{H}_y^{TE}}{{}^{I}\mathcal{H}_y^{TM}} & 1 & \boldsymbol{\vartheta}\frac{{}^{I}\mathcal{H}_y^{TE}}{{}^{I}\mathcal{H}_y^{TM}} & -1 & -\boldsymbol{\vartheta}\frac{{}^{II}\mathcal{H}_y^{TE}}{{}^{I}\mathcal{H}_y^{TM}} & -\boldsymbol{\vartheta}\frac{{}^{II}\mathcal{H}_y^{TM}}{{}^{I}\mathcal{H}_y^{TM}} & -\boldsymbol{\vartheta}\frac{{}^{II}\mathcal{H}_y^{TE}}{{}^{I}\mathcal{H}_y^{TM}} & \boldsymbol{\vartheta}\frac{{}^{II}\mathcal{H}_x^{TM}}{{}^{I}\mathcal{H}_y^{TM}} \\
0 & \boldsymbol{\vartheta}\frac{{}^{I}\mathcal{E}_x^{TM}}{{}^{II}\mathcal{E}_x^{TM}} & 0 & \boldsymbol{\vartheta}\frac{{}^{I}\mathcal{E}_x^{TM}}{{}^{II}\mathcal{E}_x^{TM}} & 0 & -1 & 0 & -1
\end{bmatrix} \cdot \begin{bmatrix} {}^{I}\vec{A}\,^{TE} \\ {}^{I}\vec{A}\,^{TM} \\ {}^{I}\vec{B}\,^{TE} \\ {}^{I}\vec{B}\,^{TM} \\ {}^{II}\vec{A}\,^{TE} \\ {}^{II}\vec{A}\,^{TM} \\ {}^{II}\vec{B}\,^{TE} \\ {}^{II}\vec{B}\,^{TM} \end{bmatrix}
$$

we split the interface matrix into two matrices $\boldsymbol{\vartheta}_{Known}$ and $\boldsymbol{\vartheta}_{Unknown}$ so that the system of equations can be rewritten as

$$
\vec{0} = \underbrace{\begin{bmatrix}
1 & 0 & -\boldsymbol{\vartheta}\frac{{}^{II}\mathcal{H}_x^{TE}}{{}^{I}\mathcal{H}_x^{TE}} & 0 \\
\boldsymbol{\vartheta}\frac{{}^{I}\mathcal{E}_y^{TE}}{{}^{II}\mathcal{E}_y^{TE}} & \boldsymbol{\vartheta}\frac{{}^{I}\mathcal{E}_y^{TM}}{{}^{II}\mathcal{E}_y^{TE}} & 1 & 0 \\
\boldsymbol{\vartheta}\frac{{}^{I}\mathcal{H}_y^{TE}}{{}^{I}\mathcal{H}_y^{TM}} & 1 & -\boldsymbol{\vartheta}\frac{{}^{II}\mathcal{H}_y^{TE}}{{}^{I}\mathcal{H}_y^{TM}} & \boldsymbol{\vartheta}\frac{{}^{II}\mathcal{H}_y^{TM}}{{}^{I}\mathcal{H}_y^{TM}} \\
0 & \boldsymbol{\vartheta}\frac{{}^{I}\mathcal{E}_x^{TM}}{{}^{II}\mathcal{E}_x^{TM}} & 0 & -1
\end{bmatrix}}_{\boldsymbol{\vartheta}_{Known}} \cdot \underbrace{\begin{bmatrix} {}^{I}\vec{A}\,^{TE} \\ {}^{I}\vec{A}\,^{TM} \\ {}^{II}\vec{B}\,^{TE} \\ {}^{II}\vec{B}\,^{TM} \end{bmatrix}}_{\vec{\alpha}_{Known}} + \underbrace{\begin{bmatrix}
1 & 0 & -\boldsymbol{\vartheta}\frac{{}^{II}\mathcal{H}_x^{TE}}{{}^{I}\mathcal{H}_x^{TE}} & 0 \\
-\boldsymbol{\vartheta}\frac{{}^{I}\mathcal{E}_y^{TE}}{{}^{II}\mathcal{E}_y^{TE}} & \boldsymbol{\vartheta}\frac{{}^{I}\mathcal{E}_y^{TM}}{{}^{II}\mathcal{E}_y^{TE}} & -1 & 0 \\
\boldsymbol{\vartheta}\frac{{}^{I}\mathcal{H}_y^{TE}}{{}^{I}\mathcal{H}_y^{TM}} & -1 & -\boldsymbol{\vartheta}\frac{{}^{II}\mathcal{H}_y^{TE}}{{}^{I}\mathcal{H}_y^{TM}} & -\boldsymbol{\vartheta}\frac{{}^{II}\mathcal{H}_y^{TM}}{{}^{I}\mathcal{H}_y^{TM}} \\
0 & \boldsymbol{\vartheta}\frac{{}^{I}\mathcal{E}_x^{TM}}{{}^{II}\mathcal{E}_x^{TM}} & 0 & -1
\end{bmatrix}}_{\boldsymbol{\vartheta}_{Unknown}} \cdot \underbrace{\begin{bmatrix} {}^{I}\vec{B}\,^{TE} \\ {}^{I}\vec{B}\,^{TM} \\ {}^{II}\vec{A}\,^{TE} \\ {}^{II}\vec{A}\,^{TM} \end{bmatrix}}_{\vec{\alpha}_{Unknown},}
$$

which in short denotes as

$$
\vec{0} = \boldsymbol{\vartheta}_{Known} \cdot \vec{\alpha}_{Known} + \boldsymbol{\vartheta}_{Unknown} \cdot \vec{\alpha}_{Unknown}.
$$

Note that $\boldsymbol{\vartheta}_{Known}$ represents the left and right side of $\boldsymbol{\vartheta}$ while $\boldsymbol{\vartheta}_{Unknown}$ represents the center part of $\boldsymbol{\vartheta}$.

The field problem then can be solved by solving the inhomogeneous system of equations

$$
-\boldsymbol{\vartheta}_{Known} \cdot \vec{\alpha}_{Known} = \boldsymbol{\vartheta}_{Unknown} \cdot \vec{\alpha}_{Unknown}. \tag{4.27}
$$

In chapter 6 of this thesis we will discuss means to solve "real-life" structures such as filters, which can be interpreted as cascades of several interfaces connected by non-zero length waveguide segments.

4.6. Special Case: Decoupled TE and TM Solution

An interesting special case occurs when at suitable interfaces the excitation is chosen in a fashion that any coupling between TE and TM Eigenmodes is irrelevant and thus can be ignored.

The system of equations (4.23) is then decoupled into two systems of equations

$$\vec{0} = \begin{bmatrix} 1 & -\vartheta \frac{{}^{II}\mathcal{H}_x^{TE}}{{}^{I}\mathcal{H}_x^{TE}} & 1 & -\vartheta \frac{{}^{II}\mathcal{H}_x^{TE}}{{}^{I}\mathcal{H}_x^{TE}} \\ \vartheta \frac{{}^{I}\mathcal{E}_y^{TE}}{{}^{II}\mathcal{E}_y^{TE}} & -1 & -\vartheta \frac{{}^{I}\mathcal{E}_y^{TE}}{{}^{II}\mathcal{E}_y^{TE}} & 1 \end{bmatrix} \cdot \begin{bmatrix} {}^{I}\vec{A}^{TE} \\ {}^{II}\vec{A}^{TE} \\ {}^{I}\vec{B}^{TE} \\ {}^{II}\vec{B}^{TE} \end{bmatrix} \tag{4.28}$$

$$\vec{0} = \begin{bmatrix} 1 & -\vartheta \frac{{}^{II}\mathcal{H}_y^{TM}}{{}^{I}\mathcal{H}_y^{TM}} & -1 & \vartheta \frac{{}^{II}\mathcal{H}_x^{TM}}{{}^{I}\mathcal{H}_y^{TM}} \\ \vartheta \frac{{}^{I}\mathcal{E}_x^{TM}}{{}^{II}\mathcal{E}_x^{TM}} & -1 & \vartheta \frac{{}^{I}\mathcal{E}_x^{TM}}{{}^{II}\mathcal{E}_x^{TM}} & -1 \end{bmatrix} \cdot \begin{bmatrix} {}^{I}\vec{A}^{TM} \\ {}^{II}\vec{A}^{TM} \\ {}^{I}\vec{B}^{TM} \\ {}^{II}\vec{B}^{TM} \end{bmatrix} \tag{4.29}$$

and the structure's solution for TE and TM Eigenmodes can be regarded as individual problems, which can be solved independently. We will encounter such a special case in chapter 7 later.

For a TE-only problem, the total number[13] of Eigenmodes per waveport becomes

$$^{I}\mathcal{M} = 2MN \qquad ^{II}\mathcal{M} = 2ST \tag{4.30}$$

while for a TM-only problem, the total number denotes

$$^{I}\mathcal{M} = 2PQ \qquad ^{II}\mathcal{M} = 2VW. \tag{4.31}$$

[13] i.e. the number of Eigenmodes for both directions

5. The Single Waveguide - Modeling Non-Zero Length Waveguides of Uniform Cross-Section

In the previous chapter, we have carried out an in-depth discussion of the Mode Matching process at the interface between two waveguides of different geometry. In order to treat more advanced structures such as filters and resonators (as we will do later in this thesis), a second type of building blocks, namely non-zero length waveguides of uniform cross-section are required.

To model such waveguides, *waveguide matrices* $\boldsymbol{\gamma}$, which are designed to use a similar amplitude ordering scheme as $\boldsymbol{\vartheta}$, are introduced in this chapter.

Figure 5.1 shows a waveguide of uniform cross-section with length l, which shall be represented under form of a waveguide matrix $\boldsymbol{\gamma}$. To set up this matrix, we have to analyse the relationships between the amplitudes at waveport I and II of the waveguide.

Let us now consider an arbitrary Eigenmode with an amplitude ${}^{I}A$ entering the waveguide at waveport I and assume z to be the direction of propagation. The Eigenmode's amplitude at waveport II then can be denoted as

$$ {}^{II}A = {}^{I}A \cdot e^{-jk_z l}. \tag{5.1} $$

The exponential term in (5.1) either describes wave propagation (i. e. a phase delay) from waveport I to waveport II or an exponential decay if the Eigenmode is evanescent, that is, an Eigenmode below cut-off. If we consider a propagating Eigenmode, the exponential term can be interpreted as $e^{-\gamma l} = e^{-j\beta l}$, that is, the propagation constant is purely imaginary and does not account for any loss present in a real waveguide. By letting $\gamma = \alpha + j\beta$ for a propagating Eigenmode, loss inside the structure under consideration can be partially modelled. The attenuation constant α can be determined using the power loss method as discussed for example in [6].

Following a similar reasoning, the amplitude ${}^{I}B$ on waveport I of an arbitrary Eigenmode entering the structure with an amplitude ${}^{II}B$ via waveport II can be denoted as

$$ {}^{I}B = {}^{II}B \cdot e^{jk_z(-l)} = {}^{II}B \cdot e^{-jk_z l}, \tag{5.2} $$

which describes wave propagation in negative z-direction.

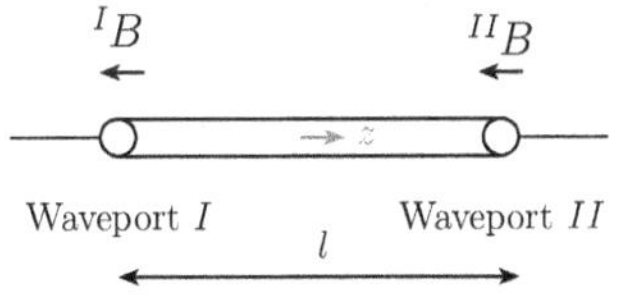

Figure 5.1.: Amplitude definitions for a waveguide of non-zero length

After some manipulations, (5.1) and (5.2) are rewritten

$$0 = {}^{I}A \cdot e^{-jk_z l} - {}^{II}A \tag{5.3}$$

and

$$0 = {}^{I}B - {}^{II}B \cdot e^{-jk_z l}. \tag{5.4}$$

By constructing such equations for all mode types relevant to the problem under consideration, a homogeneous system of equations can be assembled. Recalling the structure of the interface matrix $\boldsymbol{\vartheta}$ from (4.23), the homogeneous system of equations describing a waveguide must have a form

$$\vec{0} = \underbrace{\begin{bmatrix} \gamma^{TE} & 0 & 0 & 0 & -1 & 0 & 0 & 0 \\ 0 & \gamma^{TM} & 0 & 0 & 0 & -1 & 0 & 0 \\ 0 & 0 & 1 & 0 & 0 & 0 & -\gamma^{TE} & 0 \\ 0 & 0 & 0 & 1 & 0 & 0 & 0 & -\gamma^{TM} \end{bmatrix}}_{\gamma} \cdot \begin{bmatrix} {}^{I}\vec{A}^{TE} \\ {}^{I}\vec{A}^{TM} \\ {}^{I}\vec{B}^{TE} \\ {}^{I}\vec{B}^{TM} \\ {}^{II}\vec{A}^{TE} \\ {}^{II}\vec{A}^{TM} \\ {}^{II}\vec{B}^{TE} \\ {}^{II}\vec{B}^{TM} \end{bmatrix} \tag{5.5}$$

where γ is defined as the waveguide matrix. The sub-matrices denote

$$\gamma^{TE} = \begin{bmatrix} \ddots & & 0 \\ & e^{-jk^{TE}_{z|m,n} l} & \\ 0 & & \ddots \end{bmatrix} \tag{5.6}$$

and

$$\gamma^{TM} = \begin{bmatrix} \ddots & & 0 \\ & e^{-jk^{TM}_{z|p,q} l} & \\ 0 & & \ddots \end{bmatrix} \tag{5.7}$$

and as a consequence, the waveguide matrix consists out of two diagonals only.

While (5.5) is valid for rectangular waveguide structures as introduced in section 4.2, the presented concept of waveguide matrices can similarly be applied for problems requiring different coordinate systems or alternate mode-sets.

6. Advanced Structures - Cascading Segments and Assembling the System Matrix

In chapter 4 interface matrices $\boldsymbol{\vartheta}$ for modeling interfaces between waveguides of different geometry were introduced. Similarly, in chapter 5, waveguide matrices $\boldsymbol{\gamma}$ were introduced to model non-zero length waveguides of constant geometry. Obviously, both of these kinds of "building blocks" are required to model "real life" structures such as filters and resonators. Consequently, the question arises how to properly cascade these matrices. This topic will be covered in this chapter.

6.1. Preliminary Considerations - Using Segments and Junctions

In the following, we will refer to interfaces and waveguides in a generalised form as *segments*. The corresponding interface matrices $\boldsymbol{\vartheta}$ and waveguide matrices $\boldsymbol{\gamma}$ are compatible by design, that is, they use a similar amplitude ordering scheme as can be seen from (4.23) and (5.5). When cascading such matrices, for the structure of the resulting system of equations it is in fact irrelevant whether the cascaded matrices represent interfaces or waveguides.[1]

It is thus useful to introduce a generalised *segment matrix* $\boldsymbol{\Theta}$, which may either represent an interface matrix or a waveguide matrix in a general fashion. Before defining this matrix in greater detail, in the following section, preliminary considerations shall be given on improving the cascadability of this type of matrix. Formally, segment matrices will be introduced in section 6.2.

6.1.1. Junction Amplitudes

So far, we have discussed both waveguide interfaces as well as waveguides in terms of waveports and the waveports' corresponding amplitudes as shown in figure 6.1. Regarding the compilation of an overall system of equations describing a cascade of interfaces and waveguides, this notation is rather clumsy as it contains a considerable amount of duplicate Eigenmode amplitudes, that is,

$$^{\eta}\vec{A} = {}^{\eta-1}\vec{A} \qquad {}^{\eta}\vec{B} = {}^{\eta-1}\vec{B} \tag{6.1}$$

where $^{\eta}\vec{A}$, $^{\eta}\vec{B}$ and $^{\eta-1}\vec{A}$, $^{\eta-1}\vec{B}$ denote amplitudes at adjacent waveports η, $\eta-1$ as shown in figure 6.1. The unnecessary large number of Eigenmode amplitudes leads to an unnecessary large number of matrix columns as well as equations (i.e. rows), which are required to equate amplitudes at adjacent waveports as per (6.1).

[1]Obviously, in order to obtain an electromagnetically "meaningful" structure, only interfaces and waveguides with appropriate geometries may be cascaded. Consequently, only waveguides of similar geometry may be directly cascaded. If waveguides of different geometry shall be cascaded, an appropriate interface has to be inserted in order to account for parasitic electromagnetic effects.

We thus introduce junction amplitudes

$$\begin{aligned} {}^{\nu}\breve{A} &= {}^{\eta}\vec{A} = {}^{\eta-1}\vec{A} \\ {}^{\nu}\breve{B} &= {}^{\eta}\vec{B} = {}^{\eta-1}\vec{B} \end{aligned} \tag{6.2}$$

where η, $\eta - 1$ shall be the waveports adjacent to junction ν on the left of segment ν as is shown in figure 6.2. It should be noted that for waveport numbering regular Roman numbers $\{I, II, III, ...\}$ are used while junction and segment numbering is done using calligraphic Roman numbers $\{\mathcal{I}, \mathcal{II}, \mathcal{III}, ..., \mathcal{N}\}$ where $\mathcal{N}$ is the total number of segments.

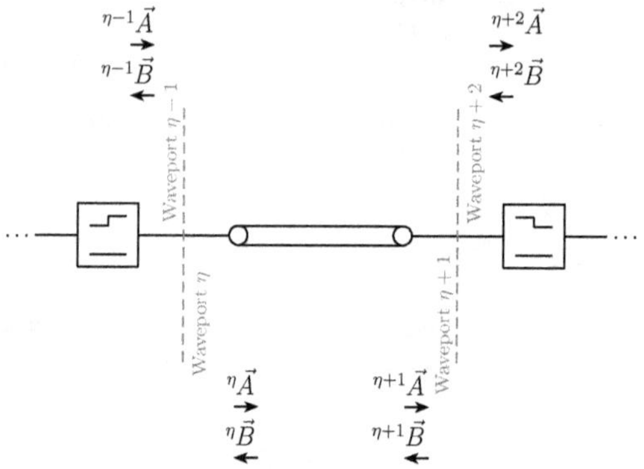

Figure 6.1.: Waveport amplitude representation of a structure

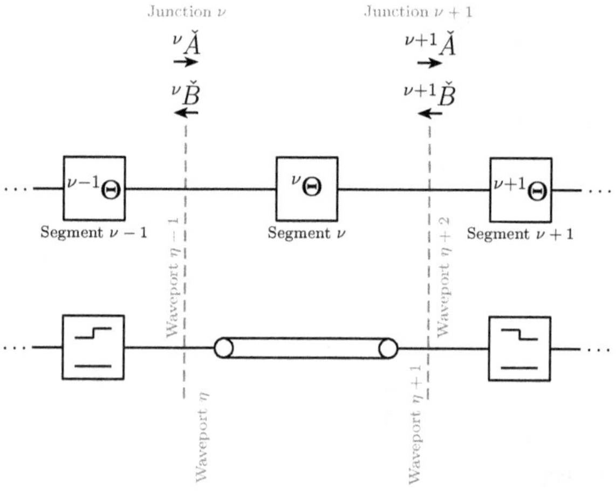

Figure 6.2.: Junction amplitude representation of a structure. ${}^{\nu}\boldsymbol{\Theta}$ denotes the segment matrix of segment ν.

6.2. The Segment Matrix $\boldsymbol{\Theta}$

As discussed in the previous section, when dealing with more advanced structures such as filters and resonators, which represent cascades of interfaces and waveguides, it is helpful to define a generalised segment matrix, which may represent both an interface or waveguide matrix and uses junction amplitudes rather than waveport amplitudes.

Consider the segment ν shown in figure 6.3. For this segment a system of equations

$$\vec{0} = {}^{\nu}\boldsymbol{\Theta} \begin{bmatrix} {}^{\nu}\breve{A}^{\,TE} \\ {}^{\nu}\breve{A}^{\,TM} \\ {}^{\nu}\breve{B}^{\,TE} \\ {}^{\nu}\breve{B}^{\,TM} \\ {}^{\nu+1}\breve{A}^{\,TE} \\ {}^{\nu+1}\breve{A}^{\,TM} \\ {}^{\nu+1}\breve{B}^{\,TE} \\ {}^{\nu+1}\breve{B}^{\,TM} \end{bmatrix} \tag{6.3}$$

may be set up where ${}^{\nu}\boldsymbol{\Theta}$ is defined as segment matrix. The segment matrix may either represent an interface matrix $\boldsymbol{\vartheta}$ or a waveguide matrix $\boldsymbol{\gamma}$ and thus its dimensions are

$${}^{\nu}\mathcal{C} = {}^{\nu}\mathcal{M} + {}^{\nu+1}\mathcal{M} \qquad {}^{\nu}\mathcal{R} = \frac{{}^{\nu}\mathcal{M}+{}^{\nu+1}\mathcal{M}}{2}, \tag{6.4}$$

which directly follows from (4.12) if we let the number of Eigenmodes on the segment's waveports to be equal to the number Eigenmodes on the adjacent junctions, that is, ${}^{\nu}\mathcal{M} = {}^{\eta}\mathcal{M} = {}^{\eta-1}\mathcal{M}$ and ${}^{\nu+1}\mathcal{M} = {}^{\eta+1}\mathcal{M} = {}^{\eta+2}\mathcal{M}$.

${}^{\nu}\breve{A}$, ${}^{\nu}\breve{B}$ denote the amplitudes at the left junction of segment ν and ${}^{\nu+1}\breve{A}$, ${}^{\nu+1}\breve{B}$ represent amplitudes at the right junction of this segment. Note that while $\boldsymbol{\vartheta}$ and $\boldsymbol{\gamma}$ were introduced in chapter 4 and 5 for waveport amplitudes, they are equally valid for the junction amplitudes introduced in the previous section.

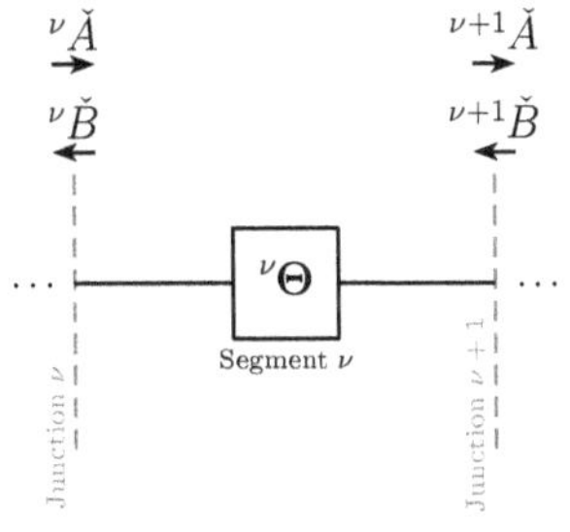

Figure 6.3.: On the definition of the segment matrix ${}^{\nu}\boldsymbol{\Theta}$

Because proper assembling of segment matrices is a crucial part of solving electromagnetic problems using the Mode Matching technique, the following detailed representation of interface matrices $\boldsymbol{\vartheta}$ and waveguide matrices $\boldsymbol{\gamma}$ under form of segment matrices is given for reference purposes. Note that only the amplitude vectors differ from (4.23) and (5.5).

The segment matrix representation of an interface matrix $\boldsymbol{\vartheta}$ denotes

$$\vec{0} = \underbrace{\begin{bmatrix} 1 & 0 & 1 & 0 & -\boldsymbol{\vartheta}^{{}^{II}\mathcal{H}_x^{TE}}_{{}^{I}\mathcal{H}_x^{TE}} & 0 & -\boldsymbol{\vartheta}^{{}^{II}\mathcal{H}_x^{TE}}_{{}^{I}\mathcal{H}_x^{TE}} & 0 \\ \boldsymbol{\vartheta}^{{}^{I}\mathcal{E}_y^{TE}}_{{}^{II}\mathcal{E}_y^{TE}} & \boldsymbol{\vartheta}^{{}^{I}\mathcal{E}_y^{TM}}_{{}^{II}\mathcal{E}_y^{TE}} & -\boldsymbol{\vartheta}^{{}^{I}\mathcal{E}_y^{TE}}_{{}^{II}\mathcal{E}_y^{TE}} & \boldsymbol{\vartheta}^{{}^{I}\mathcal{E}_y^{TM}}_{{}^{II}\mathcal{E}_y^{TE}} & -1 & -\boldsymbol{\vartheta}^{{}^{II}\mathcal{E}_y^{TM}}_{{}^{II}\mathcal{E}_y^{TE}} & 1 & -\boldsymbol{\vartheta}^{{}^{II}\mathcal{E}_y^{TM}}_{{}^{II}\mathcal{E}_y^{TE}} \\ \boldsymbol{\vartheta}^{{}^{I}\mathcal{H}_y^{TE}}_{{}^{I}\mathcal{H}_y^{TM}} & 1 & \boldsymbol{\vartheta}^{{}^{I}\mathcal{H}_y^{TE}}_{{}^{I}\mathcal{H}_y^{TM}} & -1 & -\boldsymbol{\vartheta}^{{}^{II}\mathcal{H}_y^{TE}}_{{}^{I}\mathcal{H}_y^{TM}} & -\boldsymbol{\vartheta}^{{}^{II}\mathcal{H}_y^{TM}}_{{}^{I}\mathcal{H}_y^{TM}} & -\boldsymbol{\vartheta}^{{}^{II}\mathcal{H}_y^{TE}}_{{}^{I}\mathcal{H}_y^{TM}} & \boldsymbol{\vartheta}^{{}^{II}\mathcal{H}_x^{TM}}_{{}^{I}\mathcal{H}_y^{TM}} \\ 0 & \boldsymbol{\vartheta}^{{}^{I}\mathcal{E}_x^{TM}}_{{}^{II}\mathcal{E}_x^{TM}} & 0 & \boldsymbol{\vartheta}^{{}^{I}\mathcal{E}_x^{TM}}_{{}^{II}\mathcal{E}_x^{TM}} & 0 & -1 & 0 & -1 \end{bmatrix}}_{{}^{\nu}\boldsymbol{\Theta} = \boldsymbol{\vartheta}} \cdot \begin{bmatrix} {}^{\nu}\breve{A}^{TE} \\ {}^{\nu}\breve{A}^{TM} \\ {}^{\nu}\breve{B}^{TE} \\ {}^{\nu}\breve{B}^{TM} \\ {}^{\nu+1}\breve{A}^{TE} \\ {}^{\nu+1}\breve{A}^{TM} \\ {}^{\nu+1}\breve{B}^{TE} \\ {}^{\nu+1}\breve{B}^{TM} \end{bmatrix} \tag{6.5}$$

Similarly, waveguide matrices $\boldsymbol{\gamma}$ introduced in (5.5) may be converted into segment matrices, which then denote

$$\vec{0} = \underbrace{\begin{bmatrix} \boldsymbol{\gamma}^{TE} & 0 & 0 & 0 & -1 & 0 & 0 & 0 \\ 0 & \boldsymbol{\gamma}^{TM} & 0 & 0 & 0 & -1 & 0 & 0 \\ 0 & 0 & 1 & 0 & 0 & 0 & -\boldsymbol{\gamma}^{TE} & 0 \\ 0 & 0 & 0 & 1 & 0 & 0 & 0 & -\boldsymbol{\gamma}^{TM} \end{bmatrix}}_{{}^{\nu}\boldsymbol{\Theta} = \boldsymbol{\gamma}} \cdot \begin{bmatrix} {}^{\nu}\breve{A}^{TE} \\ {}^{\nu}\breve{A}^{TM} \\ {}^{\nu}\breve{B}^{TE} \\ {}^{\nu}\breve{B}^{TM} \\ {}^{\nu+1}\breve{A}^{TE} \\ {}^{\nu+1}\breve{A}^{TM} \\ {}^{\nu+1}\breve{B}^{TE} \\ {}^{\nu+1}\breve{B}^{TM} \end{bmatrix} \tag{6.6}$$

where

$$\boldsymbol{\gamma}^{TE} = \begin{bmatrix} \ddots & & 0 \\ & e^{-jk^{TE}_{z|m,n}l} & \\ 0 & & \ddots \end{bmatrix} \tag{6.7}$$

and

$$\boldsymbol{\gamma}^{TM} = \begin{bmatrix} \ddots & & 0 \\ & e^{-jk^{TM}_{z|p,q}l} & \\ 0 & & \ddots \end{bmatrix}. \tag{6.8}$$

6.3. Cascading Segments and Assembling the System Matrix $\boldsymbol{\Psi}$

Consider the cascade of segments shown in figure 6.4 and remember that each segment either represents an interface between two waveguides of different geometry or a non-zero length waveguide of uniform cross-section. For every segment, a homogeneous system of equations can be set up as presented in chapter 4 or 5. The coefficients of each system are assembled in individual segment matrices ${}^{\nu}\boldsymbol{\Theta}$.

Obviously, the structure's overall field solution must maintain each and every equation out of these sets of equations and consequently, the equations are combined to form an overall system of equations

$$\vec{0} = \boldsymbol{\Psi} \cdot \vec{a} = \boldsymbol{\Psi} \cdot \begin{bmatrix} {}^{\mathcal{I}}\breve{A} \\ {}^{\mathcal{I}}\breve{B} \\ {}^{\mathcal{II}}\breve{A} \\ {}^{\mathcal{II}}\breve{B} \\ {}^{\mathcal{III}}\breve{A} \\ {}^{\mathcal{III}}\breve{B} \\ {}^{\mathcal{IV}}\breve{A} \\ {}^{\mathcal{IV}}\breve{B} \\ \vdots \\ {}^{\mathcal{N}+1}\breve{A} \\ {}^{\mathcal{N}+1}\breve{B} \end{bmatrix} \tag{6.9}$$

where $\boldsymbol{\Psi}$ shall be referred to as *system matrix*, which completely describes the structure and is dependent on the structure's geometry and the solution frequency only. $\vec{a}$ denotes an amplitude vector summarising all junction amplitudes. ${}^{\nu}\breve{A}$ and ${}^{\nu}\breve{B}$ comprise both amplitudes of TE and TM Eigenmodes.

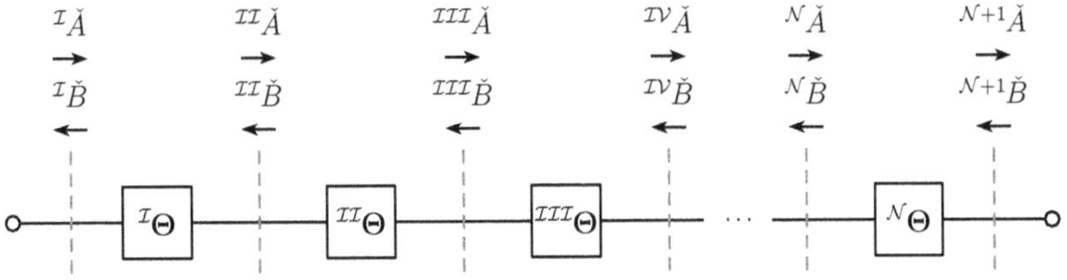

Figure 6.4.: Segment / Junction amplitude representation of an arbitrary structure

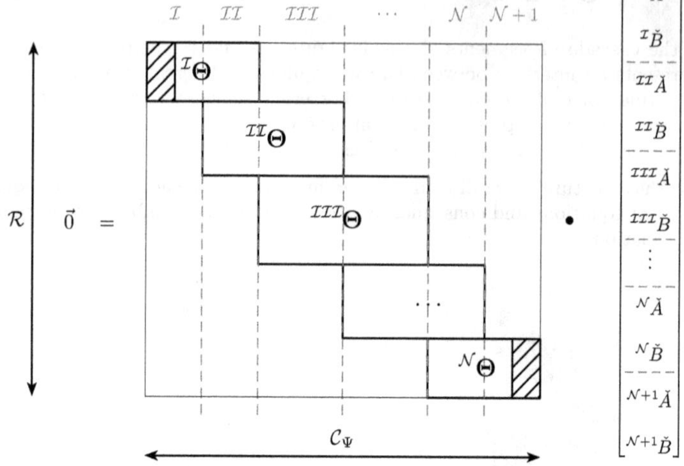

Figure 6.5.: Structure of (6.9) and the system matrix $\boldsymbol{\Psi}$

The structure of (6.9) and the system matrix $\boldsymbol{\Psi}$ is illustrated in figure 6.5.

Note that the segment matrices ${}^{\nu}\boldsymbol{\Theta}$ are arranged in a stair-like fashion in order to properly assign the system's coefficients to their corresponding junction amplitudes as indicated by red dashed lines. Crosshatched parts of the matrix contain coefficients belonging to the known Eigenmode amplitudes ${}^{\mathcal{I}}\check{A}$ and ${}^{\mathcal{N}+1}\check{B}$ exciting the structure as discussed later.

Recalling from section 6.2 that for the individual segment matrix' dimensions we have

$$ {}^{\nu}\mathcal{C} = {}^{\nu}\mathcal{M} + {}^{\nu+1}\mathcal{M} \qquad {}^{\nu}\mathcal{R} = \frac{{}^{\nu}\mathcal{M}+{}^{\nu+1}\mathcal{M}}{2}, \tag{6.10} $$

we find

$$ \mathcal{R} = \sum_{\nu=\mathcal{I}}^{\mathcal{N}} {}^{\nu}\mathcal{R} = \frac{1}{2}\sum_{\nu=\mathcal{I}}^{\mathcal{N}} {}^{\nu}\mathcal{M}+{}^{\nu+1}\mathcal{M} = \frac{1}{2}\left[\sum_{\nu=\mathcal{I}}^{\mathcal{N}} {}^{\nu}\mathcal{M} + \sum_{\nu=\mathcal{II}}^{\mathcal{N}+1} {}^{\nu}\mathcal{M}\right] = \frac{{}^{\mathcal{I}}\mathcal{M}}{2} + \sum_{\nu=\mathcal{II}}^{\mathcal{N}} {}^{\nu}\mathcal{M} + \frac{{}^{\mathcal{N}+1}\mathcal{M}}{2}. \tag{6.11} $$

Similarly, by summing up the number of amplitudes over all junctions, we find

$$ \mathcal{C}_{\Psi} = \sum_{\nu=\mathcal{I}}^{\mathcal{N}+1} {}^{\nu}\mathcal{M} \tag{6.12} $$

6.4. Solving the Structure's System of Equations Using the Reduced System Matrix $\mathbf{\Psi}_{Unknown}$

Following a similar reasoning[2] as in section 4.4, the homogeneous system of equations $\vec{0} = \mathbf{\Psi} \cdot \vec{a}$ describing the field problem imposed by the structure shown in figure 6.4 has

$$\text{DOF} = \mathcal{C}_{\Psi} - \mathcal{R}$$

degrees of freedom, that is, the problem comprises $\mathcal{Q} = \text{DOF}$ *unknowns*.

Here we assumed that $\mathbf{\Psi}$ has full rank, which is obviously true because the individual segment matrices have full rank too: Interface matrices have full rank due to the orthogonality of the Eigenmodes, while waveguide matrices have full rank by definition (cf. (6.6)).

By inserting (6.11) and (6.12) into this equation, we find that the problem's number of DOF is

$$\text{DOF} = \sum_{\nu=\mathcal{I}}^{\mathcal{N}+1} {}^{\nu}\mathcal{M} - \left[\frac{{}^{\mathcal{I}}\mathcal{M}}{2} + \sum_{\nu=\mathcal{II}}^{\mathcal{N}} {}^{\nu}\mathcal{M} + \frac{{}^{\mathcal{N}+1}\mathcal{M}}{2}\right] = \frac{{}^{\mathcal{I}}\mathcal{M}}{2} + \frac{{}^{\mathcal{N}+1}\mathcal{M}}{2}, \tag{6.13}$$

which states that exactly half the number of amplitudes on the structures outer junctions must be predefined. An obvious choice (also see section 4.4) is to predefine the amplitudes travelling into the structure, namely

$${}^{\mathcal{I}}\breve{A} = \begin{bmatrix} {}^{\mathcal{I}}\breve{A}^{TE} \\ \\ {}^{\mathcal{I}}\breve{A}^{TM} \end{bmatrix} \qquad \text{and} \qquad {}^{\mathcal{N}+1}\breve{B} = \begin{bmatrix} {}^{\mathcal{N}+1}\breve{B}^{TE} \\ \\ {}^{\mathcal{N}+1}\breve{B}^{TM} \end{bmatrix}.$$

In order to solve the system of equations (6.9), similar to section 4.5 the system matrix $\mathbf{\Psi}$ must be split into two matrices $\mathbf{\Psi}_{Known}$ and $\mathbf{\Psi}_{Unknown}$ where the former matrix contains all coefficients corresponding to known junction amplitudes and the latter one includes all coefficients associated with unknown junction amplitudes.

The system of equations (6.9) then can be rewritten as

$$\vec{0} = \mathbf{\Psi}_{Known} \cdot \vec{a}_{Known} + \mathbf{\Psi}_{Unknown} \cdot \vec{a}_{Unknown}$$

where $\vec{a}_{Known}$ and $\vec{a}_{Unknown}$ denote amplitude vectors

$$\vec{a}_{Known} = \begin{bmatrix} {}^{\mathcal{I}}\breve{A} & {}^{\mathcal{N}+1}\breve{B} \end{bmatrix}^T$$

and

$$\vec{a}_{Unknown} = \begin{bmatrix} {}^{\mathcal{I}}\breve{B} & {}^{\mathcal{II}}\breve{A} & {}^{\mathcal{II}}\breve{B} & \dots & {}^{\mathcal{N}}\breve{A} & {}^{\mathcal{N}}\breve{B} & {}^{\mathcal{N}+1}\breve{A} \end{bmatrix}^T.$$

[2] See footnote 12 in section 4.4.

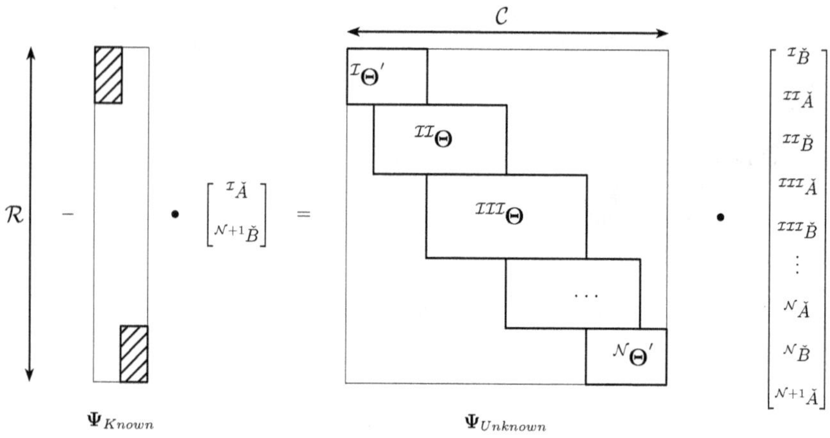

Figure 6.6.: Structure of the inhomogeneous system of equations as per (6.14)

Finally, the field problem can be solved by solving the inhomogeneous system of equations

$$\check{S} = -\mathbf{\Psi}_{Known} \cdot \vec{a}_{Known} = \mathbf{\Psi}_{Unknown} \cdot \vec{a}_{Unknown}. \tag{6.14}$$

The structure of this system of equations is illustrated in figure 6.6.[3] In the following, the matrix $\mathbf{\Psi}_{Unknown}$ shall be referred to as *reduced system matrix*.

While the total number of rows in the reduced system matrix, i.e. the number of linearly independent equations is identical to the system matrix $\mathbf{\Psi}$, that is,

$$\mathcal{R} = \frac{{}^{\mathcal{I}}\mathcal{M}}{2} + \sum_{\nu=\mathcal{II}}^{\mathcal{N}} {}^{\nu}\mathcal{M} + \frac{{}^{\mathcal{N}+1}\mathcal{M}}{2}. \tag{6.15}$$

the number of columns, i.e. the number of (unknown) junction amplitudes now similarly calculates

$$\mathcal{C} = \frac{{}^{\mathcal{I}}\mathcal{M}}{2} + \sum_{\nu=\mathcal{II}}^{\mathcal{N}} {}^{\nu}\mathcal{M} + \frac{{}^{\mathcal{N}+1}\mathcal{M}}{2}. \tag{6.16}$$

and the reduced system matrix' dimension may thus be denoted as $\mathcal{R} \times \mathcal{C}$, that is, the matrix is square. The number of *unknowns* consequently denotes $\mathcal{Q} = \mathcal{R} = \mathcal{C}$.

[3] Again, dashed parts of the matrix on the left contain coefficients corresponding to known junction amplitudes travelling into the structure and thus exciting it. These coefficients were cropped from the segment matrices ${}^{\mathcal{I}}\mathbf{\Theta}$ and ${}^{\mathcal{N}}\mathbf{\Theta}$ on the right side of (6.14), which is why the segment matrices of the first and last segment are marked with an additional apostrophe ($'$).

It should be noted that in practical applications the vector $\check{S}$ only contains the result of a multiplication of a single column of $\mathbf{\Psi}_{Known}$ with a single non-zero element contained in $\vec{a}_{Known}$ because when solving modal problems, the structure is typically excited by a single Eigenmode entering the cascade on one side. Thus, the effort for computing $\check{S}$ is negligible compared to populating and solving the inhomogeneous system of equations (6.14).

A system of equations comparable to (6.14) (as illustrated in figure 6.6) can be set up for all single-branch[4] modal problems, regardless of the actual type of waveguides used in the structure.[5]

Obviously, the size of the reduced system matrix $\mathbf{\Psi}_{Unknown}$ and thus its memory requirements as well as the computational effort to populate $\mathbf{\Psi}_{Unknown}$ and to solve the structure's inhomogeneous system of equations depends on the number of unknowns, that is, on the structure's number of segments as well as on the number of Eigenmodes considered.

In chapter 11 we will investigate these relations in greater detail.

[4] If the structure branched out at some place, a more complex system of equations would be obtained.

[5] However, it should be noted that when dealing with resonant problems, the structure's overall system of equations remains under form of an homogeneous system of equations similar to (6.9). In this system, additional boundary conditions, i.e. PEC walls, are introduced to appropriately terminate the cascade on both ends, that is, to establish relations between ${}^{\mathcal{I}}\check{A}$ and ${}^{\mathcal{I}}\check{B}$ as well as ${}^{\mathcal{N}+1}\check{A}$ and ${}^{\mathcal{N}+1}\check{B}$.

7. Investigation of a Simple Structure

In the previous chapters, we have extensively studied both Eigenmodes of waveguides as well as the concept and formalism of the Mode Matching technique used in this thesis. In the following, we will now investigate the application of the Mode Matching technique to the structure shown in figure 7.1, which comprises multiple interfaces and waveguides. Several important insights can be gained from this analysis:

Firstly, we will acquire a detailed understanding of the coupling mechanisms at the discontinuities between the waveguide segments. We can use these insights to choose the smallest possible set of Eigenmodes suitable for treating these discontinuities and thus reduce the computation time required for solving the structure.

These findings can directly be applied for solving the waveguide iris filter to be discussed in section 13.4.1. Rather than including each and every TE and TM Eigenmode in our analysis, we reduce the set of Eigenmodes taken into account to the odd-order $TE^x_{m,0}$ Eigenmodes. This obviously provides a huge computational advantage. For the tubular filters, which will be investigated in section 13.2.2, a comparable analysis can be carried out based on the very same concept.

Secondly, by analysing the structure's reduced system matrix, the reader will gain a better understanding on how the matrix-based representation of the structure in terms of the Mode Matching technique is realised.

This chapter is organised as follows: In the following section, we will firstly analyse the coupling mechanisms at the interface between the narrow and wide waveguide, which are described in terms of an interface matrix $\boldsymbol{\vartheta}$. We will see that if the structure is excited by the $TE^x_{1,0}$ Eigenmode, only odd-order $TE^x_{m,0}$ Eigenmodes need to be taken into account while for the given excitation no coupling with TM Eigenmodes occurs.

Next, we will analyse waveguide matrices $\boldsymbol{\gamma}$, which are used to model waveguide sections.

The interface and waveguide matrices can be reinterpreted as segment matrices $\boldsymbol{\Theta}$ as discussed in chapter 6.2. Finally, we will combine these segment matrices to form the structure's reduced system matrix $\boldsymbol{\Psi}_{Unknown}$.

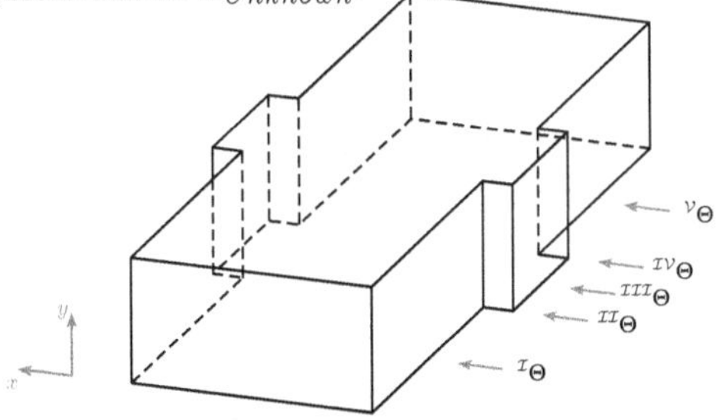

Figure 7.1.: Exemplary structure

7.1. The Discontinuities' Interface Matrices

In the following, we will investigate coupling between the waveguides' various Eigenmodes in order to determine the smallest possible set of Eigenmodes which may be used to analyse the structure assuming that the $TE^x_{1,0}$ Eigenmode is used for feeding the structure.

In general, since the structure is discontinuous in the x direction only, no coupling can occur between Eigenmodes of different y dependence. The reason for this is that the overlap integrals between Eigenmodes of different y dependence will always vanish due to orthogonality.

7.1.1. Coupling between TE and TM Eigenmodes

We will now discuss coupling between TE and TM Eigenmodes and show that no coupling occurs between these two sets of Eigenmodes.

As can be seen in figure 7.3a and 7.3b, $TE^x_{m,0}$ Eigenmodes do not exhibit a y dependence. In contrast, $TM^x_{v,1}$ Eigenmodes do have a y dependence, namely $\cos(k_y y)$ for E_y ($v > 0$ only). The $TM^x_{0,1}$ Eigenmode shown in figure 7.3c is a special case for which the field components E_y or H_x vanish so that coupling with the $TE^x_{m,0}$ Eigenmodes is impossible anyway.

In figure 7.2 it is shown that the $TE^x_{m,0}$ and $TM^x_{v,1}$ Eigenmodes have different y dependencies and thus coupling between Eigenmodes of these two sets of Eigenmodes does not occur. As a consequence, the solution for the $TE^x_{m,0}$ Eigenmodes is decoupled from the $TM^x_{v,1}$ Eigenmodes' solution (also see section 4.6).

For illustrative purposes, we will nevertheless include $TM^x_{v,1}$ Eigenmodes in our analysis, which therefore covers ${}^{I}TE^x_{m,0}$, ${}^{II}TE^x_{s,0}$, ${}^{I}TM^x_{v,1}$, and ${}^{II}TM^x_{p,1}$ Eigenmodes. To keep the matrices overseeable, we will only consider the first three Eigenmodes per mode type.[1]

In terms of the Mode Matching technique, coupling at a waveguide discontinuity is described by an interface matrix $\boldsymbol{\vartheta}$. The interface matrix for the discontinuity between waveguide I and II is depicted in figure 7.4. By careful comparison of figure 7.4 with the definition of the interface matrix' reprinted below figure 7.4 we see that sub-matrices describing coupling between $TE^x_{m,0}$ and $TM^x_{v,1}$ indeed vanish.

In the following, we thus only need to discuss the sub-matrices

$$\boldsymbol{\vartheta}^{{}^{I}\mathcal{E}_y^{TE}}_{{}^{II}\mathcal{E}_y^{TE}} \quad \boldsymbol{\vartheta}^{{}^{II}\mathcal{H}_x^{TE}}_{{}^{I}\mathcal{H}_x^{TE}} \quad \boldsymbol{\vartheta}^{{}^{I}\mathcal{E}_x^{TM}}_{{}^{II}\mathcal{E}_x^{TM}} \quad \boldsymbol{\vartheta}^{{}^{II}\mathcal{H}_y^{TM}}_{{}^{I}\mathcal{H}_y^{TM}}, \tag{7.1}$$

which describe coupling between Eigenmodes of the same type.

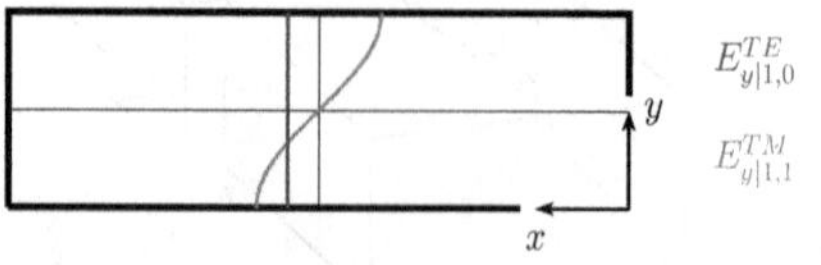

Figure 7.2.: On coupling between TE and TM Eigenmodes

[1] See (4.11), $M = P = S = V = 3$, $N = Q = T = W = 1$.

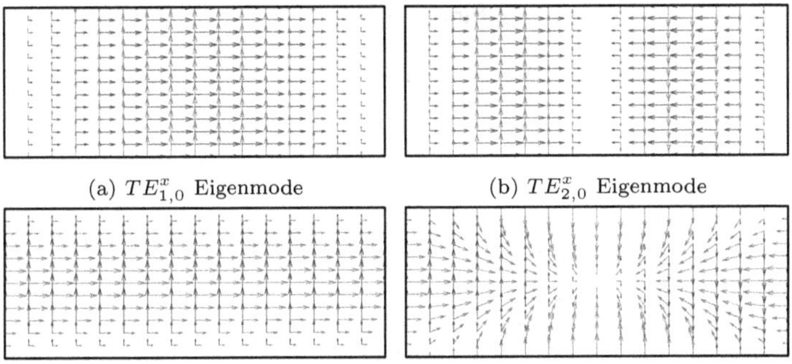

(a) $TE^x_{1,0}$ Eigenmode (b) $TE^x_{2,0}$ Eigenmode

(c) $TM^x_{0,1}$ Eigenmode (d) $TM^x_{1,1}$ Eigenmode

Figure 7.3.: Field plots of the fundamental TE and TM Eigenmodes
Electric field shown in red, magnetic field shown in blue

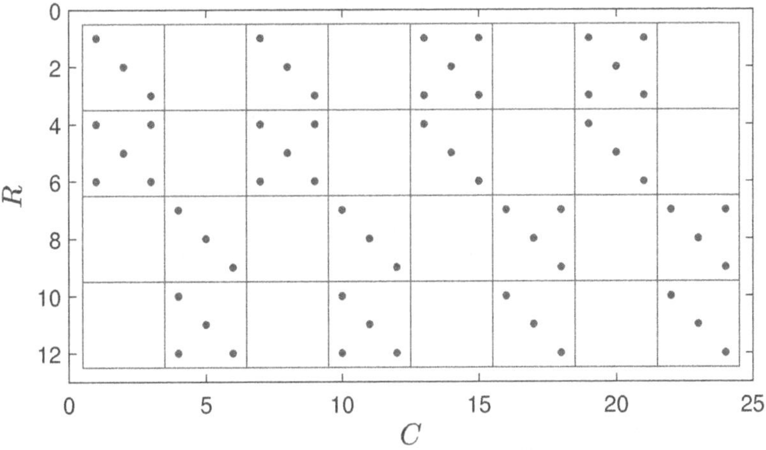

Figure 7.4.: Population pattern of the waveguide matrix applicable for the interface between waveguide I and II

$$
{}^{\nu}\Theta = \begin{bmatrix}
1 & 0 & 1 & 0 & -\vartheta^{{}^{II}\mathcal{H}_x^{TE}}_{{}^{I}\mathcal{H}_x^{TE}} & 0 & -\vartheta^{{}^{II}\mathcal{H}_x^{TE}}_{{}^{I}\mathcal{H}_x^{TE}} & 0 \\
\vartheta^{{}^{I}\mathcal{E}_y^{TE}}_{{}^{II}\mathcal{E}_y^{TE}} & \vartheta^{{}^{I}\mathcal{E}_y^{TM}}_{{}^{II}\mathcal{E}_y^{TE}} & -\vartheta^{{}^{I}\mathcal{E}_y^{TE}}_{{}^{II}\mathcal{E}_y^{TE}} & \vartheta^{{}^{I}\mathcal{E}_y^{TM}}_{{}^{II}\mathcal{E}_y^{TE}} & -1 & -\vartheta^{{}^{II}\mathcal{E}_y^{TM}}_{{}^{II}\mathcal{E}_y^{TE}} & 1 & -\vartheta^{{}^{II}\mathcal{E}_y^{TM}}_{{}^{II}\mathcal{E}_y^{TE}} \\
\vartheta^{{}^{I}\mathcal{H}_y^{TE}}_{{}^{I}\mathcal{H}_y^{TM}} & 1 & \vartheta^{{}^{I}\mathcal{H}_y^{TE}}_{{}^{I}\mathcal{H}_y^{TM}} & -1 & -\vartheta^{{}^{II}\mathcal{H}_y^{TE}}_{{}^{I}\mathcal{H}_y^{TM}} & -\vartheta^{{}^{II}\mathcal{H}_y^{TM}}_{{}^{I}\mathcal{H}_y^{TM}} & -\vartheta^{{}^{II}\mathcal{H}_y^{TE}}_{{}^{I}\mathcal{H}_y^{TM}} & \vartheta^{{}^{II}\mathcal{H}_x^{TM}}_{{}^{I}\mathcal{H}_y^{TM}} \\
0 & \vartheta^{{}^{I}\mathcal{E}_x^{TM}}_{{}^{II}\mathcal{E}_x^{TM}} & 0 & \vartheta^{{}^{I}\mathcal{E}_x^{TM}}_{{}^{II}\mathcal{E}_x^{TM}} & 0 & -1 & 0 & -1
\end{bmatrix}
$$

Structure of the segment matrix corresponding to an interface matrix - Reprint of (4.23)

7.1.2. Coupling between TE Eigenmodes

Let us now investigate coupling between TE Eigenmodes. The relevant sub-matrices

$$\boldsymbol{\vartheta}_{{}^{II}\mathcal{E}_y^{TE}}^{{}^{I}\mathcal{E}_y^{TE}} \qquad \boldsymbol{\vartheta}_{{}^{I}\mathcal{H}_x^{TE}}^{{}^{II}\mathcal{H}_x^{TE}} \tag{7.2}$$

are depicted in figure 7.5. Each entry in figure 7.5 corresponds to a dot in figure 7.4 and the notation Eigenmode 1 : Eigenmode 2 can be read as Eigenmode 1 excites / contributes to Eigenmode 2.

Consequently, rows indicate all Eigenmodes contributing to a specific Eigenmode while columns indicate all Eigenmodes used for expanding a given Eigenmode. This point becomes obvious when remembering the fact that each matrix row represents an equation which was obtained by orthogonal expansion of the continuity condition at the interface (also see (4.9)).

In order to comprehend the population patterns depicted in figure 7.5, we can exemplarily examine the expansion of the electric field component E_y from waveguide I to II. The field component E_y is depicted in figure 7.6 for the first few Eigenmodes.

By taking the overlap integrals of the depicted field components, it can be seen that overlap integrals between odd- and even-order Eigenmodes vanish while overlap integrals between odd- and odd-order Eigenmodes are non-zero. Consequently, the fundamental $TE_{1,0}^x$ Eigenmode excites higher-order odd TE Eigenmodes while higher-order even TE Eigenmodes remain unaffected. This explains the off-diagonal elements' population pattern shown in figure 7.5.

Obviously, the overlap integrals between Eigenmodes of similiar order never vanish. As a consequence, the entire main diagonal of the sub-matrix is populated.

For the expansion of H_x from II to I the exact same findings apply because E_y and H_x are related by the wave impedance of the individual Eigenmodes.

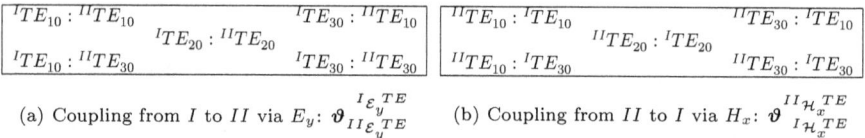

${}^{I}TE_{10} : {}^{II}TE_{10}$		${}^{I}TE_{30} : {}^{II}TE_{10}$
	${}^{I}TE_{20} : {}^{II}TE_{20}$	
${}^{I}TE_{10} : {}^{II}TE_{30}$		${}^{I}TE_{30} : {}^{II}TE_{30}$

(a) Coupling from I to II via E_y: $\boldsymbol{\vartheta}_{{}^{II}\mathcal{E}_y^{TE}}^{{}^{I}\mathcal{E}_y^{TE}}$

${}^{II}TE_{10} : {}^{I}TE_{10}$		${}^{II}TE_{30} : {}^{I}TE_{10}$
	${}^{II}TE_{20} : {}^{I}TE_{20}$	
${}^{II}TE_{10} : {}^{I}TE_{30}$		${}^{II}TE_{30} : {}^{I}TE_{30}$

(b) Coupling from II to I via H_x: $\boldsymbol{\vartheta}_{{}^{I}\mathcal{H}_x^{TE}}^{{}^{II}\mathcal{H}_x^{TE}}$

Figure 7.5.: Non-zero elements of the TE Eigenmode coupling matrices

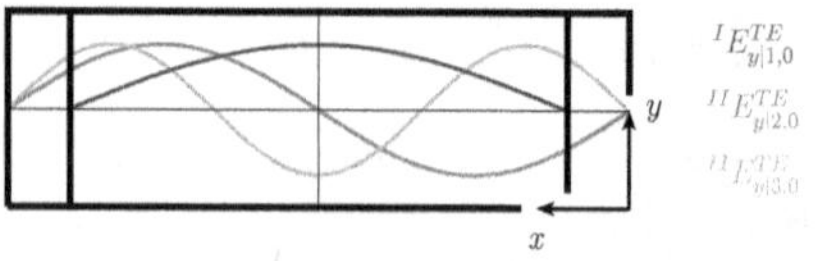

Figure 7.6.: On coupling between even and odd TE Eigenmodes via E_y

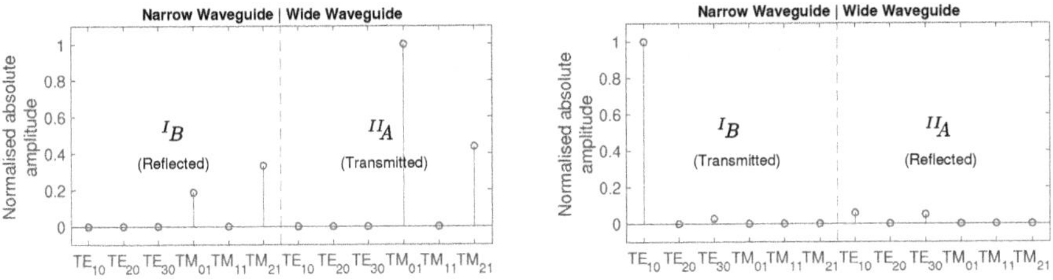

Figure 7.7.: Amplitudes of reflected and transmitted Eigenmodes when solving the waveguide interface's matrix depicted in figure 7.4 independently. Left: TE_{10} excitation on narrow waveguide. Right: TE_{10} excitation on wide waveguide.

Figure 7.7 depicts the amplitudes of the reflected and transmitted Eigenmodes obtained by individually solving the interface matrix given in figure 7.4 for a $TE_{1,0}$ excitation. As one would expect from the previous discussion, only odd-order ${}^{I}TE^{x}_{m,0}$ and ${}^{II}TE^{x}_{s,0}$ Eigenmodes are excited by the fundamental (odd-order) $TE_{1,0}$ Eigenmode while even-order TE Eigenmodes and TM Eigenmodes remain unaffected.

7.1.3. Coupling between TM Eigenmodes

In the previous sections we came to the conclusion that for the given structure and excitation TM Eigenmodes do not need to be taken into account. However, from figure 7.4 the reader may have noticed that the population scheme of the sub-matrices

$$\boldsymbol{\vartheta}_{{}^{II}\mathcal{E}_x^{TM}}^{{}^{I}\mathcal{E}_x^{TM}} \quad \boldsymbol{\vartheta}_{{}^{I}\mathcal{H}_y^{TM}}^{{}^{II}\mathcal{H}_y^{TM}} \tag{7.3}$$

is not symmetric.

For example, the expansion of E_x described by the sub-matrix shown in figure 7.8a indicates that the ${}^{I}TM^{x}_{0,1}$ Eigenmode on waveguide I excites the ${}^{II}TM^{x}_{2,1}$ Eigenmode on waveguide II. In contrast, the expansion of H_y (see figure 7.8b) does not indicate an excitation of the ${}^{I}TM^{x}_{2,1}$ Eigenmodes (or any other higher-order Eigenmode) on waveguide I by the ${}^{II}TM^{x}_{0,1}$ Eigenmode on waveguide II.

This is a suspicious discovery, which demands further investigation.

${}^{I}TM_{01} : {}^{II}TM_{01}$		
	${}^{I}TM_{11} : {}^{II}TM_{11}$	
${}^{I}TM_{01} : {}^{II}TM_{21}$		${}^{I}TM_{21} : {}^{II}TM_{21}$

(a) Coupling from I to II via E_x: $\boldsymbol{\vartheta}_{{}^{II}\mathcal{E}_x^{TM}}^{{}^{I}\mathcal{E}_x^{TM}}$

${}^{II}TM_{01} : {}^{I}TM_{01}$		${}^{II}TM_{21} : {}^{I}TM_{01}$
	${}^{II}TM_{11} : {}^{I}TM_{11}$	
		${}^{II}TM_{21} : {}^{I}TM_{21}$

(b) Coupling from II to I via H_y: $\boldsymbol{\vartheta}_{{}^{I}\mathcal{H}_y^{TM}}^{{}^{II}\mathcal{H}_y^{TM}}$

Figure 7.8.: Non-zero elements of the TM Eigenmode coupling matrices

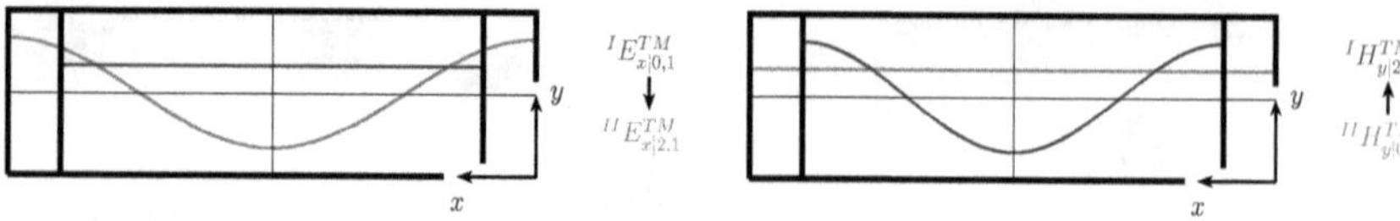

(a) Coupling via E_x

(b) No coupling via H_y

Figure 7.9.: On coupling via the TM Eigenmodes' fields

If we analyse the overlap integrals of the field distributions depicted in figure 7.9 in a similar fashion as we did in the previous section, we find that the population schemes depicted in figure 7.8 are indeed correct.

Still in terms of reciprocity, this finding seems somewhat strange as one would expect that a similar feeding from either side of the structure excites a comparable selection of parasitic higher-order Eigenmodes on both waveguides. This assumption is indeed true because the ${}^{II}TM^x_{0,1}$ on waveguide II is indirectly coupled with the ${}^{I}TM^x_{2,1}$ Eigenmode via the ${}^{I}TM^x_{0,1}$ Eigenmode on waveguide I:

The ${}^{II}TM^x_{0,1}$ Eigenmode on waveguide II excites the ${}^{I}TM^x_{0,1}$ Eigenmode on waveguide I via the magnetic field as per figure 7.8b. Then, due to figure 7.8a, the ${}^{I}TM^x_{0,1}$ Eigenmode on waveguide I excites the ${}^{II}TM^x_{2,1}$ Eigenmode on waveguide II via the electric field E_x. This Eigenmode in turn excites the ${}^{I}TM^x_{2,1}$ Eigenmode via H_y as per figure 7.8b.

7.2. The Waveguide Matrix

While the interface matrix $\boldsymbol{\vartheta}$ discussed in the previous section is used to model waveguide discontinuities, the waveguide matrix $\boldsymbol{\gamma}$ describes waveguide segments of uniform cross-section. The waveguide matrix was introduced in chapter 5 and its definition can be found in (5.5). The waveguide matrix applicable for any of the waveguide segments contained in the structure is depicted in figure 7.10. As one would expect from the waveguide matrix' definition, the matrix contains two diagonals only.

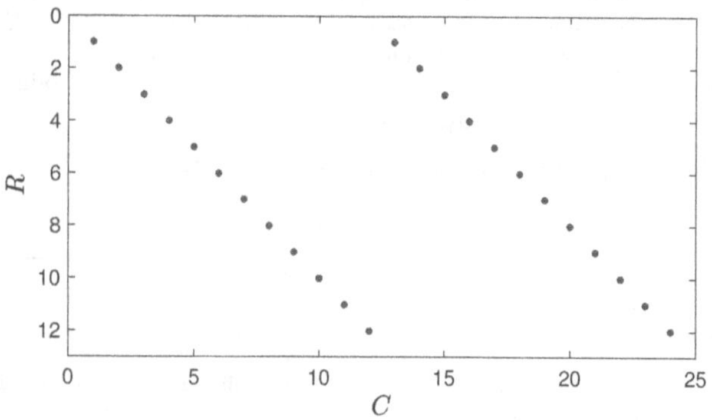

Figure 7.10.: Population pattern of the waveguide matrices

7.3. The Structure's Reduced System Matrix

Having analysed both the required interface and waveguide matrices in the previous sections, we can now study the entire structure's reduced system matrix $\mathbf{\Psi}_{Unknown}$ generated by the Mode Matching solver.

The reduced system matrix is depicted in figure 7.11. The individual segment matrices are indicated by blue boxes. It can be seen that the segment matrices are stacked in a shifted fashion in order to equate the appropriate amplitudes.[2] The red lines indicate the matrix' bandwidth.

The properties of the reduced system matrix are given in table 7.1. Because our analysis considered three Eigenmodes per Eigenmode type and direction of propagation, we have $\mathcal{M} = 12$ amplitudes per junction.[3] Because the structure contains 5 segments, the total number of unknowns is $\mathcal{Q} = 60$ and the reduced system matrix' dimensions are 60×60.

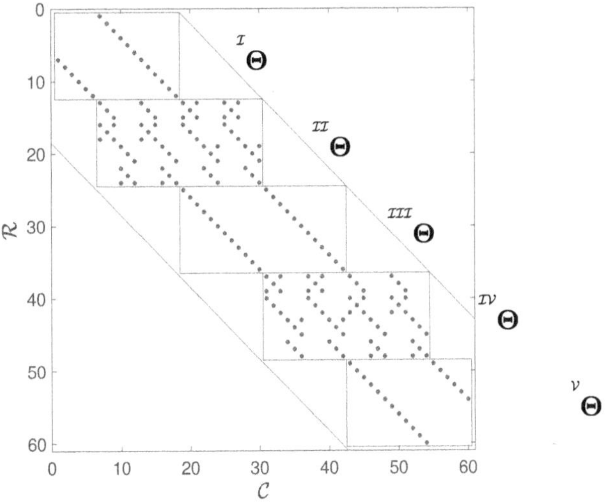

Figure 7.11.: Sparsity pattern of the reduced system matrix $\mathbf{\Psi}_{Unknown}$ for the structure shown in figure 7.1.

Amplitudes per junction $\mathcal{M}$	Segments $\mathcal{N}$	Dimensions $\mathcal{R} \times \mathcal{C}$	Unknowns $\mathcal{Q}$	Elements (see sec. 11.1) ξ	Memory (complex double)
12	5	60×60	60	3.600	56.25 kByte

Table 7.1.: Properties of the reduced system matrix $\mathbf{\Psi}_{Unknown}$ shown in figure 7.11.

[2] While the system matrix is defined for junction amplitudes, in a strict sense, the stacked structure equates waveport amplitudes as discussed in section 6.1.1.

[3] See footnote 1 on page 64 and (4.12) on page 43.

8. Post-Processing: Scattering Parameter Calculations

In the previous chapters we have developed the Mode Matching formalism required to solve the electromagnetic field problems imposed by the presented structures. Obviously, the final aim of our endeavours is to calculate scattering parameters for structures representing modal problems. The calculation of scattering parameters is the focus of this chapter. Additional post-processing steps will be discussed later in chapter 10.

8.1. Scattering Parameter Calculations

Before discussing the calculation of *scattering parameters* from the Eigenmodes' amplitudes obtained by solving the structure's inhomogeneous system of equations, we will briefly revise the definition of scattering parameters.

For an in-depth treatment of the theory of scattering parameters, refer to Kurokawa's classic paper [77] or to the excellent discussions presented in [6, 78]. In the following, we will then discuss means to calculate scattering parameters from the amplitude vectors obtained using Mode Matching.

8.1.1. Definition of Scattering Parameters

Scattering parameters describe the behaviour of an arbitrary $\mathfrak{N}$-port[1] network as is shown in figure 8.1 in terms of incident power waves[2]

$$\mathfrak{a}_i = U_i^+/\sqrt{Z_{0,i}} = \sqrt{Z_{0,i}} I_i^+ \tag{8.1}$$

and scattered power waves

$$\mathfrak{b}_i = U_i^-/\sqrt{Z_{0,i}} = \sqrt{Z_{0,i}} I_i^- \tag{8.2}$$

on all ports $i \in \mathfrak{N}$. U_i^+, I_i^+ represent incident voltage and current waves on port i. Similarly, U_i^-, I_i^- denote scattered voltage and current waves on port i.

[1] In this thesis, the term *port* is used in a scattering parameter sense, that is, it refers to the terminals of a superordinate structure. The term *waveport* is used in the context of interfaces and waveguides as introduced in the previous chapters.

[2] Note that there is an important difference in the nomenclature for Eigenmodes on waveports and power waves on ports. Since from an electromagnetic viewpoint the direction of propagation is most important, Eigenmodes propagating in positive direction of propagation are labeled A and Eigenmodes propagating in the opposite direction are labeled B. In contrast, in view of scattering parameters, incident power waves, which may propagate in positive or negative direction, are labeled $\mathfrak{a}$. Scattered power waves, which also may propagate in both directions, are labeled $\mathfrak{b}$.

$Z_{0,i}$ denotes the *characteristic impedance* of the considered Eigenmodes on port i [6], which represents the ratio of a voltage and a current wave. In contrast, the *wave impedance* denotes the ratio of an electric and magnetic field component perpendicular to the direction of propagation.

The ratio of an incident and scattered power wave is described by scattering coefficients

$$s_{ij} = \frac{\mathfrak{b}_i}{\mathfrak{a}_j}, \tag{8.3}$$

for which $\mathfrak{a}_k$ and U_k^+, I_k^+ respectively are zero for $k \neq j$ by definition [6].

If the amplitudes of incident and scattered power waves are assembled in two vectors $\vec{\mathfrak{a}}$, $\vec{\mathfrak{b}}$ and the scattering coefficients are compiled under form of a scattering matrix $\boldsymbol{S}$, the network's behaviour may be described under form of a vector-matrix equation

$$\vec{\mathfrak{b}} = \boldsymbol{S}\,\vec{\mathfrak{a}}. \tag{8.4}$$

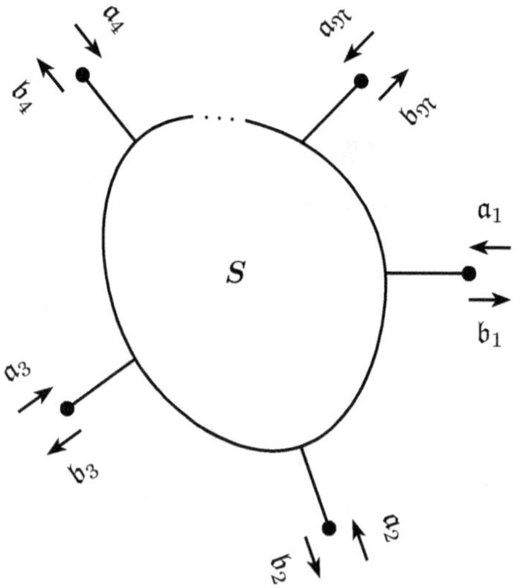

Figure 8.1.: On the definition of the scattering matrix for a $\mathfrak{N}$-port network

8.1.2. Calculating Scattering Parameters from Eigenmode Amplitudes

While scattering parameters can easily be calculated using

$$s_{ij} = \frac{U_i^- / \sqrt{Z_{0,i}}}{U_j^+ / \sqrt{Z_{0,j}}}, \tag{8.5}$$

which follows directly from (8.3) with (8.1) and (8.2), correctly determining the voltage waves and characteristic impedances requires a sound understanding of the definitions of voltage, current and characteristic impedance in the examined waveguides. Although these definitions are of crucial importance, an in-depth review of these quantities was moved to appendix C to keep this chapter as concise as possible.

If the structure comprises rectangular waveguides, voltages on the structure's ports are calculated as per (C.21) and the waveguides' characteristic impedances are obtained from (C.24). If the structure consists of coaxial waveguides, voltages are determined using (C.9) and the characteristic impedances are calculated from (C.11).

The fundamental Eigenmodes' amplitudes required for calculating port voltages are obtained from the amplitude vectors $\vec{a}_{Unknown}$ and $\vec{a}_{Known}$.

When calculating the voltage on the waveguides' ports, it is of crucial importance to observe the dependence[3] of the electric field distribution's sign on the Eigenmode's direction of propagation. Failure to do so will lead to scattering parameters with incorrect phase terms.

8.1.3. Passivity Check on Scattering Parameters

In order to validate energy conservation of the obtained scattering parameters, a passivity check can be performed by validating that

$$1 = |s_{11}|^2 + |s_{21}|^2 \qquad 1 = |s_{22}|^2 + |s_{12}|^2 \tag{8.6}$$

applies for the fundamental Eigenmode. Here we assume that all other Eigenmodes are below cut-off as well as that the structure is lossless.

While performing a passivity check in order to validate whether a simulation has converged appears to be a viable approach at first sight, in chapter 9 we will see that convergence cannot be verified by such means. This is because energy conservation is always fulfilled, independently of the number of Eigenmodes retained [74].

Still, performing a passivity check might be useful in order to detect unphysical structures (see sec. 9, page 75f) as well as to discover flaws in the implementation of the Mode Matching algorithm or the scattering parameter calculation.

[3] The change in sign is due to the fact that calculating the electric field components from a vector potential requires calculating the derivative of the corresponding vector potential, which includes a wave propagation term $e^{\mp j k_z z}$, w.r.t. z, i.e. the direction of propagation. This issue arises with all types of waveguides considered in the scope of this thesis.

9. Convergence of Modal Problems

When discussing computational electromagnetics methods such as the Mode Matching technique, *convergence* obviously is a topic of considerable importance.

In the context of Mode Matching, the decision if a simulation has converged can be reformulated as the question whether the infinite series (4.9) used for expanding the tangential fields at the interface have properly been truncated so that a stable solution is obtained. Depending on the complexity of the interface, a considerable number of infinite series have to be truncated in a proper manner and as a consequence, the appropriate choice of the number of Eigenmodes retained is a difficult one to make.

The convergence of Mode Matching solutions in conjunction with the proper truncation of the infinite series has extensively been studied in literature (see e.g. [29, 38, 39, 67]). It is a widely accepted fact that the optimal ratio of the number of Eigenmodes retained on two adjacent waveguides is identical to the waveguides' dimensions ratios [38] as we have already discussed in section 4.3.3.

While the waveguides' dimensions dictated the proper ratio of the number of Eigenmodes retained on two adjacent waveguides, the question how many Eigenmodes are actually required to properly model parasitic effects in the structure has not yet been answered.

The most "natural approach" [38] to overcome this problem is to observe the variation of the quantities which shall be obtained from the simulation when the numbers of Eigenmodes included in the analysis is increased [38, 79][1]. A comparable approach will be introduced in the following section of this chapter.

For completeness sake, it should be mentioned that when using the Mode Matching technique, the energy balance of modal problems does not suit as a means to observe convergence of the solution. Most notably, a passivity check as introduced in section 8.1.3 will always hold as long as the structure to be solved is physically meaningful.

The reason for this is the fact that in a lossless structure, energy conservation is always fulfilled, independently from the number of Eigenmodes retained [74]. This point becomes obvious when remembering that the Mode Matching technique is based on the orthogonality relation (2.60), which preserves complex power over the discontinuity [67] (also see section 2.5.1).

A different point of view [74] on energy conservation becomes obvious when considering the fact that the Mode Matching technique is based on orthogonal expansions of the waveguides' Eigenmodes, for which Parseval's theorem [48, 62] applies:

Parseval's Theorem states that the integral of the square integrated field distribution to be expanded is equal to the sum of the squared amplitudes of the basis functions, that is, the power in the field distribution to be expanded is equal to the sum of powers contained in the individual Eigenmodes used as basis functions.

[1] while at the same time underlining the importance of the proper ratio of the number of Eigenmodes retained.

Let us now introduce a convergence criterion applicable for modal problems where our aim is to determine scattering parameters. In order to assess convergence of a Mode Matching solution to such a problem we calculate

$$\begin{aligned}
\Delta s_{11} &= |s_{11}^{\mathfrak{I}} - s_{11}^{\mathfrak{I}-1}| \\
\Delta s_{21} &= |s_{21}^{\mathfrak{I}} - s_{21}^{\mathfrak{I}-1}| \\
\Delta s_{12} &= |s_{12}^{\mathfrak{I}} - s_{12}^{\mathfrak{I}-1}| \\
\Delta s_{22} &= |s_{22}^{\mathfrak{I}} - s_{22}^{\mathfrak{I}-1}|
\end{aligned} \tag{9.1}$$

where $s_{ij}^{\mathfrak{I}}$, $s_{ij}^{\mathfrak{I}-1}$ denote scattering parameters obtained from two successive Mode Matching solutions. For the iteration $\mathfrak{I}$ a larger number of Eigenmodes was retained than for the iteration $\mathfrak{I}-1$.

As the speed of convergence of the individual scattering parameters may vary considerably, we consider the largest variation between two sets of scattering parameters, that is,

$$\Delta s_{max} = \max(\Delta s_{11}, \Delta s_{21}, \Delta s_{12}, \Delta s_{22}). \tag{9.2}$$

If Δs_{max} has become smaller than a predefined threshold value, the simulation is assumed to have converged.

We will apply this convergence criterion in order to assess the Mode Matching solution for a rectangular waveguide filter in chapter 13.4.1 later on.

This convergence criterion is similar to that used by Ansys HFSS, a FEM solver, ("Max Delta S") to observe the convergence of the adaptive meshing process by inspecting the maximum difference between the scattering parameters of two mesh iterations [80].

10. Advanced Post-Processing: Equivalent Circuits

The calculation of scattering parameters for modal problems has already been studied in chapter 8 of this thesis. However, for a large number of problems, equivalent circuits for waveguide discontinuities provide additional insight, which may be of crucial help for solving the underlying engineering problem. A classic example is waveguide filter design:

On the one hand, equivalent circuits for waveguide discontinuities provide information on how the filter's lumped component representation can be translated into a waveguide filter design. Furthermore, the actual geometry of the waveguide discontinuity corresponding to the replaced lumped component's value can be calculated from the equivalent circuit. For instance, as we will see in section 13.4, in a rectangular waveguide filter design, the waveguide irises discussed in section 10.2 of this chapter can be used to realise shunt inductances.

On the other hand, equivalent circuits may be used to examine and describe the parasitic effects of waveguide discontinuities on filter designs. For example, the tubular stepped-impedance filters discussed in section 13.2 are detuned by the parasitic capacitive behaviour of the inner conductor's steps discussed in section 10.2 of this chapter.

The calculation of equivalent circuits for waveguide discontinuities will be discussed in the following section 10.1 of this chapter.

In contrast, when solving resonant problems using the Mode Matching technique as discussed in chapter 12, entirely different means are required for post-processing. Most notably, for such problems the unloaded quality factor $\mathfrak{Q}_0$ is of interest. The calculation of a cavity's unloaded Q factor will be discussed in chapter 12.

10.1. Calculation of Equivalent Circuits

Equivalent circuits of waveguide discontinuities are quite helpful to develop a better understanding of the parasitic effects caused by waveguide discontinuities. A practical way to obtain a discontinuity's equivalent circuit is to firstly calculate the impedance matrix $\boldsymbol{Z}$ or the admittance matrix $\boldsymbol{Y}$ from the scattering matrix $\boldsymbol{S}$ introduced in chapter 8. In a next step, $\boldsymbol{Z}$ or $\boldsymbol{Y}$ are mapped to π- or T-topology equivalent circuits.

Unfortunately, textbooks like e.g. [6] only provide schemes to calculate $\boldsymbol{Z}$ from $\boldsymbol{S}$ if all port impedances are equivalent. Obviously, typical Mode Matching problems often cannot fulfil this requirement as a matter of principle. In subsection 10.1.1 a scheme which accounts for different port impedances is developed. This derivation is following [81].

Finally, equivalent circuits for two exemplary discontinuities are discussed in chapter 10.2.

10.1.1. Calculating the Impedance Matrix from the Scattering Matrix

Consider the $\mathfrak{N}$-port network shown in figure 10.1. The total voltage on port i can be denoted as the sum of the incident and scattered voltage wave

$$U_i = U_i^+ + U_i^- \tag{10.1}$$

while the total current calculates as difference between the incident and scattered current wave

$$I_i = I_i^+ - I_i^- . \tag{10.2}$$

The impedance matrix $\boldsymbol{Z}$ applicable for such a network is defined [6]

$$\vec{U} = \boldsymbol{Z}\,\vec{I}. \tag{10.3}$$

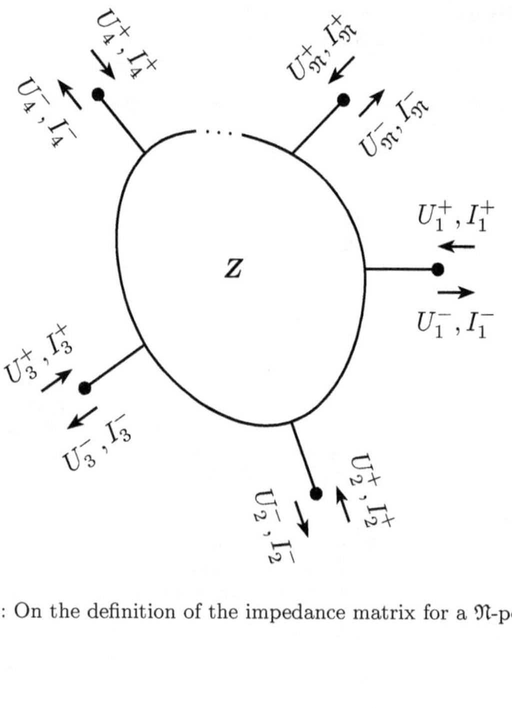

Figure 10.1.: On the definition of the impedance matrix for a $\mathfrak{N}$-port network

The following derivation (eq. 10.4 - 10.8) is due to [81]: Denoting the port voltage and current in terms of power waves gives

$$U_i = \sqrt{Z_{0,i}}\,(\mathfrak{a}_i + \mathfrak{b}_i) \quad \text{and} \quad I_i = 1/\sqrt{Z_{0,i}}\,(\mathfrak{a}_i - \mathfrak{b}_i)\,, \tag{10.4}$$

which follows directly from (8.1) and (8.2). Using (10.4), we can now rewrite (10.3) as

$$\begin{bmatrix} \sqrt{Z_{0,1}}\,(\mathfrak{a}_1 + \mathfrak{b}_1) \\ \vdots \\ \sqrt{Z_{0,\mathfrak{N}}}\,(\mathfrak{a}_\mathfrak{N} + \mathfrak{b}_\mathfrak{N}) \end{bmatrix} = \boldsymbol{Z} \begin{bmatrix} 1/\sqrt{Z_{0,1}}\,(\mathfrak{a}_1 - \mathfrak{b}_1) \\ \vdots \\ 1/\sqrt{Z_{0,\mathfrak{N}}}\,(\mathfrak{a}_\mathfrak{N} - \mathfrak{b}_\mathfrak{N}) \end{bmatrix}.$$

This equation may be rewritten using a newly defined *characteristic impedance matrix*[1]

$$\boldsymbol{Z}_0 = \begin{bmatrix} Z_{0,1} & \cdots & 0 \\ \vdots & \ddots & \vdots \\ 0 & \cdots & Z_{0,\mathfrak{N}} \end{bmatrix} \tag{10.5}$$

where $Z_{0,1}$ to $Z_{0,\mathfrak{N}}$ denote the characteristic impedances of the Eigenmodes considered on the networks' $\mathfrak{N}$ ports as

$$\sqrt{\boldsymbol{Z}_0}\,\left(\vec{\mathfrak{a}} + \vec{\mathfrak{b}}\right) = \boldsymbol{Z}\,\left(\sqrt{\boldsymbol{Z}_0}\right)^{-1}\left(\vec{\mathfrak{a}} - \vec{\mathfrak{b}}\right). \tag{10.6}$$

We now rearrange this equation so that all terms containing the vector of scattered power waves $\vec{\mathfrak{b}}$ are located on the left side while all terms containing the vector of incident power waves $\vec{\mathfrak{a}}$ are on the right side, which gives

$$\left(\sqrt{\boldsymbol{Z}_0} + \boldsymbol{Z}\left(\sqrt{\boldsymbol{Z}_0}\right)^{-1}\right)\vec{\mathfrak{b}} = \left(\boldsymbol{Z}\left(\sqrt{\boldsymbol{Z}_0}\right)^{-1} - \sqrt{\boldsymbol{Z}_0}\right)\vec{\mathfrak{a}}.$$

Next, we multiply by the inverse matrix of $\left(\sqrt{\boldsymbol{Z}_0} + \boldsymbol{Z}\,(\sqrt{\boldsymbol{Z}_0})^{-1}\right)$ and obtain a form

$$\vec{\mathfrak{b}} = \left(\sqrt{\boldsymbol{Z}_0} + \boldsymbol{Z}\left(\sqrt{\boldsymbol{Z}_0}\right)^{-1}\right)^{-1}\left(\boldsymbol{Z}\left(\sqrt{\boldsymbol{Z}_0}\right)^{-1} - \sqrt{\boldsymbol{Z}_0}\right)\vec{\mathfrak{a}},$$

from which we find by comparison with $\vec{\mathfrak{b}} = \boldsymbol{S}\,\vec{\mathfrak{a}}$ (8.4) that

$$\boldsymbol{S} = \left(\sqrt{\boldsymbol{Z}_0} + \boldsymbol{Z}\left(\sqrt{\boldsymbol{Z}_0}\right)^{-1}\right)^{-1}\left(\boldsymbol{Z}\left(\sqrt{\boldsymbol{Z}_0}\right)^{-1} - \sqrt{\boldsymbol{Z}_0}\right). \tag{10.7}$$

[1] Because $\boldsymbol{Z}_0$ is a diagonal matrix,

$$\sqrt{\boldsymbol{Z}_0} = \begin{bmatrix} \sqrt{Z_{0,1}} & \cdots & 0 \\ \vdots & \ddots & \vdots \\ 0 & \cdots & \sqrt{Z_{0,\mathfrak{N}}} \end{bmatrix} \quad \text{and} \quad \boldsymbol{Z}_0^{-1} = \begin{bmatrix} 1/Z_{0,1} & \cdots & 0 \\ \vdots & \ddots & \vdots \\ 0 & \cdots & 1/Z_{0,\mathfrak{N}} \end{bmatrix}.$$

In the following, this expression for the scattering matrix $\boldsymbol{S}$ needs to be rearranged for the impedance matrix $\boldsymbol{Z}$, giving

$$
\begin{aligned}
\left(\sqrt{\boldsymbol{Z}_0} + \boldsymbol{Z}\left(\sqrt{\boldsymbol{Z}_0}\right)^{-1}\right)\boldsymbol{S} &= \left(\boldsymbol{Z}\left(\sqrt{\boldsymbol{Z}_0}\right)^{-1} - \sqrt{\boldsymbol{Z}_0}\right) \\
\sqrt{\boldsymbol{Z}_0}\boldsymbol{S} + \boldsymbol{Z}\left(\sqrt{\boldsymbol{Z}_0}\right)^{-1}\boldsymbol{S} &= \boldsymbol{Z}\left(\sqrt{\boldsymbol{Z}_0}\right)^{-1} - \sqrt{\boldsymbol{Z}_0}. \\
\boldsymbol{Z}\left(\sqrt{\boldsymbol{Z}_0}\right)^{-1} - \boldsymbol{Z}\left(\sqrt{\boldsymbol{Z}_0}\right)^{-1}\boldsymbol{S} &= \sqrt{\boldsymbol{Z}_0}\boldsymbol{S} + \sqrt{\boldsymbol{Z}_0} \\
\boldsymbol{Z}\left(\sqrt{\boldsymbol{Z}_0}\right)^{-1}(\boldsymbol{1} - \boldsymbol{S}) &= \sqrt{\boldsymbol{Z}_0}\,(\boldsymbol{1} + \boldsymbol{S})
\end{aligned}
$$

and finally

$$\boldsymbol{Z} = \sqrt{\boldsymbol{Z}_0}\,(\boldsymbol{1} + \boldsymbol{S})\,(\boldsymbol{1} - \boldsymbol{S})\,\sqrt{\boldsymbol{Z}_0}, \tag{10.8}$$

which is the desired expression for the impedance matrix.

If a π-topology equivalent circuit is employed, it is necessary to convert the impedance matrix into an admittance matrix. This can be done using expressions given in [6, tab. 4.2 on p. 187].

10.1.2. Mapping the Impedance or Admittance Matrix to Equivalent Circuits

Finally, in order to obtain the equivalent circuit of the discontinuity, the coefficients of the impedance or admittance matrix need to be mapped to the component values of the applicable π- or T-topology equivalent circuit. Again we will only summarise the results required to obtain the equivalent circuits' component values. The interested reader is referred to [6] for an in-depth discussion of the subject.

Assuming that the circuit under consideration is reciprocal, the mapping of the impedance matrix to a T-topology equivalent circuit is shown in figure 10.2. The corresponding π-topology equivalent circuit and appropriate expressions for its components are given in [6, fig. 4.13, p. 188].

Note that in order to obtain meaningful component values for the equivalent circuit, some engineering judgement is required in order to select the most appropriate topology.

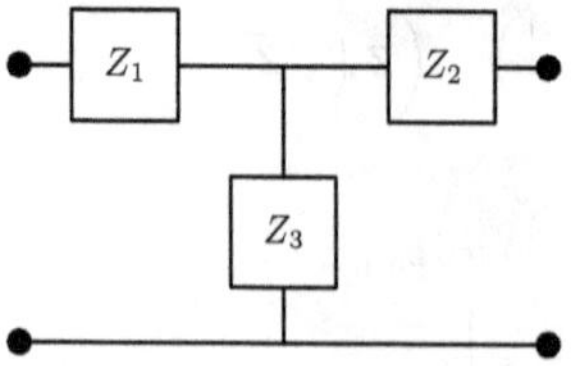

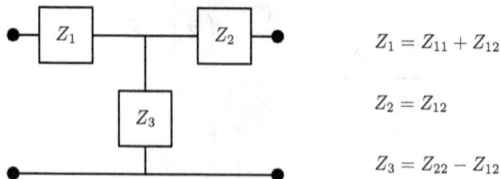

$Z_1 = Z_{11} + Z_{12}$

$Z_2 = Z_{12}$

$Z_3 = Z_{22} - Z_{12}$

Figure 10.2.: Reciprocal T-topology equivalent circuit

10.2. Equivalent Circuits for Exemplary Waveguide Discontinuities

We now conclude our discussion of equivalent circuits by reviewing the equivalent circuits of two waveguide discontinuities, namely of a symmetric iris in an rectangular waveguide and a step of the inner conductor in a coaxial waveguide. Let us briefly discuss the expected behaviour of the discontinuities before comparing results obtained using Mode Matching with analytical solutions from Marcuvitz' Waveguide Handbook [17]:

As will be shown in section B.1.1, a symmetric *iris* in a rectangular waveguide as is depicted in figure 10.4 can be treated taking only $TE_{m,0}$ Eigenmodes into account. In section B.1.1 we will see that these Eigenmodes can be denoted as

$$\begin{aligned} E^{TE}_{y|m,0} &= \pm\, j\, k_{z|m,0}\, A^{TE}_{m,0} \sin\left(k_{x|m} x\right) e^{\mp j k_{z|m,0} z} \\ H^{TE}_{x|m,0} &= -\, j \frac{k^2_{z|m,0}}{\omega\mu}\, A^{TE}_{m,0} \sin\left(k_{x|m} x\right) e^{\mp j k_{z|m,0} z}. \end{aligned} \tag{10.9}$$

In contrast, a step of the inner conductor in a coaxial waveguide as is shown in figure 10.6. can be treated taking only the fundamental TEM mode and rotationally symmetric TM_m Eigenmodes into account. These Eigenmodes denote

$$\begin{aligned} E^{TM}_{\rho|m} &= \mp \frac{j k_{\rho|m} k_{z,m}}{j\omega\varepsilon} \left[N_0(k_{\rho|m} a)\, J_0'(k_{\rho|m}\rho) - J_0(k_{\rho|m} a)\, N_0'(k_{\rho|m}\rho)\right] e^{\mp j k_{z|m} z} \\ H^{TM}_{\phi|m} &= -\quad k_{\rho|m} \left[N_0(k_{\rho|m} a)\, J_0'(k_{\rho|m}\rho) - J_0(k_{\rho|m} a)\, N_0'(k_{\rho|m}\rho)\right] e^{\mp j k_{z|m} z} \end{aligned} \tag{10.10}$$

as is shown in section B.3.1. J_0 denotes zero-order Bessel functions of first kind and N_0 denotes zero-order Bessel functions of second kind (also: Neumann functions).

By applying $Z_0^{TE} = \frac{-E_y}{H_x}$ to (10.9) we find that the wave impedance of a TE Eigenmode calculates as [48]

$$Z^{TE}_{wave} = \frac{\omega\mu}{k_z} \tag{10.11}$$

while the wave impedance of a TM wave is obtained from (10.10) as [48]

$$Z^{TM}_{wave} = \frac{k_z}{\omega\varepsilon}. \tag{10.12}$$

Note that these results are of general nature, that is, (10.11) and (10.12) apply regardless of the coordinate system.

In order to determine the value of the wave impedances, it is necessary to calculate the wavenumber in the direction of propagation k_z. The wavenumber k_z is defined[2] [48, p. 150]

$$k_z = \begin{cases} \sqrt{k^2 - k_c^2} & k > k_c \quad \text{(Propagating)} \\ \frac{1}{j}\sqrt{k_c^2 - k^2} & k < k_c \quad \text{(Non-propagating)} \end{cases} \tag{10.13}$$

where k denotes the free space wave number and k_c represents the cut-off wavenumber [48], whose calculation differs for the various coordinate systems.

However, k_z can be expressed in terms of the cut-off frequency f_c as

$$k_z = \begin{cases} k\sqrt{1 - \frac{f_c^2}{f^2}} & f > f_c \quad \text{(Propagating)} \\ \frac{1}{j}k\sqrt{\frac{f_c^2}{f^2} - 1} & f < f_c \quad \text{(Non-propagating).} \end{cases} \tag{10.14}$$

Inserting (10.14) into (10.11) and (10.12) allows us to examine the frequency dependence of the wave impedance for TE and TM Eigenmodes: As is shown in figure 10.3, if $f/f_c > 1$, the Eigenmode under consideration is propagating, that is, Z_{wave} is real-valued. If however $f/f_c < 1$, the Eigenmode is below cut-off, i.e. non-propagating, and Z_{wave} becomes purely imaginary. In this context, the interested reader is also referred to the discussion of power transport presented in appendix C.

For a TE Eigenmode under cut-off, Z_{wave} becomes positive imaginary, that is, the magnetic field and respectively the current "lag" the electric field and voltage, which is inductive behaviour. For a TM Eigenmode under cut-off, Z_{wave} becomes negative imaginary. The magnetic field and current "lead" the electric field and voltage, which is capacitive behaviour.

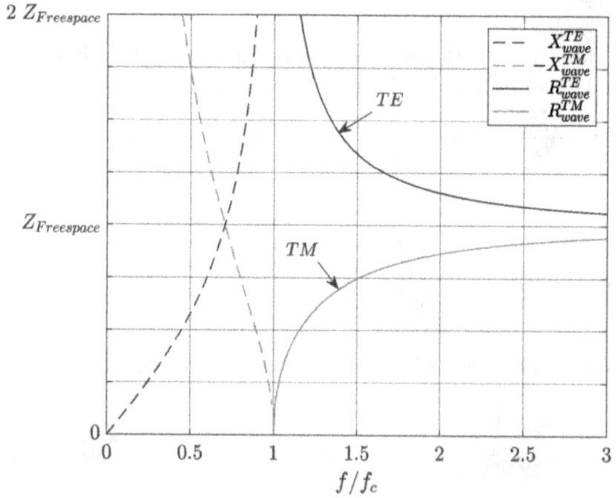

Figure 10.3.: Wave impedance of TE and TM Eigenmodes. This figure is valid regardless of the used coordinate system. A comparable plot can be found in [48].

[2] This definition results from $jk_z = j\beta$ for $k > k_c$ and $jk_z = \alpha$ for $k_c > k$ [48, p. 150].

As already mentioned earlier, at a symmetric iris in a rectangular waveguide as is shown in figure 10.4, only higher-order $TE_{m,0}$ Eigenmodes can be excited. Since TE Eigenmodes exhibit inductive behaviour, the discontinuity can be modelled using a T equivalent circuit with a shunt inductance. The two series impedances can be neglected as their value is close to zero. The obtained equivalent circuit is depicted in figure 10.5b.

In figure 10.5a, the iris' inductance obtained by analysing Mode Matching results as described in this section is compared to the analytical solution presented in [17, sec. 5.2, (1b)]. Both methods show outstanding agreement.

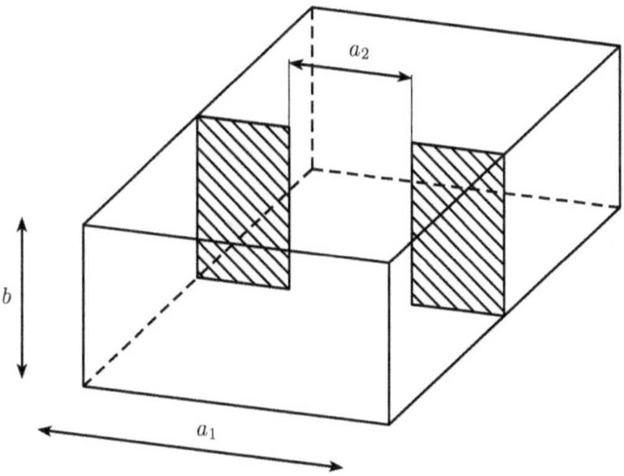

Figure 10.4.: Symmetric iris in a rectangular waveguide

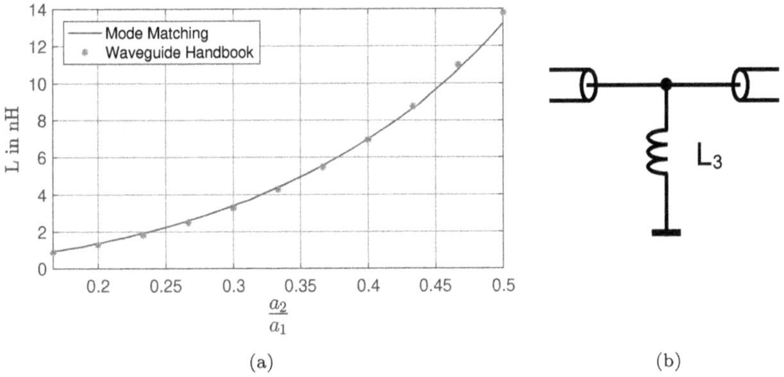

Figure 10.5.: Parasitic inductance of a symmetric iris in a rectangular waveguide $a_1/b = 2$, $f = 3$ GHz

At a step of the inner conductor in a rotationally symmetric coaxial waveguide as shown in figure 10.6, only rotationally symmetric TM_m Eigenmodes may be excited. These Eigenmodes show capacitive behaviour. Consequently, the discontinuity is modelled as T equivalent circuit with a shunt capacitance; again series impedances are neglected due to their value being close to zero. The equivalent circuit is shown in figure 10.7b.

As is illustrated in figure 10.7a, the capacitance calculated from Mode Matching results as discussed in this section shows excellent agreement with Marcuvitz' analytical solution, which is discussed in greater detail in appendix D.

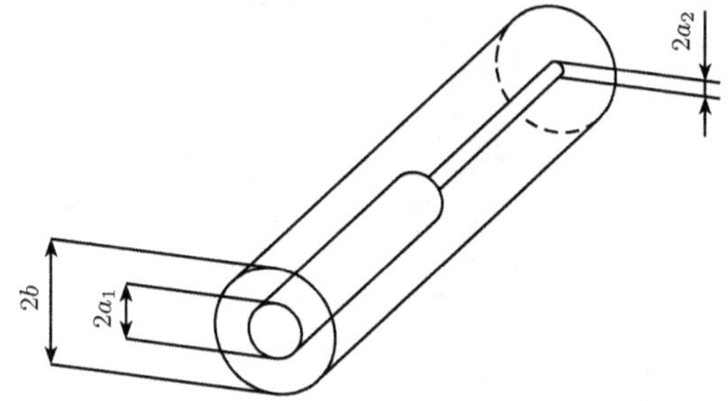

Figure 10.6.: Inner conductor step in a coaxial waveguide

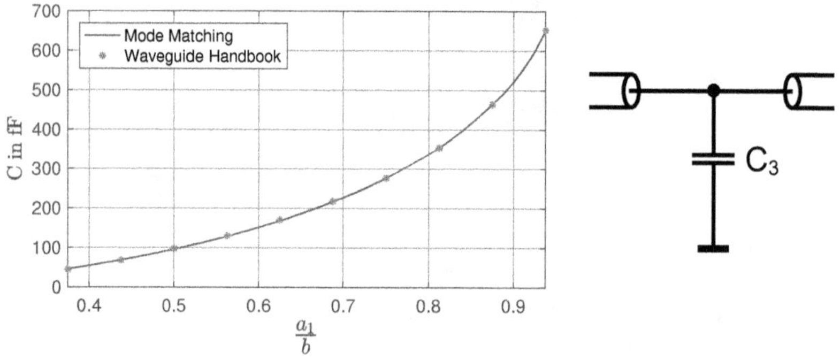

(a) (b)

Figure 10.7.: Parasitic capacitance of an inner conductor step in a coaxial waveguide $a_2/b = 1/8$, $f = 3$ GHz

11. Numerical Aspects of the Mode Matching Technique

In the previous chapters, the Mode Matching formalism and a method for cascading multiple interfaces and waveguides were developed in great detail. This chapter shall give an overview of numerical aspects of the Mode Matching technique as well as of the practical insights gained during the implementation of the method.

When solving field problems using the Mode Matching technique, three issues should be considered, namely:

- Memory requirements of the reduced system matrix $\mathbf{\Psi}_{Unknown}$
- Solving the structure's inhomogeneous system of equations $\check{S} = \mathbf{\Psi}_{Unknown} \cdot \vec{a}_{Unknown}$
- Populating the reduced system matrix, especially solving the overlap integrals of the field distributions at an interface between two adjacent waveguides of different geometry.

In order to get an impression of the relevance of these issues, we now discuss the solution process of a waveguide iris filter with $\mathcal{Q} = 4200$ unknowns, i.e. a structure with 21 segments analysed using $\mathcal{M} = 200$ Eigenmodes per junction. Note that we used an identical number of Eigenmodes for each junction in order to keep calculations simple. All calculations were carried out using standard Matlab routines on an Intel i7 ("Haswell") CPU running at 2.2 GHz.

Firstly, let us consider the size of the reduced system matrix. If the matrix is stored in a regular fashion as described in section 11.1, it requires about 270 MByte of memory if its coefficients are represented as complex numbers with double precision.[1]

The fact that only a limited number of diagonals adjacent to the matrix' main diagonal contains non-zero elements can be exploited by using the compact storage form discussed in section 11.2. This leads to a reduction of the matrix' memory requirements[2] to below 5 MByte.

Although this is an impressive improvement, a critical reader may argue that both matrix sizes may easily be handled by today's computers.

However, if we consider the process of solving the structure's inhomogeneous system of equations, we find that solving the system of equations based on the regularly stored matrix takes about two seconds.

[1] see section 11.1

[2] This is an even better memory reduction than one would obtain from the discussion in the following sections. The reason for this is twofold: Firstly, the presented analysis does not take into account that the matrix bandwidth may become smaller if some Eigenmodes do not couple at the interfaces. Secondly, Matlab's sparsity algorithm may exploit the block nature of the reduced system matrix rather than only exploiting the limited matrix bandwidth.

In contrast, a LU decomposition, a method which may be optimised to operate on matrices stored under compact form as discussed in section 11.3.1 and in greater detail in appendix E solves the system of equations in under 0.1 seconds.

Even at this point, the critical reader may still not be convinced that the optimisations proposed in this chapter are worth the extra effort because even two seconds appear to be an acceptable waiting time.

However, several aspects indeed justify the optimisation of both the storage scheme as well as the method for solving the structure's system of equations:

- Firstly, the structures considered so far are comparably small. For example, higher-order filters might lead to considerably larger systems of equations.
- Secondly, for example the approximation of milling radii at waveguide irises as well as of horn antenna tapers may drastically increase the number of segments included in the structure.
- Finally, a filter design and optimisation tool would greatly benefit from the discussed accelerations because the tool's responsivity to design changes would greatly improve.

Taking these aspects into account, the extra effort for accelerating the solution process by means to be discussed in the following sections is definitely justified.

For completeness sake, section 11.3.2 briefly discusses more advanced (iterative) techniques for solving systems of equations. However, all structures investigated by the author so far were too small to benefit from using such solvers. For example, the system of equations describing the waveguide iris filter discussed above was solved using Matlab's BiCGStab routine in about 0.15 seconds with a residual of less than 10^{-6} including previous preconditioning using an incomplete LU decomposition [82].

So far, we have only considered the computational effort for solving the structure's system of equations. However, populating the reduced system matrix is a computationally expensive task because a considerable amount of overlap integrals of the various Eigenmodes has to be calculated. If the reduced system matrix is populated without any further optimisations, this process may easily take several minutes to complete.

In contrast, if redundancies in the structure are exploited and the frequency independent calculation of the overlap integrals is carried out once only as described in section 11.3.2, the computational effort for populating the reduced system matrix may drastically be reduced. For the waveguide iris filter discussed above, the population process takes about 3.4 seconds. Note that in addition to the previously mentioned approaches, this solver calculates all overlap integrals analytically.

As already pointed out, in the following sections a brief overview of the numerical aspects required for the discussed accelerations will be given. Finally, some thoughts on parallelisation conclude this chapter.

11.1. Dimensions of the Reduced System Matrix

As already discussed in chapter 6.4, the dimensions of the reduced system matrix $\mathbf{\Psi}_{Unknown}$ depicted in figure 11.1 are $\mathcal{R} \times \mathcal{C}$ where

$$\mathcal{R} = \frac{{}^{\mathcal{I}}\mathcal{M}}{2} + \sum_{\nu=\mathcal{II}}^{\mathcal{N}} {}^{\nu}\mathcal{M} + \frac{{}^{\mathcal{N}+1}\mathcal{M}}{2} \qquad \mathcal{C} = \frac{{}^{\mathcal{I}}\mathcal{M}}{2} + \sum_{\nu=\mathcal{II}}^{\mathcal{N}} {}^{\nu}\mathcal{M} + \frac{{}^{\mathcal{N}+1}\mathcal{M}}{2}. \tag{11.1}$$

If ${}^{\nu}\mathcal{M}$ is assumed to be roughly equal for all interfaces, we find that the number of rows and columns of $\mathbf{\Psi}_{Unknown}$ linearly depends on both the number of interfaces $\mathcal{N}$ as well as on the number of Eigenmodes per junction $\mathcal{M}$.[3]

Remembering that the number of unknowns is $\mathcal{Q} = \mathcal{R} = \mathcal{C}$, the matrix' total number of *elements* is

$$\xi = \mathcal{Q}^2 \tag{11.2}$$

and consequently, the number of matrix elements increases quadratically both with the number of Eigenmodes per junction ($\mathcal{M}$) as well as with the number of junctions ($\mathcal{N}$) if we assume that ${}^{\nu}\mathcal{M}$ roughly equal for all interfaces.

If each complex coefficient is stored with double precision, each matrix element requires $2{\times}64\ \mathrm{bit}/8 = 16$ Bytes and thus $\mathbf{\Psi}_{Unknown}$ requires $16 \cdot \xi$ Bytes of memory.

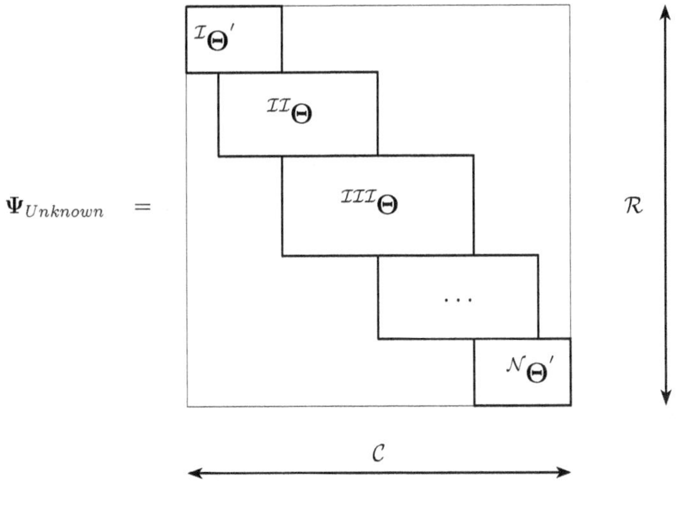

Figure 11.1.: Structure of the reduced system matrix $\mathbf{\Psi}_{Unknown}$

[3] The sums approximately represent a multiplication by $\mathcal{N}$.

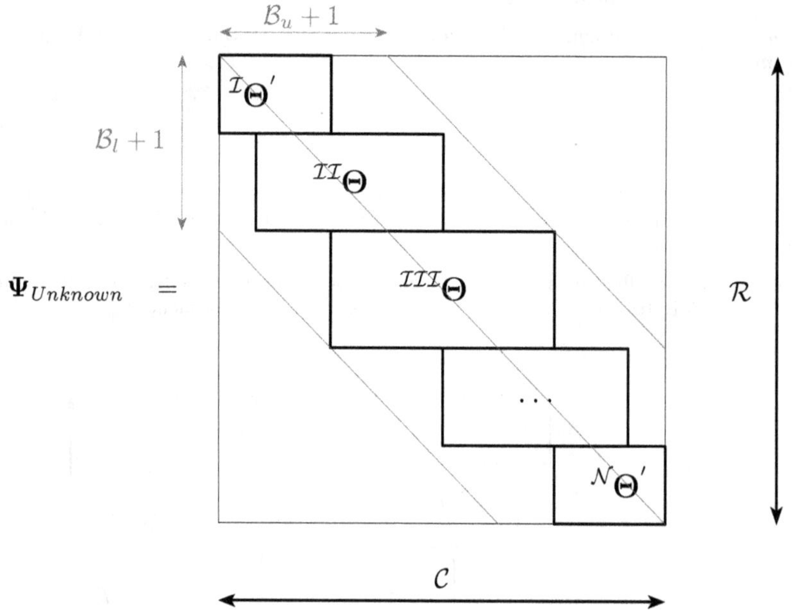

Figure 11.2.: Dimensions and sparsity of the reduced system matrix $\Psi_{Unknown}$

11.2. Sparsity and Banded Structure of the Reduced System Matrix

Regarding the reduced system matrix $\boldsymbol{\Psi}_{Unknown}$, there are two important facts to notice: Firstly, $\boldsymbol{\Psi}_{Unknown}$ is a *sparse matrix*, that is, a considerable amount of the matrix's total number of elements ξ is zero [83]. Secondly, $\boldsymbol{\Psi}_{Unknown}$ is band-limited, which means that only a small number of diagonals adjacent to the main diagonal are populated [83].

In order to assess the matrix' degree of band-limitation, an upper matrix bandwidth $\mathcal{B}_u$ is defined as the number of non-zero diagonals above the main diagonal of the matrix and similarly, a lower matrix bandwidth $\mathcal{B}_l$ is defined as the number of non-zero diagonals below the main diagonal [84]. The overall matrix bandwidth $\mathcal{B}$ is defined as the total number of non-zero diagonals, that is,

$$\mathcal{B} = \mathcal{B}_u + \mathcal{B}_l + 1 \tag{11.3}$$

where +1 is added to account for the main diagonal [85].

Under the assumption that the widest segment matrix is not located at the end of the structure, by careful inspection of figure 11.2 we find

$$\mathcal{B}_u = \max\left\{\frac{{}^{\nu}\mathcal{M}}{2} + {}^{\nu+1}\mathcal{M}\right\} - 1 \qquad \text{for } \mathcal{I} \leq \nu < \mathcal{N} \tag{11.4}$$

and similarly under the assumption that the widest segment matrix is not located at the beginning of the structure

$$\mathcal{B}_l = \max\left\{{}^{\nu}\mathcal{M} + \frac{{}^{\nu+1}\mathcal{M}}{2}\right\} - 1 \qquad \text{for } \mathcal{I} < \nu \leq \mathcal{N}\ . \tag{11.5}$$

Having analysed the band-limited nature of the reduced system matrix's structure, we can now exploit this property in order to store the matrix using a compact storage scheme under form of a $\mathcal{Q} \times \mathcal{B}$ matrix (cf. e.g. [86]) with

$$\xi_{Compact} = \mathcal{Q}\mathcal{B} \tag{11.6}$$

elements.

If we assume the number of Eigenmodes for each junction to be roughly identical, from (11.1) it can be seen that the number of rows and thus the number of unknowns $\mathcal{Q}$ grows linearly with the number of segments $\mathcal{N}$ involved. At the same time, from (11.4) and (11.5) it becomes obvious that the matrix bandwidth is independent of $\mathcal{N}$.

In other words, this means that the matrix's number of coefficients scales linearly with the size of the structure. This is an important advantage of the compact storage form over the regular, that is, dense, storage form where ξ grows with the square of $\mathcal{N}$.

11.3. Computational Effort for Solving the Structure's System of Equations

After having seen how the band-limited structure of the reduced system matrix $\Psi_{Unknown}$ can be exploited to reduce its memory requirements, the obvious question is if and how the band-limited structure of the matrix may be used to minimise the computational effort for solving the structure's system of equations.

It is well known that standard Gaussian elimination for a system of equations with a coefficient matrix $\mathbb{K}^{\mathcal{Q}\times\mathcal{Q}}$, i.e. $\mathcal{Q}$ unknowns, is a $\mathcal{O}(\mathcal{Q}^3)$ process[4] [86,87], that is, the required number of floating point operations (flops) to solve the system of equations scales with the third power of the number of unknowns.

In section 11.1 we have seen that for the reduced system matrix the number of unknowns $\mathcal{Q}$ scales both linearly with the number of interfaces $\mathcal{N}$ and the number of Eigenmodes per junction $\mathcal{M}$ if we assume the number of Eigenmodes on the structure's junctions to be roughly equivalent. As a consequence, the computational effort for solving a structure's system of equations scales with the third power of both terms, i.e.

$$\mathcal{O}(\mathcal{N}^3) \quad \text{and} \quad \mathcal{O}(\mathcal{M}^3). \tag{11.7}$$

11.3.1. LU Decomposition for Dense and Sparse Banded Matrices

A technique much more common for solving small and medium sized problems [83] than standard Gaussian elimination is *LU decomposition* [83,86,88], which is presented in greater detail in appendix E.

While LU decomposition of dense matrices is a $\mathcal{O}(\mathcal{Q}^3)$ process similar to standard Gaussian elimination, a sparse banded matrix with an upper bandwidth $\mathcal{B}_u$ and a lower bandwidth $\mathcal{B}_l$ can be decomposed with a computational effort of [88]

$$2\mathcal{B}_l(\mathcal{B}_l + \mathcal{B}_u)\mathcal{Q}. \tag{11.8}$$

This assumes additional row pivoting to be used [88]. As a consequence, solving a system of equations becomes an

$$\mathcal{O}(\mathcal{Q}\,\mathcal{B}_l^2) \qquad \mathcal{O}(\mathcal{Q}\,\mathcal{B}_u) \tag{11.9}$$

process.

Let us assess the impact of these findings on decomposing the reduced system matrix under the assumption that the number of amplitudes on all junctions is roughly equivalent.

On the one hand, since $\mathcal{O}(\mathcal{Q}\,\mathcal{B}_l^2)$ applies, the computational effort for solving the structure's system of equations still scales with the third power of the number of Eigenmodes $\mathcal{M}$ per junction as $\mathcal{M}$ linearly governs the matrix' bandwidths $\mathcal{B}_u$ and $\mathcal{B}_l$ as well as the number of unknowns $\mathcal{Q}$. However, fortunately for most structures very good results can be achieved using a comparably low number of Eigenmodes $\mathcal{M}$, thus keeping the computational effort down to an acceptable level.

On the other hand, as a consequence of $\mathcal{O}(\mathcal{Q}\,\mathcal{B}_l^2)$, the computational effort scales linearly with the number of unknowns $\mathcal{Q}$ and because of (11.1) with the number of interfaces $\mathcal{N}$.

[4] The $\mathcal{O}(\mathcal{Q}^x)$ notation implies that the process' operation count is a function of terms of order $\mathcal{Q}^x$ and lower [87].

11.3.2. Outlook on Advanced Techniques for Solving Systems of Equations

While all structures presented in this thesis were solved either by Matlab's function **mldivide** (aka backslash operator) or its LU decomposition routine **lu** in a satisfying manner, an outlook on other, more advanced solving techniques appears appropriate. For further reading on the subject, the interested reader is referred to [86] and [88], which together give an excellent introduction.

An interesting question to examine is if and how segment matrices can be manipulated in order to obtain a hermetian, positive definite reduced system matrix. If the reduced system matrix could be put under such form, *Cholesky decomposition* [83, 86, 88], which strives to decompose a matrix $\boldsymbol{A}$ as

$$\boldsymbol{A} = \boldsymbol{L}\boldsymbol{L}^T \tag{11.10}$$

where $\boldsymbol{L}$ is a lower triangular matrix could be employed. Cholesky decomposition is twice as fast [88] as LU decomposition and extremely stable without pivoting applied [86].

Applying iterative methods for solving the system of equations of larger structures would also make an interesting field of research: Although all structures discussed in the scope of this thesis were not too large for direct solvers nor did they require iterative methods in view of computation time, iterative solvers nevertheless represent an interesting subject.

One of the most commonly used iterative techniques [88] to solve large sparse systems are conjugate gradient methods (CG methods) [86, 88]. The idea behind the basic conjugate gradient method is to minimize a function

$$\mathfrak{f}(\vec{x}) = \frac{1}{2}\vec{x}\,\mathbf{A}\,\vec{x} - \vec{b}\,\vec{x}, \tag{11.11}$$

which is minimal if

$$\nabla\mathfrak{f} = \mathbf{A}\,\vec{x} - \vec{b} = \vec{0}, \tag{11.12}$$

that is, $\mathbf{A}\,\vec{x} = \vec{b}$ is satisfied [86].

It should be noted that this method theoretically represents a direct method as the system's solution is found after a finite number of iterations. However, in a practical application, it is avoided to calculate the full number of iterations as the method provides an acceptable approximation of the exact solution after a small number of iterations. Thus, the CG method typically is regarded as an iterative method [88].

Similar to the Cholesky decomposition, the basic CG method requires a symmetric, positive definite matrix [88]. This again underlines the importance of putting the reduced system matrix under such form.

Alternatively, the biconjugate gradient method (BiCG) could be used, which does not necessarily require a symmetric, positive definite matrix [88]. Note that this method is no longer directly based on the principle of function minimisation discussed previously [86].

Finally, biconjugate stabilised gradient methods (BiCGStab) should be mentioned, which additionally strive to minimise oscillations in the residual, which occur when using the BiCG method [88].

11.4. Populating the Reduced System Matrix

In the previous section, we have extensively discussed the required effort for solving the system of equations corresponding to a given structure under consideration. However, in practical applications, often the computational effort for populating the reduced system matrix describing the structure is considerably larger than the effort for solving the structure's system of equations, especially if the solver exploits the sparsity of the reduced system matrix as discussed in the previous section.

The reason for this is that in order to populate the reduced system matrix, the overlap integrals contained in the segment matrices corresponding to waveguide interfaces must be calculated numerically. This is obviously a process which is costly in terms of its computational effort.

Besides the obvious alternative to directly implement the analytical solution of the overlap integrals[5], there are also other approaches which accelerate matrix population by either exploiting the fact that common structures often contain duplicate segments or by re-using results of computational costly but frequency independent calculations.

11.4.1. Interface Reusing without and with Flipping

In most structures which are analysed using the Mode Matching technique a particular interface between two waveguides of different geometry occurs more than once. An obvious possibility to minimise the effort required for populating the reduced system matrix is to reuse a segment matrix calculated for an interface's first occurrence for all subsequent appearances. In doing so, recalculating the segment matrices for each single interface by solving the corresponding overlap integrals is avoided.

This approach becomes especially beneficial for analysing filters because several common filter polynomials lead to symmetric structures [89, 90], that is, the filter's physical realisation (which is the structure to be analysed using the Mode Matching technique) is symmetric with respect to a centre plane between the filter's input and output [90]. Consequently only half the number of segment matrices (or less) corresponding to waveguide interfaces need to be calculated.

Let ${}^{\nu_1}\boldsymbol{\Theta}$ and ${}^{\nu_2}\boldsymbol{\Theta}$ be segment matrices corresponding to identical waveguide interfaces of similar orientation. During population of the reduced system matrix, we may exploit the fact that

$$ {}^{\nu_1}\boldsymbol{\Theta} = {}^{\nu_2}\boldsymbol{\Theta} \tag{11.13} $$

to efficiently populate the matrix's elements by simple copying. Obviously, in order to enable interface reusing, careful bookkeeping of the segment matrices' storage locations in the reduced system matrix' array is mandatory.

The previously introduced technique shall be called *interface reusing without flipping*.

[5] Analytical calculation of the overlap integrals is more difficult than one would expect in the first place. The reason for this is that if the transverse wavenumbers of two Eigenmodes are identical, additional limit values have to be considered.

Interface reusing without flipping only allows to reuse segment matrices for waveguide interfaces which occur with the same orientation. However, structures to be analysed often contain waveguide interfaces which have duplicates with both similar and opposite orientation. For example, in a waveguide iris filter, at each iris, two similar waveguide interfaces with opposite orientations occur.

By performing *interface reusing with flipping*, the required segment matrix of a waveguide interface with opposite orientation can be derived from the segment matrix of an already processed waveguide interface (with "regular" orientation) without performing any costly calculations e.g. for recalculating overlap integrals again.

In order to do so, by careful inspection of the definition of the segment matrix given in (6.3) we find that interface reusing with flipping can easily be carried out during the copying process if simple swapping of the left and right part of the segment matrix obtained earlier is performed.

From a mathematical point of view the process of column swapping can be denoted for two identical interfaces of opposite orientation with segment matrices ${}^{\nu_1}\boldsymbol{\Theta}$, ${}^{\nu_2}\boldsymbol{\Theta}$ as matrix multiplication

$$ {}^{\nu_2}\boldsymbol{\Theta} = {}^{\nu_1}\boldsymbol{\Theta} \cdot {}^{\nu_1}\boldsymbol{\mathcal{P}} \tag{11.14} $$

where

$$ {}^{\nu_1}\boldsymbol{\mathcal{P}} = \begin{bmatrix} \mathbf{0}_{{}^{\nu_1}\mathcal{M} \times {}^{\nu_1+1}\mathcal{M}} & \mathbf{1}_{{}^{\nu_1}\mathcal{M} \times {}^{\nu_1}\mathcal{M}} \\ \mathbf{1}_{{}^{\nu_1+1}\mathcal{M} \times {}^{\nu_1+1}\mathcal{M}} & \mathbf{0}_{{}^{\nu_1}\mathcal{M} \times {}^{\nu_1+1}\mathcal{M}} \end{bmatrix} \tag{11.15} $$

is a *permutation matrix*[6] for permuting the segment matrix of the v_1th segment. $\mathbf{1}$ denotes unity matrices and $\mathbf{0}$ represents matrices containing zeros only. The matrices' dimensions are given in the subscript.

Interface reusing without and with flipping is illustrated in figure 11.3 on the following page.

[6]A permutation matrix is used to manipulate the order of rows and columns of a matrix by proper matrix multiplication. As a matter of principle, each and every column and row of the permutation matrix may only contain one unity element while all other elements vanish [88].

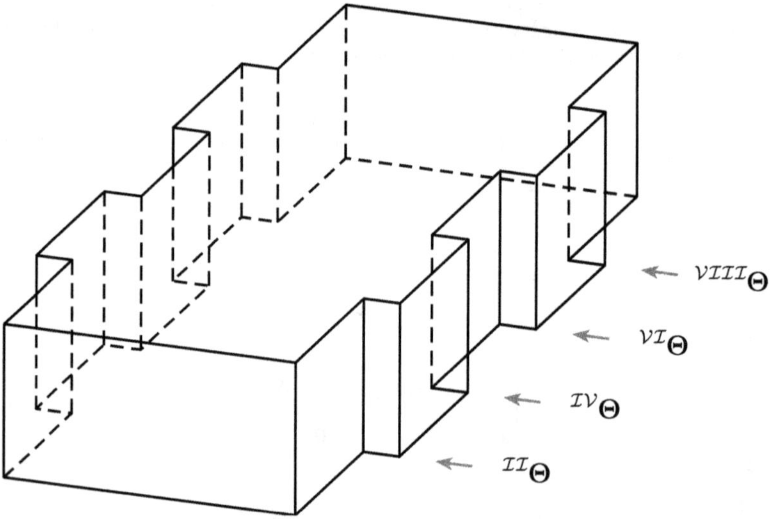

Figure 11.3.: Interface reusing without and with flipping for an exemplary structure. The segment matrices ${}^{\mathcal{II}}\boldsymbol{\Theta}$ and ${}^{\mathcal{VI}}\boldsymbol{\Theta}$ are equal, that is, the segment matrices can be reused without flipping.
For the segment matrices ${}^{\mathcal{IV}}\boldsymbol{\Theta}$, ${}^{\mathcal{VIII}}\boldsymbol{\Theta}$ we have ${}^{\mathcal{IV}}\boldsymbol{\Theta} = {}^{\mathcal{VIII}}\boldsymbol{\Theta} = {}^{\mathcal{II}}\boldsymbol{\Theta}\,{}^{\mathcal{II}}\boldsymbol{\mathcal{P}}$, that is, the segment matrix ${}^{\mathcal{II}}\boldsymbol{\Theta}$ may be reused if column swapping is performed.

11.4.2. Separation of Factorisation and Integration during the Calculation of Overlap Integrals

As we have seen in section 6.3, in order to populate the reduced system matrix of a given structure, the segment matrices corresponding to the structure's interfaces must be calculated. In order to do so, the segment matrices' sub-matrices

$$\vartheta_{{}^{II}\mathcal{E}_x^{TM}}^{{}^{I}\mathcal{E}_x^{TM}} \quad \vartheta_{{}^{II}\mathcal{E}_y^{TE}}^{{}^{I}\mathcal{E}_y^{TE}} \quad \vartheta_{{}^{II}\mathcal{E}_y^{TE}}^{{}^{I}\mathcal{E}_y^{TM}} \quad \vartheta_{{}^{I}\mathcal{H}_x^{TE}}^{{}^{II}\mathcal{H}_x^{TE}} \quad \vartheta_{{}^{I}\mathcal{H}_y^{TM}}^{{}^{II}\mathcal{H}_y^{TE}} \quad \vartheta_{{}^{I}\mathcal{H}_y^{TM}}^{{}^{II}\mathcal{H}_y^{TM}} \quad \vartheta_{{}^{II}\mathcal{E}_y^{TM}}^{{}^{II}\mathcal{E}_y^{TE}} \quad \vartheta_{{}^{I}\mathcal{H}_y^{TE}}^{{}^{I}\mathcal{H}_y^{TM}} \tag{11.16}$$

each containing coefficients as per (4.17) must be calculated, that is, each coefficient is a quotient of two inner products, which represent overlap integrals as discussed in chapter 3. While equations in (4.17) apply for rectangular coordinates, similar equations may be derived for other coordinate systems.

Recall from section 4.3.1 that $\mathcal{E}$ and $\mathcal{H}$ in (4.17) represent the unsigned field distributions of the Eigenmodes of the appropriate waveguide.

For the following discussion of the separation of factorisation an integration during calculation of the inner products, we consider arbitrary unsigned field distributions ${}^{\eta}\mathcal{F}$ and ${}^{\eta+1}\mathcal{F}$. By careful inspection of the Eigenmodes of the appropriate waveguides derived later in this thesis, it can be seen that the unsigned field distribution of each and every Eigenmode can be split into two factors

$$\mathcal{F} = \mathcal{F}_{pre} \cdot \mathcal{F}_{shape}. \tag{11.17}$$

$\mathcal{F}_{pre}$ is a prefactor, which may contain all applicable wavenumbers, e.g. k, k_x, k_y, k_z, if rectangular coordinates are considered as well as ε, μ, the imaginary unit j and the angular frequency ω. Consequently, $\mathcal{F}_{pre}$ is frequency dependent.

In contrast, $\mathcal{F}_{shape}$ contains *shape functions* determining the actual shape of the Eigenmodes' field distribution, i.e. sine and cosine functions if rectangular waveguides are considered or Bessel functions if circular structures are examined.

Obviously, these shape functions must be chosen so that the field distributions maintain the waveguide's boundary conditions at each and every frequency and consequently $\mathcal{F}_{shape}$ is frequency independent. In other words, the arguments of the shape functions are scaled by the frequency independent wavenumbers for directions perpendicular to the direction of propagation in order to ensure that the waveguide's boundary conditions are maintained.

Assuming z to be the direction of propagation, only k_x and k_y may be included in the shape functions if rectangular coordinates are considered. Similarly, if (rotationally symmetric) circular structures are considered, only k_ρ may occur in the shape functions.

Let us now return to calculating the reduced system matrix' coefficients as per (4.17). Using the linearity of the inner product [70], an arbitrary coefficient included in (11.16) can be calculated as

$$\vartheta_{{}^{\eta}\mathcal{F}}^{{}^{\eta+1}\mathcal{F}} = \frac{\left\langle {}^{\eta+1}\mathcal{F} | {}^{\eta}\mathcal{F} \right\rangle}{\left\langle {}^{\eta}\mathcal{F} | {}^{\eta}\mathcal{F} \right\rangle} = \underbrace{\frac{{}^{\eta+1}\mathcal{F}_{pre}^{*}\, {}^{\eta}\mathcal{F}_{pre}}{{}^{\eta}\mathcal{F}_{pre}^{*}\, {}^{\eta}\mathcal{F}_{pre}}}_{\text{Frequency Dependent}} \cdot \underbrace{\frac{\left\langle {}^{\eta+1}\mathcal{F}_{shape} | {}^{\eta}\mathcal{F}_{shape} \right\rangle_{\eta}}{\left\langle {}^{\eta}\mathcal{F}_{shape} | {}^{\eta}\mathcal{F}_{shape} \right\rangle_{\eta}}}_{\text{Frequency Independent}}. \tag{11.18}$$

This is an important outcome as (11.18) shows that the computational expensive task of numerically solving overlap integrals in order to determine inner products only needs to be performed once rather than for each individual point included in the simulation's frequency sweep.

For each frequency point, the pre-calculated overlap integrals only need to be scaled by a frequency-dependent factor, which is obtained by simple addition, multiplication and division.

An unfortunate exception where the separation of factorisation and integration during the calculation of the inner products is impossible are partially filled waveguides as discussed e.g. in [48] for rectangular waveguides and in [68] for partially filled coaxial waveguides.

In contrast to standard waveguides, wavenumbers for directions perpendicular to the direction of propagation are calculated from the waveguide's characteristic equation and thus become frequency dependent. Consequently, at each interface adjacent to such a waveguide, the overlap integrals need to be individually calculated for every frequency point in the simulation's frequency sweep.

11.5. Some Thoughts on Parallelisation

This chapter shall be closed by sketching some possible approaches to parallelising the Mode Matching technique. Let us assume that a manager-worker-architecture [91] is used.

An obvious approach is to calculate several frequency points of the simulation's frequency sweep in parallel, that is, each worker individually calculates a solution for the frequency assigned to it, possibly using precalculated inner products of shape functions as described in section 11.4.2.

Another approach is to distribute the population of the reduced system matrix over several workers: Either, a single worker could calculate all overlap integrals contained in a segment matrix corresponding to a single waveguide interface or, alternatively, individual workers could even be assigned to sub-matrices of a single segment matrices as per (11.16). In every case, after a worker has finished its calculations, it would store the obtained coefficients in the memory range allocated for the (sub-)matrix by the manager prior to job deployment.

Finally, the obtained overall system of equations is solved by an appropriate solver. For the sake of completeness, it should be mentioned that there are also approaches to parallelise solving linear systems of equations [85].

Like for many other parallelisation problems, a reasonable tradeoff between parallelisation and additional communication overhead between the manager and workers needs to be found. Parallelisation of the Mode Matching technique could make an interesting field of research provided that sufficiently large problems need to be solved and thus, the additional communication overhead is small compared to the computational effort of solving the problem.

12. Extension of the Mode Matching Technique for Electromagnetic Cavities

In the previous chapters, we have introduced means to solve modal problems using the Mode Matching technique. As we will see in chapter 15, solving resonant problems is of similar interest. In the following chapter, we will thus extend the Mode Matching formalism to *electromagnetic cavities*.

Firstly, we will briefly revise the rectangular cavity's boundary value problem, whose solutions are the cavity's Eigenmodes and their related Eigenvalues, namely k^2, which yield the cavity's resonant frequencies. Next, in section 12.2 we will then introduce additional boundary conditions in order to describe resonant problems in a Mode Matching sense, inspect the cavity's system matrix and discuss means to solve the underlying homogeneous system of equations. Finally, in section 12.3 we will discuss post-processing for resonant problems, namely the calculation of the energy stored inside the cavity as well as the power dissipated due to dielectric and ohmic losses. From these quantities, we eventually obtain the cavity's unloaded quality factor $\mathfrak{Q}_0$.

12.1. The Electromagnetic Cavity's Boundary Value Problem

Consider the cavity depicted in figure 12.1. In order to determine the field distribution inside the cavity's volume, it is necessary to solve the cavity's boundary value problem defined by the scalar vector potential's differential equation for source-free regions

$$\nabla^2\psi + k^2\psi = 0 \tag{12.1}$$

already introduced in section 2.1.1. The problem's boundary conditions depend on the direction to which the mode-sets are chosen to be transverse[1] to (see section 2.1.2, p. 17) and shall not be discussed in greater detail here.

In every case, (12.1) is solved by separation of variables and subsequent solution of the obtained ordinary differential equations subject to the applicable boundary conditions. The cavity's Eigenvalues k^2 follow from the separation condition [48]

$$k^2 = k_x^2 + k_y^2 + k_z^2 \tag{12.2}$$

introduced in a slightly different form using a cut-off wavenumber k_c in the context of the rectangular waveguide's Eigenmodes in section 2.2.1. From $k^2 = \omega_0^2\varepsilon\mu$ the cavity's resonant frequencies ω_0 are readily available.

It is interesting to note the conceptual similarity of solving cavities, which represent three-dimensional resonators, and the solution process for cylindrical waveguides discussed in chapter 2, which can be interpreted as two-dimensional resonators.

[1] This depends on the orientation of the vector potentials $\mathfrak{A}$ and $\mathfrak{F}$.

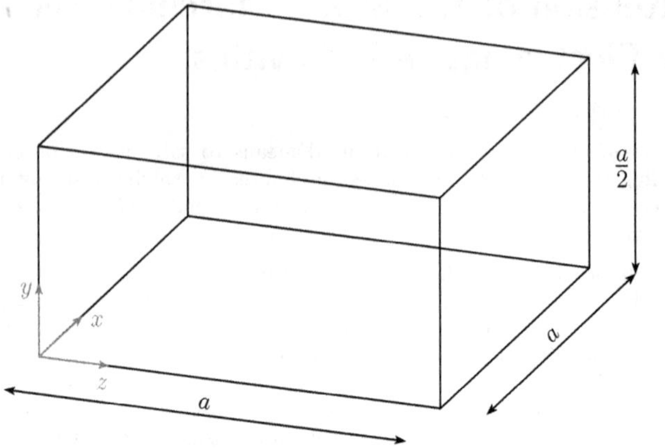

Figure 12.1.: An empty rectangular cavity

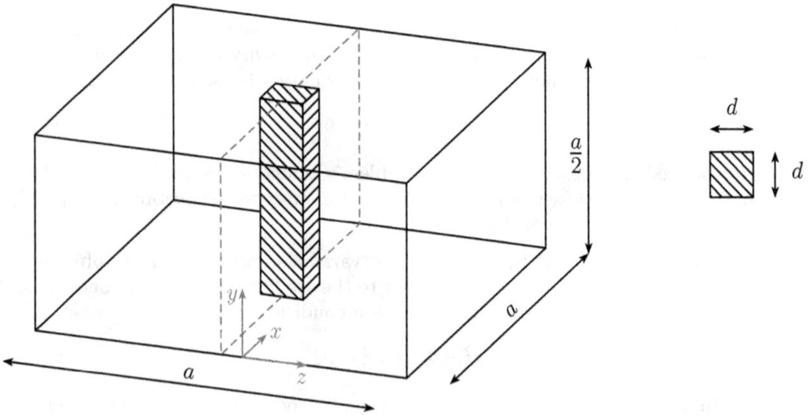

Figure 12.2.: A rectangular cavity perturbed by a centered square dielectric slab

12.2. Mode Matching Formalism for Cavities

Now consider the *perturbed cavity* depicted in figure 12.2. While approximate solutions for such cavities can be obtained using the *cavity perturbation theory* [48] originally introduced by Bethe and Schwinger [92], an exact solution by simple means as presented in the previous section is not possible. However, fortunately, it is possible to obtain quasi-analytical solutions for such cavities using the Mode Matching technique.

In order to do so, we re-interpret the perturbed cavity as a structure composed out of empty and partially filled rectangular waveguides and treat the waveguides' interfaces using the Mode Matching technique. The perturbed cavity's segment matrix representation is depicted in figure 12.4 on the following page.

In order to model the PEC boundary conditions which terminate the structure in z direction by providing a relation between the forward and backward travelling Eigenmodes on junction $\mathcal{I}$ and $\mathcal{VI}$, an additional PEC matrix was used. This matrix will be introduced in the following section.

In the remainder of this thesis, we will only be concerned with the perturbed fundamental TE Eigenmode of the rectangular cavity. Under the assumption that the general form of this Eigenmode remains unchanged, we can exploit the cavity's symmetries to obtain a smaller segment matrix representation of the structure:

The fundamental Eigenmode's magnetic field is normal to the plane indicated by dashed red lines in figure 12.2 and 12.3. Similarly, there is no electric field normal to this plane. Consequently, this plane can be interpreted as a perfectly magnetically conducting (PMC) symmetry plane.

By exploiting this PMC symmetry plane, the cavity's segment matrix representation can be halved as is shown in figure 12.5. The required PMC matrix will be introduced in the following section as well.

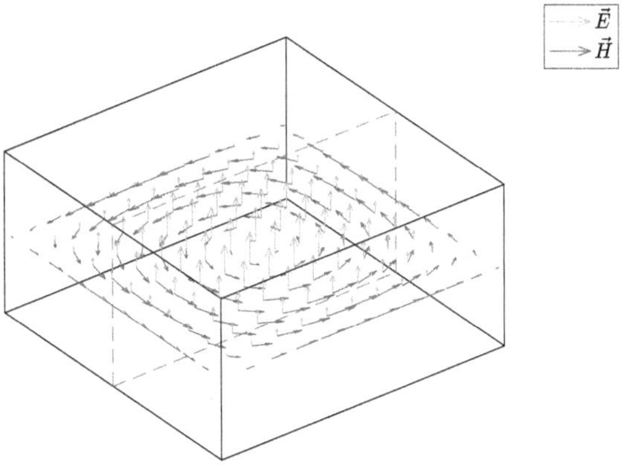

Figure 12.3.: Field distribution of the unperturbed rectangular cavity shown in figure 12.1. PMC symmetry plane indicated by red dashed lines.

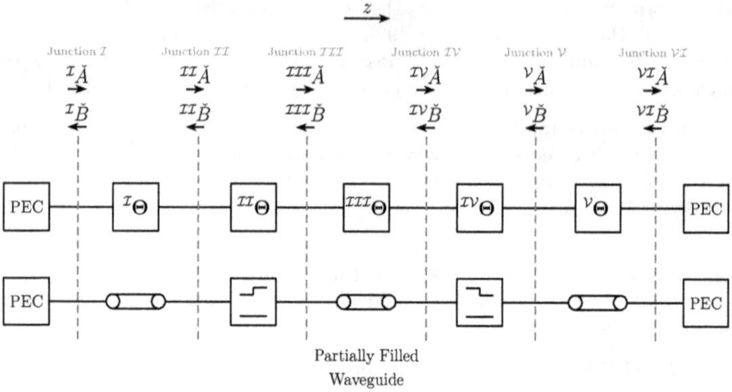

Figure 12.4.: Segment matrix representation of the perturbed cavity

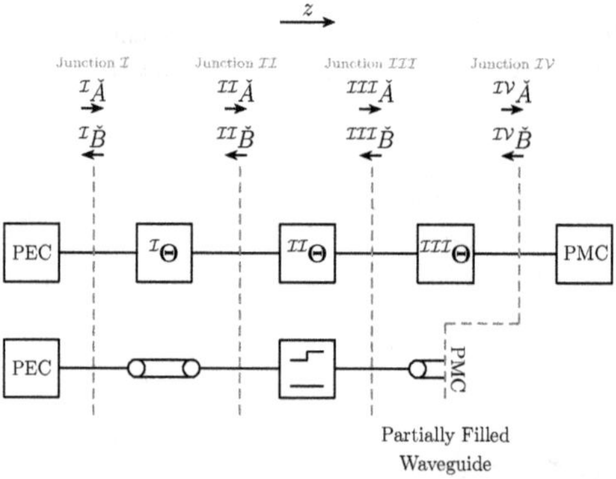

Figure 12.5.: Halved segment matrix representation of the perturbed cavity with symmetries exploited

12.2.1. Additional Boundary Conditions

PEC Matrix

At the PEC wall terminating the structure in z direction, the tangential electric field components, that is, the waveguide's transverse field components must vanish. As the Eigenmodes are orthogonal, each Eigenmode travelling towards the PEC wall can only be cancelled out by the corresponding Eigenmode travelling away from the PEC wall.

In the following, we limit our analysis to TE Eigenmodes. By careful analysis of the Ey component of the $TE_{m,n}$ Eigenmode given in appendix B.1.1 we find that due to the derivation of the scalar vector potential for z the sign of E_y is inverted if the Eigenmode's direction of propagation is reversed. Consequently at junction ν, we have

$$ {}^{\nu}\breve{A}^{TE} = {}^{\nu}\breve{B}^{TE}, \tag{12.3} $$

which may be denoted under form of a ${}^{\nu}\mathcal{M}/2 \times {}^{\nu}\mathcal{M}$ *PEC matrix*

$$ \boldsymbol{\beta}_{PEC} = \begin{bmatrix} \ddots & & \ddots & \\ & 1 & & -1 \\ & & \ddots & & \ddots \end{bmatrix} \tag{12.4} $$

where ${}^{\nu}\mathcal{M}$ denotes the total number of amplitudes on the νth junction.

PMC Matrix

At a PMC wall, tangential magnetic fields, i.e. the Eigenmodes' transverse magnetic field components have to be zero. Following a similar reasoning as for the PEC boundary condition, at junction ν we have

$$ {}^{\nu}\breve{A}^{TE} = -{}^{\nu}\breve{B}^{TE} \tag{12.5} $$

because the sign of the TE Eigenmodes' field component H_x does not change. This again may be denoted under form of a ${}^{\nu}\mathcal{M}/2 \times {}^{\nu}\mathcal{M}$ *PMC matrix*

$$ \boldsymbol{\beta}_{PMC} = \begin{bmatrix} \ddots & & \ddots & \\ & 1 & & 1 \\ & & \ddots & & \ddots \end{bmatrix}. \tag{12.6} $$

12.2.2. Solution Process

In the previous section, we have modeled the perturbed cavity depicted in figure 12.2 under form of a structure composed out of waveguide and interface matrices as well as the newly introduced PEC and PMC matrices. The corresponding system matrix $\boldsymbol{\Psi}_R$ depicted in figure 12.6 is square[2] because in contrast to modal problems, the PEC and PMC boundary conditions provide additional equations to relate the forward and backward travelling Eigenmodes on junction $\mathcal{I}$ and $\mathcal{IV}$.

As a matter of fact, we have put the cavity's field problem under form of a homogeneous system of equations

$$\vec{0} = \boldsymbol{\Psi}_R \, \vec{x}. \tag{12.7}$$

This system of equations has non-trivial solutions ($\vec{x} = \vec{x}_0 \cdot \Pi$) only if $\det(\boldsymbol{\Psi}_R) = 0$, which is true for a discrete spectrum of frequencies only. By numerically finding frequencies for which the determinant of $\boldsymbol{\Psi}_R$ vanishes we find the cavity's resonant frequencies ω_0. This process corresponds to determining the empty cavity's Eigenvalues $k^2 = \omega^2 \varepsilon \mu$ discussed at the beginning of this chapter.

After the desired resonant frequency ω_0 has been selected, the fields inside the cavity can be obtained by summing up the contributions of the individual Eigenmodes to the total field distribution. The Eigenmodes' amplitudes are obtained from the solution vector $\vec{x}$.

The remaining question now is how to obtain the basis vector $\vec{x}_0$ of the solution space. The solution space of a homogeneous system of equations $\boldsymbol{A} \cdot \vec{x} = \vec{0}$ is equal to the null-space of the matrix $\boldsymbol{A}$ [93], whose orthonormal basis can be constructed by singular value decomposition of $\boldsymbol{A}$ [86]. Thus, by performing singular value decomposition of $\boldsymbol{\Psi}_R$, the solution space's basis vector $\vec{x}_0$ can be obtained.

From a physical point of view, the fact that $\vec{x}$ can be scaled arbitrarily means that the energy stored inside the cavity can be chosen freely.

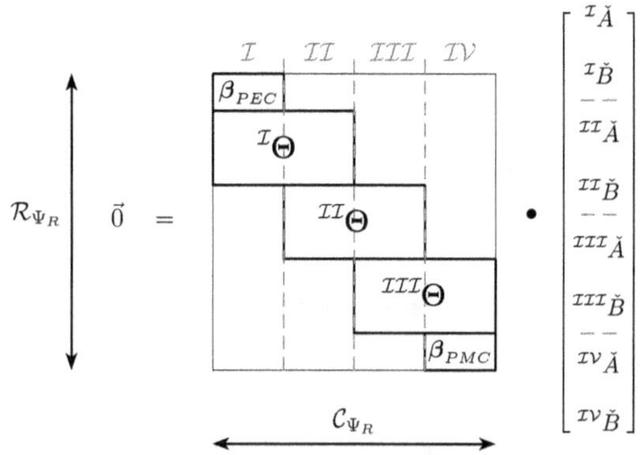

Figure 12.6.: System matrix $\boldsymbol{\Psi}_R$ of the perturbed cavity's halved segment matrix representation (cf. figure 12.5)

[2] This can also be seen from the segment matrix' dimensions as per (6.4) and the dimensions of the PEC and PMC matrix given in the previous section. Here, we assumed ${}^{\mathcal{I}}M = {}^{\mathcal{II}}M = {}^{\mathcal{III}}M$.

12.3. Post-Processing for Resonant Problems

In the previous section, we have determined the resonant frequencies ω_0 of the perturbed cavity depicted in figure 12.2. For lossy cavities, the corresponding Eigenmodes' *unloaded quality factor* [48, 78]

$$\mathfrak{Q}_0 = \frac{\omega_0 \, \mathfrak{W}}{\mathfrak{P}_l} \tag{12.8}$$

is another important figure, which represents the ratio of the energy $\mathfrak{W}$ stored inside the cavity and the power $\mathfrak{P}_l$ dissipated in it during one cycle. $\mathfrak{P}_l$ includes both the power $\mathfrak{P}_c$ dissipated in the cavity's walls due to ohmic losses as well as the power $\mathfrak{P}_d$ dissipated in the dielectric.

The Mode Matching formalism presented in this thesis is designed for loss-free waveguides only. However, under the assumption that the cavity's losses are sufficiently small, we can assume the cavity's total fields to remain unchanged in the presence of losses. This approach represents a perturbation method [6, 48].

After the cavity's total fields have been calculated by summing up the field contributions of all Eigenmodes included in the analysis, the *energy stored inside the cavity electrically* can be calculated as [48]

$$\mathfrak{W}_e = \frac{\varepsilon'}{2} \iiint\limits_{\mathcal{V}} |\vec{E}|^2 dV \tag{12.9}$$

while the *energy stored magnetically* [48] is

$$\mathfrak{W}_m = \frac{\mu'}{2} \iiint\limits_{\mathcal{V}} |\vec{H}|^2 dV. \tag{12.10}$$

Note that at resonance, we have $\mathfrak{W}_e = \mathfrak{W}_m$ [78].

The *power dissipated due to ohmic loss* can be calculated [48]

$$\mathfrak{P}_c = R_s \iint\limits_{Walls} |H_{tan}|^2 dA \tag{12.11}$$

where R_s denotes the *surface resistance* $R_s = \sqrt{\frac{\omega_0 \mu}{2\sigma}}$ and for the *power dissipated due to dielectric loss* we find [48]

$$\mathfrak{P}_d = \omega_0 \varepsilon'' \iiint\limits_{\mathcal{V}_S} |\vec{E}|^2 \, dV \tag{12.12}$$

where $\mathcal{V}_s$ denotes the volume of the dielectric perturbing the cavity. Note that all expressions given above assume effective phasor notation as is used in [48].

This chapter concludes our discussion of the Mode Matching technique. In the following part, after having studied various filter structures, we will also investigate perturbed cavity resonators. The Mode Matching solvers used for obtaining the presented results are based on the methodology introduced in this chapter.

12.1 Post-Processing for Resonant Problems

Part III.

Applications and Results

13. Microwave Filters

Microwave filters are an important building block of modern communication and radar systems [6]. Typical applications include rejection of out-of-band noise at the RX side [94], suppression of harmonics generated by the system's nonlinear components at the TX side [94], RX-TX decoupling e.g. in antenna duplexers [94] as well as channel separation [95].

In this chapter, we will see that the Mode Matching technique is often by far superior to space-segmentation techniques both in terms of computational speed and accuracy when it comes to performing full-wave simulation of *waveguide filters*[1]. In order to illustrate this point, we will compare the process of solving three exemplary filters using the Mode Matching solver developed in the scope of this thesis and a commercial FEM solver.

The results presented in the following are embedded in a concise discussion of waveguide filter theory in order to illustrate the close interrelation between the electromagnetic analysis of filter structures and the fine art of microwave filter design. In order to enable the interested reader to gain deeper insights in filter theory, in this chapter, footnotes provide additional citations to refer the reader to the appropriate literature.

The remainder of this chapter is structured as follows: In section 13.1 fundamental concepts of filter theory, namely filter polynomials, prototype filters, the lowpass-to-bandpass transformation and direct-coupled filters will be reviewed.

Next, in section 13.2 the theory of stepped-impedance lowpass filters will be discussed and we will investigate the process of solving an 11th order tubular stepped-impedance lowpass filter using a Mode Matching solver presented in this thesis.

Subsequently, we will move on to the realisation of direct-coupled bandpass filters. Firstly, in section 13.3 a tubular realisation will be discussed and a 4th order bandpass filter will be analysed. Secondly, we will investigate the realisation of direct-coupled filters in rectangular waveguide technology in section 13.4 and perform full-wave analysis of a 4th order bandpass filter for the K_a band.

13.1. Filter Design by the Insertion Loss Method

The central idea behind designing filters using the *insertion loss method* is to describe a filter's insertion loss (also: power loss) as

$$\mathcal{P}_{LR} = \frac{\text{Power available from Source}}{\text{Power delivered to load}} = 1 + \frac{M(\omega^2)}{N(\omega^2)} \tag{13.1}$$

where $M(\omega^2)$, $N(\omega^2)$ are real polynomials in ω^2 [6, 78]. Note that $\mathcal{P}_{LR} = {}^1/|s_{21}|^2$ if the filter is matched [6].

[1] Here, coaxial waveguides are explicitly included.

The power loss ratios which are most commonly used result from *Butterworth* or *Chebyshev polynomials*. The former polynomials provide a maximally flat passband while the latter provide an equal-ripple passband [6, 78]. The corresponding power loss characteristics are shown in figure 13.1.

13.1.1. Lowpass Prototype Filters

It is well known [6, 96] that filters with the power loss ratios shown in figure 13.1 can be realised using the *lowpass prototype filter* shown in figure 13.2. The *order* N of such a filter is equal to the number of reactive elements [6]. The element values[2] $g_0, \ldots, g_{N+1}$ of such a prototype filter can be calculated using equations given in literature [96]:

For Butterworth (maximally flat) filters (4.05-1) in [96] can be used. Similarly, the element values for Chebyshev (equiripple) filters can be obtained from (4.05-2) in [96]. For both filter types, tabulated values can be found in table 4.05-1 and 4.05-2 in [96]. The element values g_i are scaled for a filter impedance of $\mathfrak{R}_0 = 1\ \Omega$ and a cut-off frequency $\omega_{c,P} = 1$.

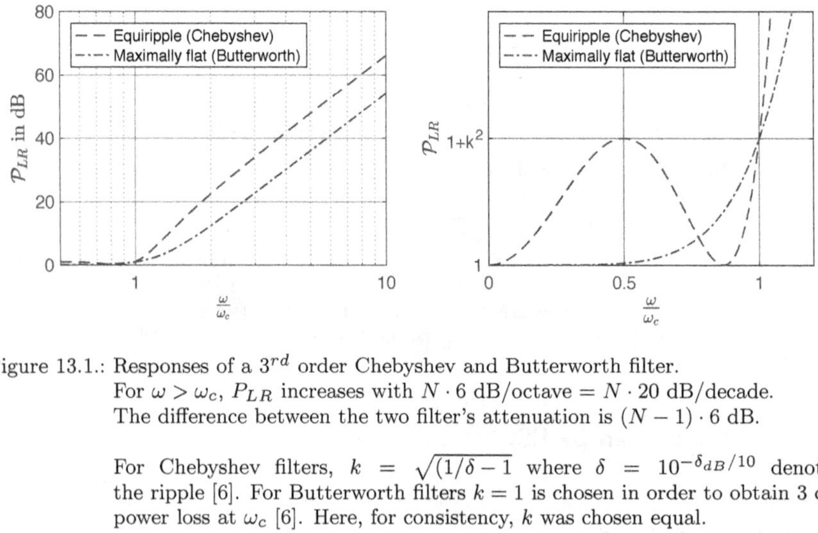

Figure 13.1.: Responses of a 3rd order Chebyshev and Butterworth filter.
For $\omega > \omega_c$, P_{LR} increases with $N \cdot 6$ dB/octave $= N \cdot 20$ dB/decade.
The difference between the two filter's attenuation is $(N-1) \cdot 6$ dB.

For Chebyshev filters, $k = \sqrt{(1/\delta - 1}$ where $\delta = 10^{-\delta_{dB}/10}$ denotes the ripple [6]. For Butterworth filters $k = 1$ is chosen in order to obtain 3 dB power loss at ω_c [6]. Here, for consistency, k was chosen equal.

[2] also see [96, p. 95f]

Rescaling for different filter impedances $\mathfrak{R}_0 = \mathfrak{G}_0^{-1}$ and cut-off frequencies ω_c can be performed by (4.04-2) to (4.04-4) in [96], which can be rewritten [6] for the scaling convention given on the previous page as

$$L_i = \frac{\mathfrak{R}_0}{\omega_c} L_{P,i} \quad C_i = \frac{1}{\mathfrak{R}_0 \omega_c} C_{P,i} \quad R_i = \mathfrak{R}_0 R_{P,i} \quad G_i = \mathfrak{G}_0 G_{P,i} \tag{13.2}$$

where $L_{P,i} = g_i$, $C_{P,i} = g_i$, $R_{P,i} = g_i$ and $G_{P,i} = g_i$, that is, the index P indicates normalised prototype elements.

After rescaling the prototype filter to the desired impedance environment, say $\mathfrak{R}_0 = 50\,\Omega$, stepped-impedance lowpass filters can directly be derived from the rescaled prototype filter's topology depicted in figure 13.3. In section 13.2.1, we will briefly revise the design of stepped impedance filters. After that, the Mode Matching solution for a coaxial (tubular) stepped-impedance filter will be presented.

In contrast to lowpass filters, the design of bandpass filters requires additional steps. In the following section, we briefly revise the lowpass to bandpass transformation of prototype filters. As the topologies obtained from this transformation are unfavourable for waveguide filter implementations, additional measures to convert these topologies to more suitable ones are required. This leads us to the theory of direct-coupled filters discussed in section 13.1.3. In section 13.3 and 13.4 we will study filter designs based on this concept.

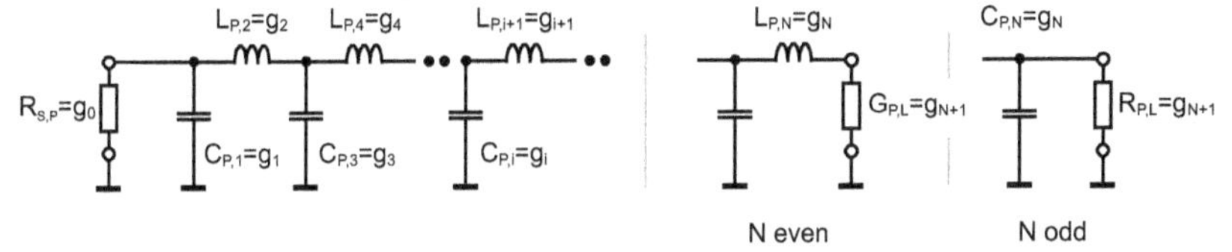

Figure 13.2.: Lowpass prototype filter with a cut-off frequency $\omega_{c,P} = 1$ scaled for $\mathfrak{R}_0 = 1\,\Omega$

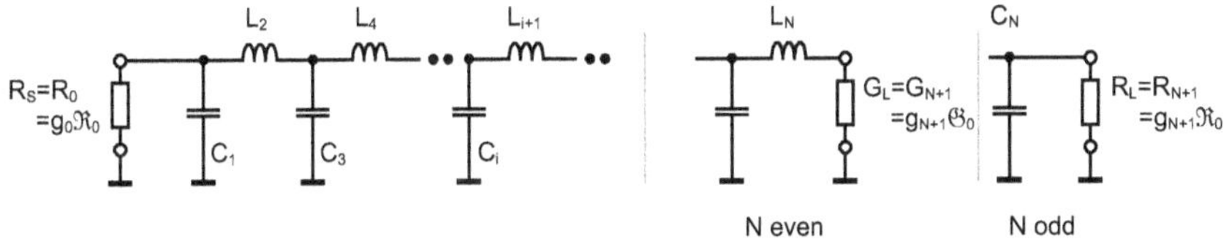

Figure 13.3.: Rescaled Lowpass prototype filter with a cut-off frequency ω_c

13.1.2. Lowpass to Bandpass Transformation

Let us firstly rescale the prototype filter depicted in figure 13.2 to the desired filter impedance $\mathfrak{R}_0$. The element values then calculate [6][3]

$$L_{p,i} = \mathfrak{R}_0 L_{P,i} \quad C_{p,i} = \frac{1}{\mathfrak{R}_0} C_{P,i} \quad R_i = \mathfrak{R}_0 R_{P,i} \quad G_i = \mathfrak{G}_0 G_{P,i} \tag{13.3}$$

where $L_{P,i} = g_i$, $C_{P,i} = g_i$, $R_{P,i} = g_i$ and $G_{P,i} = g_i$.

Next, a lowpass-to-bandpass transformation [6][4]

$$\omega \leftarrow \underbrace{\frac{1}{\mathfrak{w}} \left(\frac{\omega}{\omega_0} - \frac{\omega_0}{\omega} \right)}_{\Omega} \tag{13.4}$$

is performed where $\mathfrak{w} = {}^{BW}/_{\omega_0}$ denotes the relative bandwidth and ω_0 the center frequency of the passband.[5] Ω denotes the detuning of the filter [89].

Under this transformation[6], inductances become series resonators [6] as can be seen from

$$jX_i = j\omega L_{p,i} \rightarrow jX_i = j\frac{L_{P,i}\mathfrak{R}_0}{\mathfrak{w}} \underbrace{\left(\frac{\omega}{\omega_0} - \frac{\omega_0}{\omega} \right)}_{\Omega} = j\underbrace{\frac{L_{P,i}\mathfrak{R}_0}{\mathfrak{w}\,\omega_0}}_{L_{B,i}}\omega + \frac{1}{j\underbrace{\frac{\mathfrak{w}}{L_{P,i}\mathfrak{R}_0\omega_0}}_{C_{B,i}}\omega} \tag{13.5}$$

while capacitances are transformed into parallel resonators [6] as becomes obvious from

$$jB_i = j\omega C_{p,i} \rightarrow jB_i = j\frac{C_{P,i}}{\mathfrak{w}\mathfrak{R}_0} \underbrace{\left(\frac{\omega}{\omega_0} - \frac{\omega_0}{\omega} \right)}_{\Omega} = j\underbrace{\frac{C_{P,i}}{\mathfrak{w}\,\omega_0\mathfrak{R}_0}}_{C_{B,i}}\omega + \frac{1}{j\underbrace{\frac{\mathfrak{w}\mathfrak{R}_0}{C_{P,i}\omega_0}}_{L_{B,i}}\omega} \tag{13.6}$$

and we obtain the *bandpass prototype filter* depicted in figure 13.4. Note that the resonant frequency of all series and parallel resonators is equal to the center frequency ω_0 of the filter's passband [6][7].

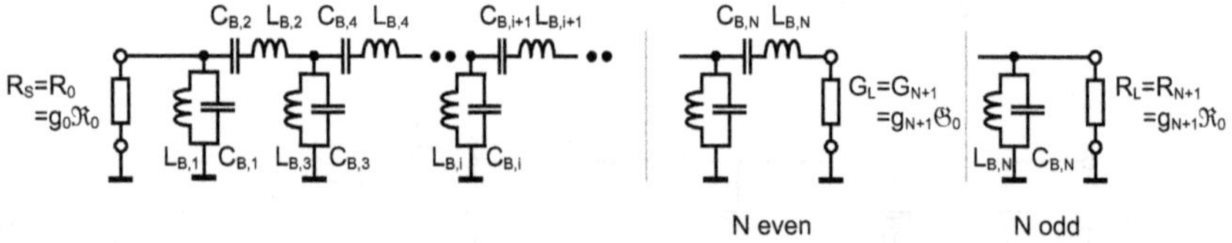

Figure 13.4.: Bandpass filter with a center frequency ω_0 and a relative bandwidth $\mathfrak{w}$

[3] see [6, p. 398]

[4] see [6, p. 401]

[5] It should be noted that under certain circumstances, this transformation may lead to a strong distortion of the filter's transfer function. Alternate design procedures, such as the mapping proposed in [97], are available.

[6] see [6, p. 403]

[7] see [6, p. 403]

A point often left out in literature on microwave filters is the order of the bandpass filter. Strictly speaking, the *order* of a filter is equal to the number of the *transfer function*'s *poles* [98][8]. For the prototype filter depicted in figure 13.2 the number of poles and thus the filter order N is equal to the number of reactive elements.

By performing the lowpass-to-bandpass transformation, the transfer function's number of poles is doubled [98][9]. This can easily be visualised in the s-plane: The lowpass-to-bandpass transformation shifts the lowpass prototype filter's poles upwards along the imaginary axis $j\omega$. At the same time, additional poles have to be added at $-j\omega$ in order to obtain a real-valued, i.e. physical meaningful impulse response [98][10].

In agreement with [99][11] we will define the bandpass filter's order to be equal to the order of the parent prototype, thus disregarding the doubling of the poles described above. To the author's best knowledge, this appears to be the definition implied by most authors of literature on microwave filters.

Obviously, when estimating the filter characteristics from the filter's order, the discussion given above has to be kept in mind: A Nth order low-pass filter and a Nth order bandpass filter exhibit the same slope $N \cdot 20$ dB/decade or $N \cdot 6$ dB/octave, although the bandpass filter has twice the number of poles.

13.1.3. Direct-Coupled Bandpass Filters and Impedance Inverters

The bandpass filter depicted in figure 13.4 contains both series and shunt resonators, which is often undesirable when realising a filter using a particular type of waveguide [6]. In order to overcome this issue, *impedance inverters* can be employed to transform the structure of this filter into a topology containing series resonators (but no shunt resonators) and impedance inverters only [96].[12]

The general concept of impedance inverters is illustrated in figure 13.5. The various realisations of impedance inverters usually contain elements with negative values, which have to be absorbed in positive-valued adjacent elements of the same type [96]. Practical realisations of impedance inverters will be discussed later in the context of the individual filters. For an extensive discussion of impedance inverters see [96], chapters 4.12 and 8.03.

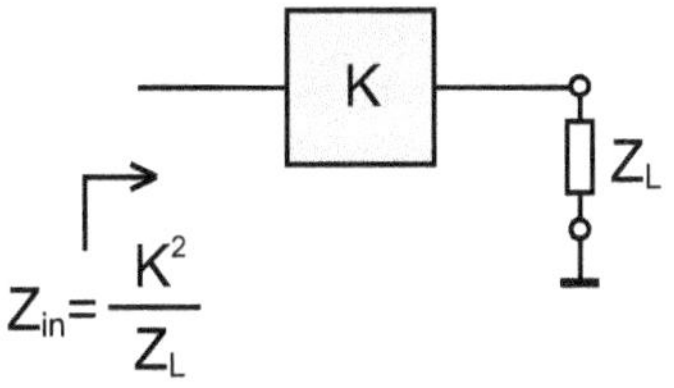

Figure 13.5.: Impedance inverter, based on [96], sec. 4.12

[8] see [98, p. 125]

[9] see [98, p. 97]

[10] see ibid.

[11] see [99, p. 145]

[12] An alternate approach is to use admittance inverters, which leads to a structure containing parallel resonators only [96, p. 431f]. In the scope of this thesis, this concept will not be investigated.

Let us now investigate the conversion of the filter topology depicted in figure 13.6a into a shunt resonator-free topology. The filter shown in Fig. 13.6a shall not be rescaled to any particular filter impedance, that is, the filter impedance is 1 Ω.

By comparing the normalised input impedance[13] of the filter shown in figure 13.6a with the normalised input impedance of the shunt resonator-free topology depicted in figure 13.6b it can be seen that the two topologies are indeed equivalent, provided the inverter constants of the impedance inverters are chosen properly.[14]

For the input impedance of the filter topology shown in figure 13.6a we find

$$Z = \cfrac{1}{jB_1 + \cfrac{1}{jX_2 + \cfrac{1}{jB_3 + \cfrac{1}{jX_4 + \ldots}}}}$$

which can be rewritten as

$$\mathfrak{z} = \cfrac{1}{jB_1 Z_0 + \cfrac{1}{\frac{jX_2}{Z_0} + \cfrac{1}{jB_3 Z_0 + \cfrac{1}{\frac{jX_4}{Z_0} + \ldots}}}}, \tag{13.7}$$

by renormalising Z to $Z_0 = R_{P,S} = R_{P,0} = g_0$.

The input impedance of the filter topology depicted in figure 13.6b can be written as

$$Z' = \cfrac{K_{0,1}^2}{jX_1' + \cfrac{K_{1,2}^2}{jX_2' + \cfrac{K_{2,3}^2}{jX_3' + \cfrac{K_{3,4}^2}{jX_4' + \ldots}}}},$$

which can be rearranged as

$$\mathfrak{z}' = \cfrac{1}{\frac{Z_0'}{K_{0,1}^2} jX_1' + \cfrac{1}{\frac{K_{0,1}^2}{K_{1,2}^2 Z_0'} jX_2' + \cfrac{1}{\frac{K_{1,2}^2 Z_0'}{K_{0,1}^2 K_{2,3}^2} jX_3' + \cfrac{1}{\frac{K_{0,1}^2 K_{2,3}^2}{K_{1,2}^2 K_{3,4}^2 Z_0'} jX_4' + \ldots}}}} \tag{13.8}$$

if Z' is normalised to $Z_0' = R_S = R_0 = g_0 \mathfrak{R}_0$.

[13]This becomes obvious from the definition

$$\Gamma = \frac{Z - Z_0}{Z + Z_0} = \frac{\frac{Z}{Z_0} - 1}{\frac{Z}{Z_0} + 1} = \frac{\mathfrak{z} - 1}{\mathfrak{z} + 1}$$

of the reflection coefficient.

[14]Comparable derivations can be found e.g. in [89].

By comparing (13.7) and (13.8), it becomes obvious that the two filter topologies depicted in figure 13.6a and 13.6b are indeed equivalent. In order to properly determine the inverter constants, we compare the colored coefficients in (13.7) and (13.8), which yields

$$jB_1 Z_0 = \frac{Z_0'}{K_{0,1}^2} jX_1' \quad \rightarrow \quad K_{0,1}^2 \quad = \quad \frac{X_1'}{B_1}\frac{Z_0'}{Z_0} \tag{13.9}$$

to be discussed later and

$$\begin{aligned}
\frac{jX_2}{Z_0} = \frac{K_{0,1}^2}{K_{1,2}^2 Z_0'} jX_2' \quad &\rightarrow \quad K_{1,2}^2 &&= \quad K_{0,1}^2 \quad \frac{X_2'}{X_2}\frac{Z_0}{Z_0'} \\
& &&= \quad \frac{X_1' X_2'}{B_1 X_2} \\
jB_3 Z_0 = \frac{K_{1,2}^2 Z_0'}{K_{0,1}^2 K_{2,3}^2} jX_3' \quad &\rightarrow \quad K_{2,3}^2 &&= \quad \frac{K_{1,2}^2}{K_{0,1}^2} \quad \frac{X_3'}{B_3}\frac{Z_0'}{Z_0} \\
& &&= \quad \frac{X_2' X_3'}{X_2 B_3} \\
\frac{jX_4}{Z_0} = \frac{K_{0,1}^2 K_{2,3}^2}{K_{1,2}^2 K_{3,4}^2 Z_0'} jX_4' \quad &\rightarrow \quad K_{3,4}^2 &&= \quad \frac{K_{0,1}^2 K_{2,3}^2}{K_{1,2}^2} \quad \frac{X_4'}{X_4}\frac{Z_0}{Z_0'} \\
& &&= \quad \frac{X_3' X_4'}{B_3 X_4} \\
& &&\dots
\end{aligned} \tag{13.10}$$

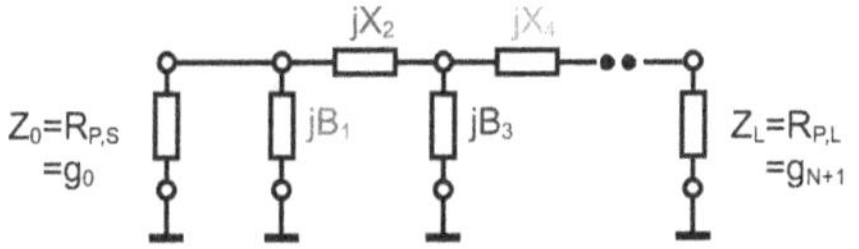

(a) Filter topology comprising both series and shunt resonators

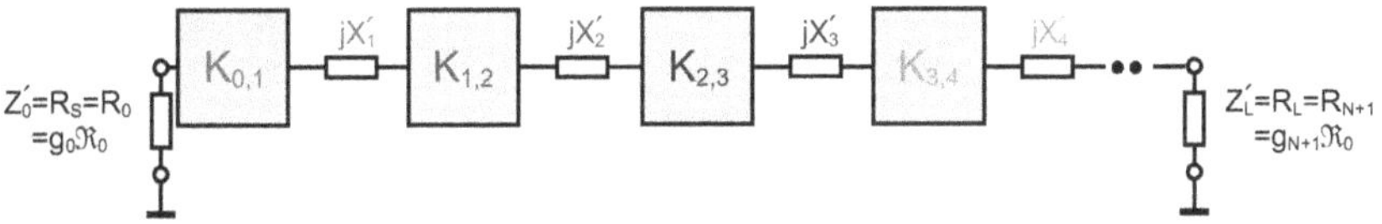

(b) Direct-coupled filter topology comprising series resonators only

Figure 13.6.: Filter topologies

Inverter Constants of the Inner Impedance Inverters $K^2_{i,i+1}$

In a bandpass filter design, the series impedances jX_i correspond to series resonators while the shunt admittances jB_i are equivalent to parallel resonators as shown in figure 13.4.

Since the filter shown in figure 13.6a is not rescaled to any particular filter impedance, the element values of the reactive elements denote

$$jX_i = j\frac{L_{P,i}}{\mathfrak{w}}\Omega \qquad jB_i = j\frac{C_{P,i}}{\mathfrak{w}}\Omega$$

where $L_{P,i} = g_i$ and $C_{P,i} = g_i$. These expressions are similar to (13.5) and (13.6) except for the impedance rescaling.

If we insert the above expressions into (13.10) and denote the impedances of the series resonators[15] in figure 13.6b as $X_i' = \omega_0 L_{S,i}\Omega$, (13.10) is rewritten as

$$\begin{aligned}
K^2_{12} &= \frac{X_1'X_2'}{B_1X_2} = \frac{\omega_0 L_{S,1}\Omega\,\omega_0 L_{S,2}\Omega}{\frac{C_{P,1}}{\mathfrak{w}}\Omega\,\frac{L_{P,2}}{\mathfrak{w}}\Omega} = \mathfrak{w}^2\frac{\chi_1\chi_2}{g_1g_2}\\
K^2_{23} &= \frac{X_2'X_3'}{X_2B_3} = \frac{\omega_0 L_{S,2}\Omega\,\omega_0 L_{S,3}\Omega}{\frac{L_{P,2}}{\mathfrak{w}}\Omega\,\frac{C_{P,3}}{\mathfrak{w}}\Omega} = \mathfrak{w}^2\frac{\chi_2\chi_3}{g_2g_3}\\
K^2_{34} &= \frac{X_3'X_4'}{B_3X_4} = \frac{\omega_0 L_{S,3}\Omega\,\omega_0 L_{S,4}\Omega}{\frac{C_{P,3}}{\mathfrak{w}}\Omega\,\frac{L_{P,4}}{\mathfrak{w}}\Omega} = \mathfrak{w}^2\frac{\chi_3\chi_4}{g_3g_4}\\
&\dots
\end{aligned} \tag{13.11}$$

where χ_i denotes the *reactance slope* of the series resonators[16] in figure 13.6b.

From (13.11) we may generalise the expression for the inverter constants of the "inner" impedance inverters $K_{i,i+1}$ as

$$K^2_{i,i+1} = \mathfrak{w}^2\frac{\chi_i\chi_{i+1}}{g_ig_{i+1}}. \tag{13.13}$$

---

[15] The impedance of a series resonator calculates [89]

$$Z = j\omega L + \frac{1}{j\omega C} = j\left[\omega L - \frac{1}{\omega C}\right] \overset{\omega_0=1/\sqrt{LC}}{=} j\left[\omega - \frac{\omega_0^2}{\omega}\right]L = j\omega_0\underbrace{\left[\frac{\omega}{\omega_0} - \frac{\omega_0}{\omega}\right]}_{\Omega}L = j\omega_0\Omega L.$$

[16] The impedance slope is defined [96, p. 430]

$$\chi = \frac{\omega_0}{2}\frac{dX}{d\omega}\bigg|_{\omega=\omega_0} \tag{13.12}$$

and consequently, for a simple series resonator we find [96, p. 430]

$$\begin{aligned}
\chi &= \frac{\omega_0}{2}\frac{d}{d\omega}\left[\omega L - \frac{1}{\omega C}\right]\bigg|_{\omega=\omega_0} = \frac{\omega_0}{2}\left[L + \frac{1}{\omega^2 C}\right]\bigg|_{\omega=\omega_0}\\
&= \frac{\omega_0}{2}\left[L + \frac{1}{\omega_0^2 C}\right] \underbrace{=}_{\text{using } \omega_0^2=1/LC} \frac{\omega_0}{2}[L + L] = \omega_0 L.
\end{aligned}$$

Inverter Constant of the First Impedance Inverter $K^2_{0,1}$

Similarly, the inverter constant of the first impedance inverter $K_{0,1}$ can be calculated from (13.9), which yields

$$K^2_{0,1} = \frac{X'_1}{B_1}\frac{Z'_0}{Z_0} = \frac{\omega_0 L_{S,1}\Omega}{\frac{g_1}{\mathfrak{w}}\Omega}\frac{Z'_0}{Z_0} = \mathfrak{w}\frac{\chi_1 R_S}{g_0 g_1} \tag{13.14}$$

where $Z_0 = R_{P,S} = g_0$ and $Z'_0 = R_S$.

Inverter Constant of the Last Impedance Inverter $K^2_{N,N+1}$

In order to determine the inverter constant of the last impedance inverter $K_{N,N+1}$ we now consider simplified versions of the filters shown in figure 13.7, which are odd-order filters. For these filters, (13.7) and (13.8) denote

$$\mathfrak{z} = \cfrac{1}{jB_1 Z_0 + \cfrac{1}{\frac{jX_2}{Z_0} + \cfrac{1}{jB_3 Z_0 + \cfrac{1}{\frac{Z_L}{Z_0}}}}} \quad \text{and} \quad \mathfrak{z}' = \cfrac{1}{\frac{Z'_0}{K^2_{01}} jX'_1 + \cfrac{1}{\frac{K^2_{01}}{K^2_{12} Z'_0} jX'_2 + \cfrac{1}{\frac{K^2_{12} Z'_0}{K^2_{01} K^2_{23}} jX'_3 + \cfrac{1}{\frac{K^2_{01} K^2_{23}}{K^2_{12} K^2_{34} Z'_0} Z'_L}}}}$$

and we find

$$\frac{Z_L}{Z_0} = \frac{K^2_{01} K^2_{23}}{K^2_{12} K^2_{34} Z'_0} Z'_L \quad \rightarrow \quad K^2_{34} = \frac{K^2_{01} K^2_{23}}{K^2_{12}} \frac{Z'_L}{Z_L} \frac{Z_0}{Z'_0},$$

which similar to the expression for K^2_{34} in (13.10) can be rewritten as

$$K^2_{34} = \frac{X'_3}{B_3}\frac{Z'_L}{Z_L} = \frac{\omega_0 L_{S,3}\Omega}{\frac{C_{P,3}}{\mathfrak{w}}\Omega}\frac{Z'_L}{Z_L} = \mathfrak{w}\frac{\chi_3 R_L}{g_3 g_4} \tag{13.15}$$

where $Z_L = R_{P,L} = g_4$ and $Z'_L = R_L$.

As can be seen from (13.10), this expression may be generalised for an odd-order filter with N resonators as

$$K^2_{N,N+1} = \mathfrak{w}\frac{\chi_N R_L}{g_N g_{N+1}}. \tag{13.16}$$

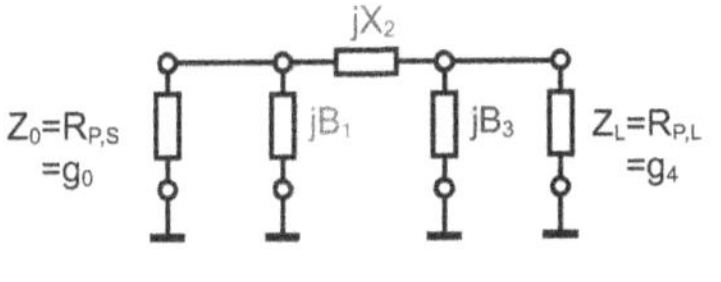

(a) Simplified version of the filter shown in figure 13.6a

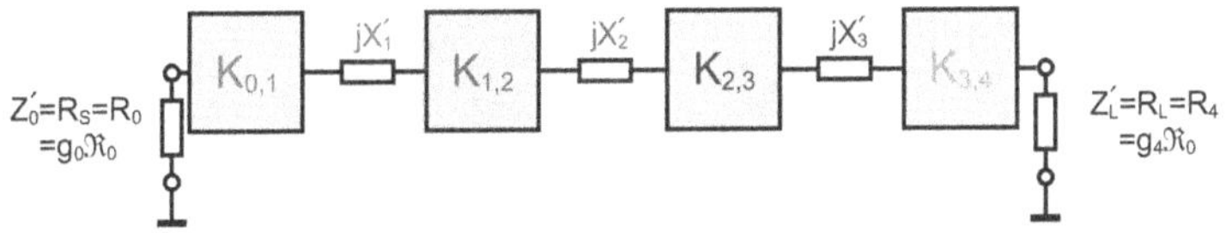

(b) Simplified version of the filter shown in figure 13.6b

Figure 13.7.: Simplified filter topologies

Summary

The filter topology introduced on the previous pages is called *direct-coupled filter* because the impedance inverters, which provide coupling between the series resonators, are connected to the resonators at distinct points. A general representation of the direct-coupled filter topology is depicted in figure 13.8. The inverter constants calculate

$$K_{0,1} = \sqrt{\mathfrak{w}\frac{\chi_1 R_S}{g_0 g_1}} \qquad K_{i,i+1} = \mathfrak{w}\sqrt{\frac{\chi_i \chi_{i+1}}{g_i g_{i+1}}} \qquad K_{N,N+1} = \sqrt{\mathfrak{w}\frac{\chi_N R_L}{g_N g_{N+1}}} \tag{13.17}$$

as derived on the previous pages.

The theory of direct-coupled filters was developed by Cohn in [97] and can also be found in Mathaei, Young and Jones' classic book [96]. More precisely, (13.17) can be found in [96, fig. 8.02-3, p. 432].

A central point of Cohn's analysis of direct-coupled filters is that resonators are described independently of their actual representation using their resonant frequency ω_0 and their reactance slope χ (series resonators) or susceptance slope β (parallel resonators) [96][17].

While the resonant frequency ω_0 is known from the lowpass-to-bandpass transformation, the reactance slopes $\chi_1, .., \chi_N$ may be chosen "arbitrarily to be of any size corresponding to convenient resonator designs" [96][18].

The shape of the filter's response is then realised by finding the *inverter constants* using (13.17) [96][19]. In other words, the shape of the filter's response is only determined by the coupling of the resonators described by the inverter constants.

Finally, note that the frequency sensitivity of the real impedance inverters limits the accuracy of the above analysis to relative bandwidths of approx. 20 % [96, 97].

We now conclude our review of the theory of direct-coupled bandpass filters. In sections 13.3 and 13.4 we will derive filter topologies from the general structure of direct-coupled resonators shown in figure 13.8 whose resonators and impedance inverters are well-suited for implementation in a particular type of waveguide. More precisely, in section 13.3, a direct-coupled resonator filter based on foreshortened coaxial transmission line resonators [100] will be examined. Subsequently in section 13.4, a rectangular waveguide iris filter will be shown.

However, firstly, stepped-impedance lowpass filters shall be examined in the following.

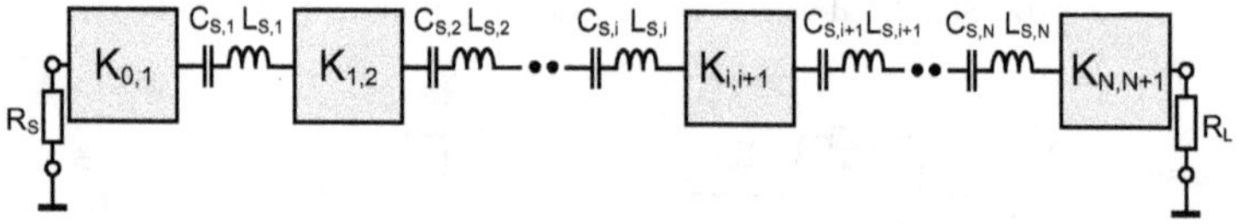

Figure 13.8.: General direct-coupled bandpass filter

[17] see [96, p. 430]
[18] see [96, p. 431]
[19] see [96, p. 431]

13.2. Tubular Filters: Stepped-Impedance Lowpass Filter

13.2.1. Theory of Stepped-Impedance Lowpass Filters

From the rescaled prototype filter shown in figure 13.3, *stepped-impedance lowpass filters* can directly be derived. The reason for this is that shunt capacitances can be realised as short transmission lines of low characteristic impedance Z_l while series inductances can be implemented as short transmission lines of high characteristic impedance Z_h [6]. In doing so, the prototype filter's element values are represented by the length of their equivalent transmission line sections. The electric length of inductive sections calculates [6][20]

$$\beta l_i = \frac{L_{P,i}\mathfrak{R}_0}{Z_h} \tag{13.18}$$

while the electric length of capacitive sections can be determined using [6][21]

$$\beta l_i = \frac{C_{P,i}Z_l}{\mathfrak{R}_0}. \tag{13.19}$$

$C_{P,i}$ and $L_{P,i}$ correspond to the prototype filter's coefficients g_i while $\mathfrak{R}_0$ denotes the filter impedance.

Stepped-impedance filters can be realised using various technologies, among which coaxial and microstrip realisations are probably the two most commonly used approaches. In the scope of this thesis, only tubular stepped-impedance filters will be considered.

An exemplary realisation[22] of a tubular stepped-impedance coaxial waveguide filter is depicted in figure 13.9a. As already discussed in section 10.2, inner conductor steps exhibit capacitive behaviour. As a consequence, additional parasitic capacitances have to be added to the filter's equivalent circuit as is shown in figure 13.10a (red).

The filter's scattering parameters calculated from the equivalent circuit shown in figure 13.9a (top) are depicted in figure 13.9b (red line). Note that these scattering parameters ignore the parasitic capacitances present in the filter design. As can be seen by comparing the filter's Mode Matching solution (blue line in figure 13.9b) with the analytical solution (red line), the filter is skewed (150 MHz less bandwidth at -3 dB) by the parasitic capacitances.

In order to get a better impression of the parasitic behaviour of the inner conductor's steps, it is interesting to compare the element values of the filter's capacitances C_1 and C_3 with the structure's parasitic capacitances $C_{par,1}$ and $C_{par,2}$. While $C_1 = C_3 = 3$ pF, C_{P1} and C_{P2} are in the region of several hundred fF as can be seen in figure 13.10b. The values given in this figure were obtained using Mode Matching as discussed in chapter 10.

Obviously, the element values of the parasitic capacitances can be used to correct the filter. In order to do so, we recalculate the filter's capacitances as $C_1' = C_1 - C_{par,1} - C_{par,2}$ and $C_3' = C_3 - C_{par,1} - C_{par,2}$ and modify the individual line segments' lengths accordingly.

As can be seen from figure 13.9b, the Mode Matching solution for the corrected filter (black line) shows much better agreement with the response calculated based on the equivalent circuit (red line).

[20] see [6, p. 413f]
[21] see [6, p. 413f]
[22] $Z_h = 120$ Ohm, $Z_l = 12$ Ohm, $Z_{Feed} = 50$ Ohm. Diameter of outer conductor: 16 mm.

However, it should be noted that at higher frequencies the filter's actual response obtained using Mode Matching still differs from the expected response. The reason for this is the imperfect representation of the lumped elements by transmission lines, whose electric length increases to higher frequencies so that propagation effects become more relevant.[23]

Consequently, while equivalent circuits may be useful for correcting a filter design, they do not replace a full wave solution, e.g. by Mode Matching, especially if for example stop-band attenuation is critical. Moreover, it should be noted that more advanced waveguide discontinuities may require higher-order modelling of the parasitics' frequency behaviour.

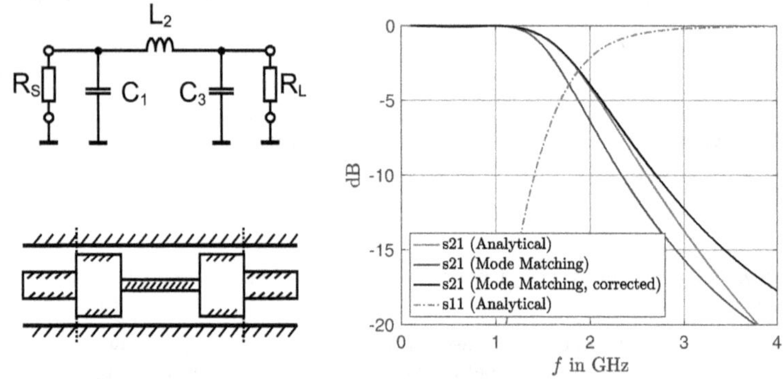

(a) Rescaled prototype (top), coaxial realisation (bottom)

(b) Scattering Parameters

Figure 13.9.: Coaxial stepped-impedance lowpass filter
Third-order Chebyshev ($N = 3$), $\omega_C = 1$ GHz

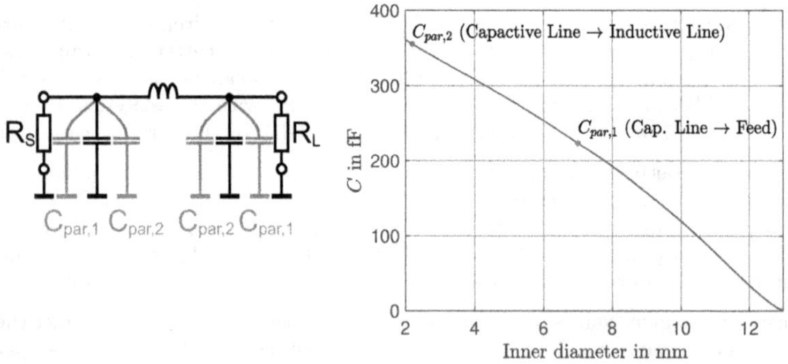

(a) Equivalent circuit

(b) Element values of the parasitic capacitances obtained by Mode Matching

Figure 13.10.: Parasitic capacitances resulting from the filter's coaxial realisation at 1 GHz

[23]In contrast, it is interesting to note that the element values of the parasitic capacitances change only in the region of a few fF over the frequency range of figure 13.9b.

13.2.2. Results: Solving an 11th Order Tubular Stepped-Impedance Filter

In the previous section, we reviewed the theory of stepped-impedance filters and investigated the impact of parasitic effects originating in the filter's coaxial waveguide realisation. In this section, we will now present Mode Matching results for a more complex filter which was intended to be used in mobile telephone jammers. The filter has a cut-off frequency of 1 GHz and was designed to provide a stopband attenuation better than 60 dB above 1580 MHz[24]. The latter requirement is necessary in order to achieve sufficient harmonics suppression for the 800 MHz LTE and 900 MHz GSM downlinks to comply with regulatory requirements as well as to avoid triggering of GSM1800 or UMTS detectors.

The filter was realised as an 11th order stepped-impedance filter. For the given design, the prototype filter shown in figure 13.11 was used. Note that this prototype filter is dual to the prototype filter shown in figure 13.2 and provides the very same response [96]. The filter geometry to be analysed was kindly provided by SF Microwave GmbH. Figure 13.12 shows the filter. A photograph of the filter's inner conductor is depicted in figure 13.13.

Let us firstly exploit the rotationally symmetric nature of the filter to choose the smallest set of Eigenmodes appropriate for the structure so that computation time is kept to a minimum.

Since the filter is rotationally symmetric, only rotationally symmetric Eigenmodes can be excited by the TEM Eigenmode used for feeding the filter. In contrast, TM and TE Eigenmodes, which are not rotationally symmetric, may not be excited because their overlap integrals with the TEM Eigenmode always vanish.

As the filter is excited by a TEM Eigenmode, coupling with higher-order Eigenmodes can obviously occur only via E_ρ and H_ϕ. However, as can be seen from

$$E_\rho^{TE} = -\frac{1}{\rho}\frac{\partial\psi}{\partial\phi} \qquad H_\phi^{TE} = \frac{1}{j\omega\mu\rho}\frac{\partial^2\psi}{\partial\phi\,\partial z}, \tag{13.20}$$

which follows directly from (2.20) and (2.22), rotationally symmetric TE Eigenmodes do not incorporate these field components and thus may not be excited.

Consequently, a Mode Matching solution can be performed based on the rotationally symmetric $TM^z_{m,0}$ mode-set given in appendix B.3.1 and the fundamental TEM Eigenmode given in appendix B.3.2.

The scattering parameters of the filter under consideration are depicted in figure 13.14. Results obtained using the Mode Matching solver discussed in this thesis are shown in blue while scattering parameters calculated using Ansys HFSS are shown as dashed red lines. Both methods show outstanding agreement.

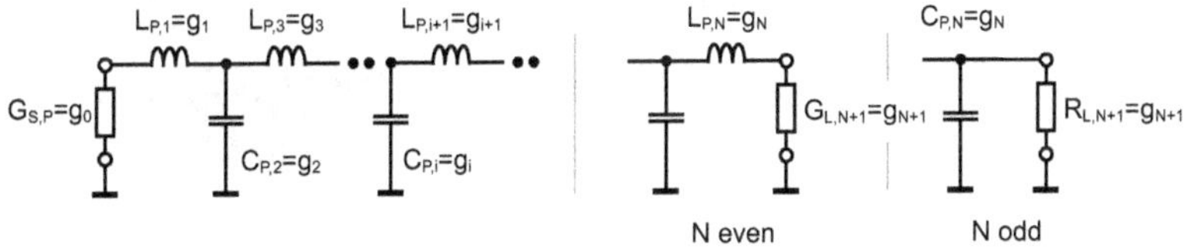

Figure 13.11.: Lowpass prototype filter with a cut-off frequency $\omega_{pc} = 1$ scaled for $g_0 = 1$ Dual structure to figure 13.2

[24]First harmonic of the LTE band starting at 790 MHz.

The Mode Matching solver required approximately 50 seconds for calculating a frequency sweep comprising 300 points with $V = P = 10$ Eigenmodes per waveguide and direction of propagation.[25] In contrast, Ansys HFSS requires 30 Minutes for solving the same sweep with $\Delta s_{max} = 0.01$. The mesh consisted out of approx. 17,000 elements.

If the desired accuracy for the HFSS simulation is reduced to $\Delta s_{max} = 0.1$, solving the structure takes about 10 minutes. However, the scattering parameters calculated with $\Delta s_{max} = 0.1$ show some deviations in the second passband as well as at high attenuations in the stopband compared to the two consistent solutions discussed above. For completeness sake, it should be mentioned that using heavy parallelisation,[26] Ansys HFSS can solve the filter with $\Delta s_{max} = 0.01$ in about 5 minutes.

All calculations were carried on a machine equipped with two Intel Xeon E5-2630 ("Sandy Bridge") CPUs (each providing 6 cores and 12 threads) and 64 GByte RAM.

Due to the significantly smaller computation time of the Mode Matching technique we can conclude that for the given tubular filter our Mode Matching solver is by far superior to Ansys HFSS' FEM approach.

A comparison between the filter's measured and simulated scattering parameters is shown in figure 13.15. Up to approx. 2 GHz s_{21} shows excellent agreement. Above 2 GHz, some variations can be observed, most notably, the measured filter's second passband lies about 200 MHz lower than one would expect from the Mode Matching solution. It is most likely that this deviation is due to the additional dielectric disks[27] shown in figure 13.13 added at a later stage in the design process. Also, mechanical inaccuracies and/or the transitions between the inner conductor and the filter's N jacks could be the cause for these deviations. The same reasoning applies for the deviations of s_{11} around 900 MHz.

Finally, an enlarged version of the filter's passband is shown in figure 13.16. In terms of the filter's cut-off frequency, all three curves show good agreement. The remaining differences are either due to the imperfect representation of the real filter's geometry in our simulation tools or are caused by the fact that both simulations do not incorporate ohmic losses. From figure 13.16 it can be seen that a reasonable worst case estimate in terms of s_{21} for the effect of neglecting ohmic losses is a value below 0.5 dB.

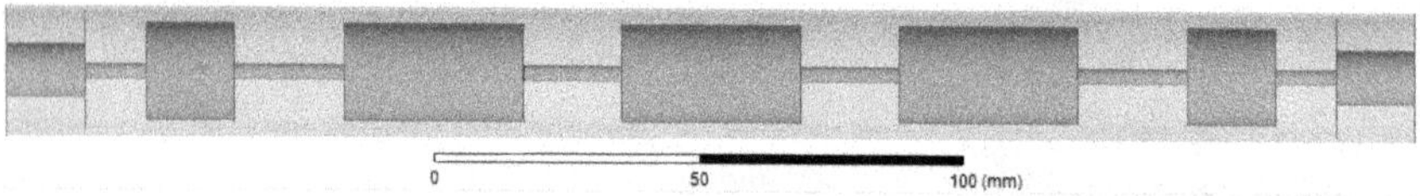

Figure 13.12.: 11th order coaxial stepped-impedance lowpass filter
The filter's transmission lines have characteristic impedances $Z_l = 12$ Ohms or $Z_h = 120$ Ohms. Filter geometry courtesy of SF Microwave GmbH

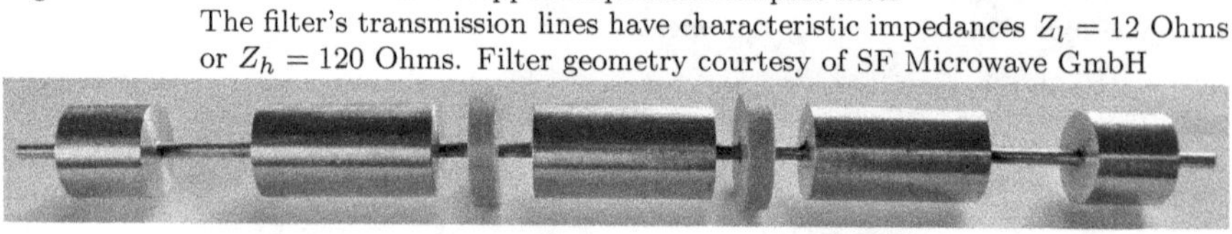

Figure 13.13.: Inner conductor of an 11th order coaxial stepped-impedance lowpass filter
Photograph courtesy of SF Microwave GmbH

[25] We thus have $\mathcal{M}$=20 Eigenmodes per junction.

[26] Adaptive passes: 20 threads; frequency sweep: 5 frequency points parallel, each using 4 threads.

[27] The dielectric disks are required for centering the inner conductor.

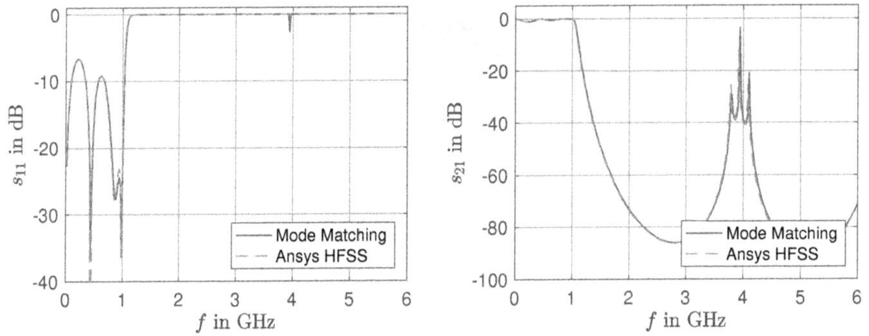

Figure 13.14.: Simulated scattering parameters of the 11th order lowpass filter

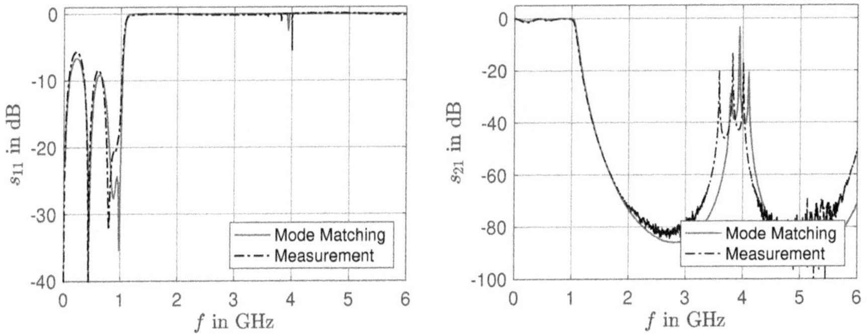

Figure 13.15.: Measured and simulated scattering parameters of the 11th order lowpass filter. Measurement data courtesy of SF Microwave GmbH

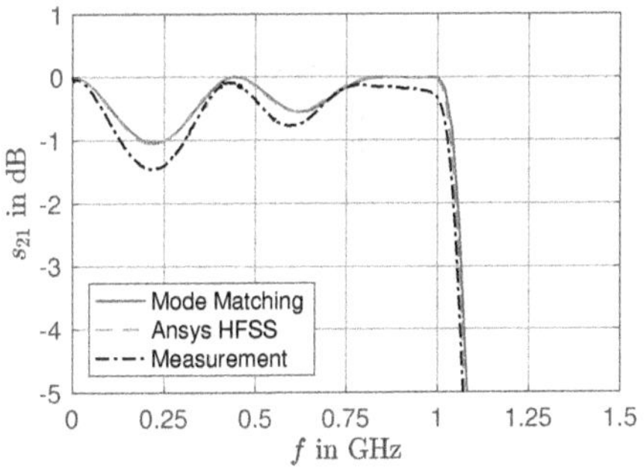

Figure 13.16.: Measured and simulated scattering parameters of the 11th order lowpass filter. Enlarged representation of the filter's passband. The ripple below 750 MHz is due to the fact that the filter was optimised to exhibit low loss between 790 - 960 MHz. Measurement data courtesy of SF Microwave GmbH

13.3. Tubular filters: Direct-Coupled Bandpass Filters

While the stepped-impedance lowpass filter presented in the previous section maintains the design criteria discussed earlier, its mechanical realisation is quite expensive as a large number of inner conductor sections with different lengths and diameters is required. In addition, after the filter was manufactured, modifications hardly can be carried out.

In order to overcome these issues, a direct-coupled bandpass filter was designed and realised by SF Microwave GmbH as an alternative to the stepped-impedance lowpass filter discussed earlier. This filter and its solution process using Mode Matching will be discussed in the following.

13.3.1. Theoretical Aspects of Tubular Direct-Coupled Bandpass Filters

In section 13.1.3 we have seen that bandpass filters can be implemented in a fashion well-suited for realisation as waveguide filters using the topology shown in figure 13.18a.

Common direct-coupled bandpass filters realised as waveguide filters usually use half-wave *transmission line resonators* to mimic the resonators' behaviour. Typical examples of such filter realisations are capacitive-gap-coupled microstrip filters[28] [6, 96] as well as the shunt inductance-coupled waveguide iris filters [96] to be discussed in the following section.

In order to reduce the filter's total length as well as to avoid higher-order passbands at multiples of the "fundamental" passband's center frequency[29] [100], here, a different approach was chosen.

In the given design, each series resonator was replaced by a *foreshortened transmission line resonator* [100] as is shown in figure 13.18b.[30] These resonators consist of a transmission line with a high characteristic impedance Z_H and an electric length of $\lambda/8$ as well as two series capacitances added at each end of the transmission line [100]. The two additional capacitances are required to virtually increase the transmission line's electric length so that the resonator's resonant frequency remains unchanged [100].

By splitting the series resonators' inherent capacitance $C_{S,i}$ into two capacitances $C'_{S,i}$ at each side of the inductance $L_{S,i}$, we can combine the resonators' inherent capacitance with the additional capacitances resulting from the foreshortening of the resonators to form the capacitances C_T as shown in figure 13.18b [100].

Since the resonant frequency of all resonators is equal to the filter's passband center frequency and because the reactance slopes of the series resonators may arbitrarily be chosen to be equivalent, C_T and χ are equal for all resonators comprised in the design.

The element value of the capacitances C_T and the foreshortened resonators' reactance slope χ required for dimensioning the impedance inverters can be calculated using [100][31]

$$C_T \approx \frac{2.414}{\omega_0 Z_H} \qquad \chi \approx 0.8743 \cdot Z_H. \tag{13.21}$$

[28] Strictly speaking, these filters use the dual topology of figure 13.18a, which is based on parallel shunt resonators and admittance inverters.

[29] Higher-order passbands at multiples of the "fundamental" passband's center frequency are highly undesirable in a filter used for harmonics suppression.

[30] In "ordinary" tubular bandpass filter designs, C-L-C-π-shunt-resonators (cf. figure 3 in [100]) are used, which are realised by means of discrete components. Capacitances are realised as short transmission lines with a low characteristic impedance while inductances are realised as wire-wound inductors. This approach enables the realisation of resonators which are short compared to the wavelength and thus also can be interpreted as "foreshortened".

[31] see [100, (1)]

Having determined the reactance slope χ of the foreshortened transmission line resonators, we can now examine possible realisations of the impedance inverters.

As the final design goal was a coaxial realisation of the filter, impedance inverters using shunt inductances generally cannot be used. This is because in a coaxial design, shunt inductances cannot (easily) be realised by waveguide discontinuities as can be seen from figure 5.07-1 in [96] or section 5.27 in Marcuvitz' waveguide handbook [17].

A well-known *shunt-capacitance impedance inverter* topology is shown in figure 13.17a. The *inverter constant* of this inverter calculates [96][32]

$$K_{i,i+1} = \frac{1}{\omega_0 C_{K,i,i+1}} \tag{13.22}$$

and it should be noticed that the series capacitances $-C_{K,i,i+1}$ are negative and thus need to be absorbed in the adjacent resonators' capacitances [96].

From the expressions (13.17) for the inverter constants given previously it can be seen that small bandwidths lead to smaller coupling coefficients. As a consequence, for narrow bandwidth filters realised using the inverter circuit shown in figure 13.17a, a large capacitance $C_{K,i,i+1}$ would be required as can be seen from (13.22). The realisation of such large $C_{K,i,i+1}$ leads to two problems:

As we have seen during our discussion of stepped-impedance filters, shunt capacitances can be realised as transmission line sections of low characteristic impedance (see (13.19)). If however the required capacitance becomes large, the corresponding transmission line length may exceed the maximum permissible length below which propagation effects on the line may be neglected.

Moreover, because the negative series capacitances $-C_{K,i,i+1}$ need to be absorbed in adjacent positive series capacitances, $C_{K,i,i+1}$ may not exceed the following series resonator's capacitance (here C_T).

In order to avoid this issue, the impedance inverter shown in figure 13.17b was introduced by Schöbel [100]. An additional advantage of this inverter topology is the fact that the capacitance $C_{I,i}$ can be arbitrarily defined, thus introducing an additional degree of freedom to the design, which may be used to improve the manufacturability [100].

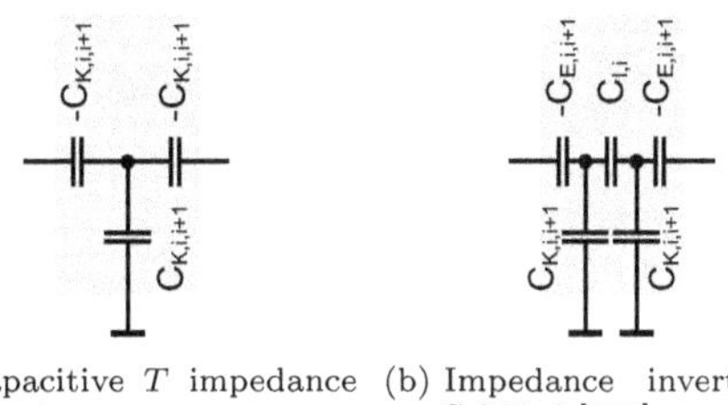

(a) Capacitive T impedance inverter.
See [96, fig. 8.03-1]

(b) Impedance inverter by Schöbel [100].

Figure 13.17.: Possible realisations of the impedance inverter $K_{i,i+1}$

[32] see [96, p. 436]

The filter topology using the inverter shown in figure 13.17b is depicted in figure 13.18c. Note that the impedance inverters' negative capacitances $-C_{E,i,i+1}$ are absorbed in the resonators' capacitances C_T, thus forming the capacitances $C_{A,i}$ or $C_{B,i}$.

The coaxial realisation of the Filter topology depicted in figure 13.18c is given in figure 13.18d. The shunt capacitances $C_{K,i,i+1}$ are realised as coaxial transmission lines with a low characteristic impedance while the series capacitances $C_{A,i}$, $C_{B,i}$ and $C_{I,i}$ are implemented as a dielectric disks, which in fact represent circular waveguide segments. The inner conductor step connecting the resonator's high impedance transmission line to the impedance inverter is not included in the equivalent circuit and has to be accounted for by performing full-wave simulation of the filter.

Finally, two aspects should be pointed out which are important for the mechanical realisation. Firstly, the inner conductor "dumbbells" representing the filter's resonators are all equal because the same C_T and χ applies. Secondly, using the inverter topology's additional degree of freedom, all dielectric disks were designed to have similar thickness.

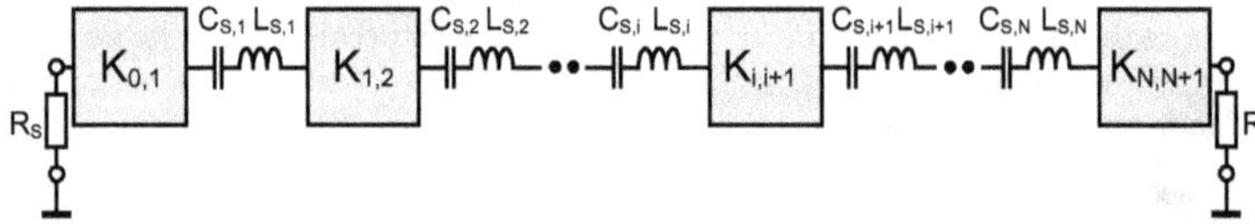

(a) Generalised direct-coupled bandpass filter

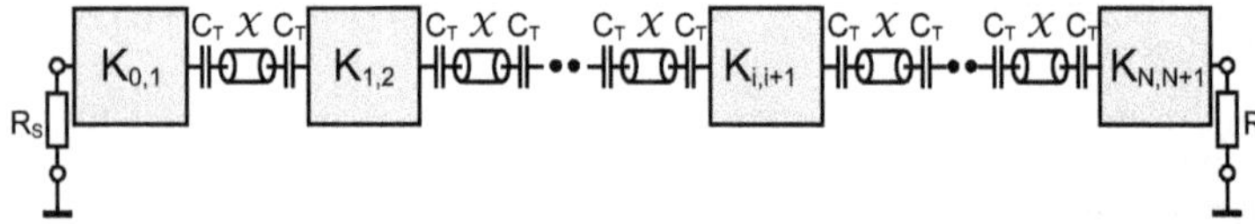

(b) Direct-coupled bandpass filter using foreshortened transmission line resonators, based on [100]

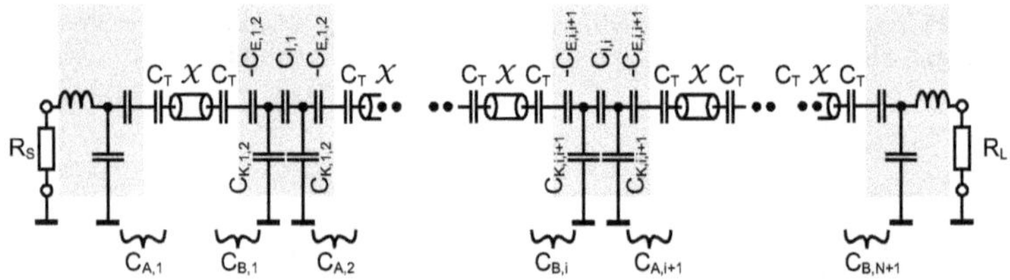

(c) Direct-coupled bandpass filter using the novel impedance inverter topology as well as foreshortened transmission line resonators, based on [100]

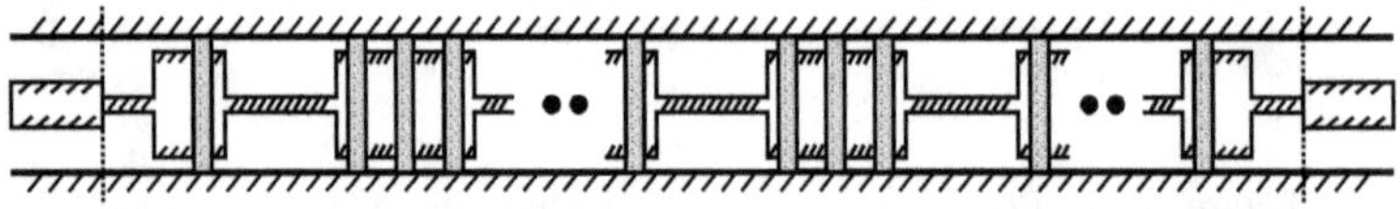

(d) Coaxial waveguide realisation of the filter shown in figure 13.18c [100]

Figure 13.18.: Tubular direct-coupled bandpass filter

13.3.2. Results: Solving a 4th Order Tubular Direct-Coupled Bandpass Filter

In section 13.2.2, a low-pass filter to be used in mobile telephone jammers was discussed. With the same application in mind, an alternate tubular direct-coupled bandpass filter was designed and realised by SF Microwave GmbH in order to avoid the issues of the stepped-impedance filter design discussed at the beginning of this section.

The bandpass filter was designed to have a passband center frequency of 875 MHz and a bandwidth of 300 MHz. The filter's passband thus includes both the LTE800 and GSM900 downlinks and provides some margin for the passband to shift due to imperfections of the real filter. Again, above 1580 MHz, stopband attenuation was required to be better than 60 dB in order to comply with regulatory requirements as well as to avoid triggering of GSM1800 and UMTS detectors.

Based on the filter topology depicted in figure 13.18c, a direct-coupled bandpass filter using the coaxial waveguide realisation depicted in figure 13.18d was realised from a 4th order Chebyshev filter with a maximum ripple of 0.04 dB. Figure 13.19 depicts the filter. A photograph of the filter's inner conductor is shown in figure 13.20. The filter geometry was kindly provided by SF Microwave GmbH.

Following a similar reasoning as in section 13.2.2, a Mode Matching solution of the structure shown in figure 13.19 can be carried out based on the fundamental TEM Eigenmode[33] and the rotationally symmetric $TM^z_{m,0}$ mode-set[34] for the coaxial waveguide and the rotationally symmetric $TM^z_{m,0}$ Eigenmodes of the circular waveguide given in section B.4.1.

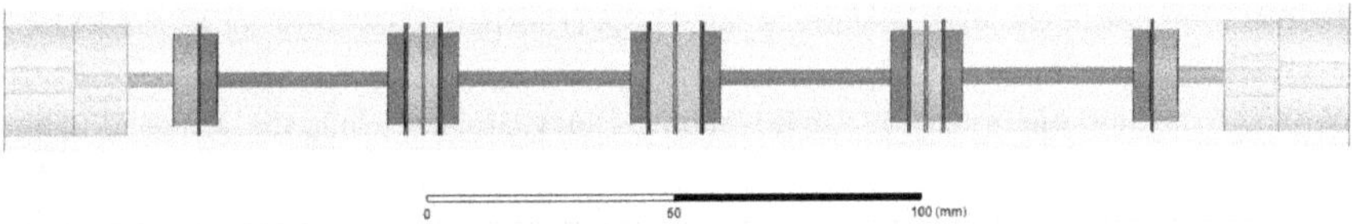

Figure 13.19.: 4th order tubular direct-coupled bandpass filter
Note that all resonators have the same length and all dielectric disks have the same thickness.

Filter geometry courtesy of SF Microwave GmbH

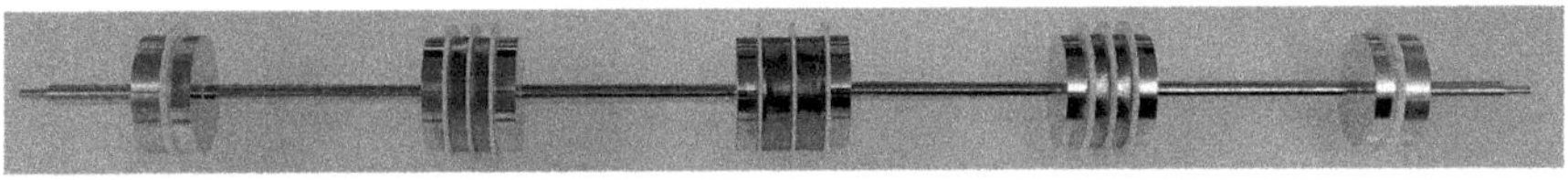

Figure 13.20.: Inner conductor of a 4th order tubular direct-coupled bandpass filter

Photograph courtesy of SF Microwave GmbH

[33] see appendix B.3.2
[34] see appendix B.3.1

Figure 13.21 shows a comparison of the filter's scattering parameters obtained using the Mode Matching solver (blue lines) developed in the process of preparing this thesis and Ansys HFSS (red dashed lines). The results obtained using the two methods again show outstanding agreement.

The Mode Matching solution was calculated using $V = P = 10$ Eigenmodes per waveguide. Solving a 166 point frequency sweep[35] took about 55 seconds to complete.

On the contrary, Ansys HFSS required about 2.5 hours for solving the same sweep with $\Delta s_{max} = 0.01$. For this solution, a mesh with approximately 70,000 elements was used. If Δs_{max} is reduced to 0.1, the solution time can be decreased to approximately 20 minutes. For the sake of completeness, note that using heavy[36] parallelisation, Ansys HFSS solves the given frequency sweep in 23 Minutes with $\Delta s_{max} = 0.01$. The technical data of the machine used for our simulations was given in section 13.2.2.

Taking these results into account, we can again conclude that for solving tubular filters, our Mode Matching approach is superior to Ansys HFSS' FEM approach.

The filter's measured scattering parameters are compared to the Mode Matching solution in figure 13.22. The measured scattering parameters are shown as black dash-dotted lines while the Mode Matching solution is shown as blue solid line. The two curves show excellent agreement over the entire bandwidth.

Finally, in figure 13.23 an enlarged representation of the filter's passband is given. Even in this enlarged representation the Mode Matching solution and the scattering parameters obtained using Ansys HFSS show almost perfect agreement. The filter's measured scattering parameters are also in good agreement, however, the filter's lower transistion band was shifted upwards by roughly 50 MHz. This is most likely due to mechanical imperfections of the real filter. Still, the filter maintains the design criteria discussed earlier as the LTE800 downlink starting at 790 MHz is fully included.

We now conclude our study of solving tubular filter designs using the Mode Matching technique and move on to direct-coupled bandpass filters realised by means of rectangular waveguides. While the underlying filter design concepts are exactly the same, the electromagnetic treatment of such filter is obviously somewhat different due to the different waveguides involved.

[35] This unusual point count is due to the fact that the filter's passbands were swept with a stepwidth of 10 MHz while for the stopbands 100 MHz were chosen.

[36] See footnote 26

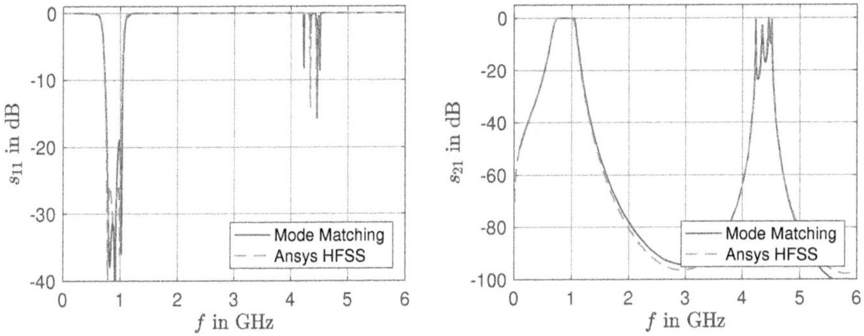

Figure 13.21.: Simulated scattering parameters of the 4th order tubular direct-coupled bandpass filter

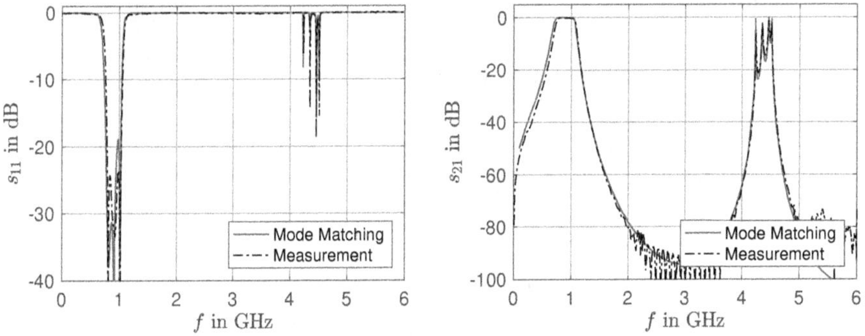

Figure 13.22.: Measured and simulated scattering parameters of the 4th order tubular direct-coupled bandpass filter
Measurement data courtesy of SF Microwave GmbH

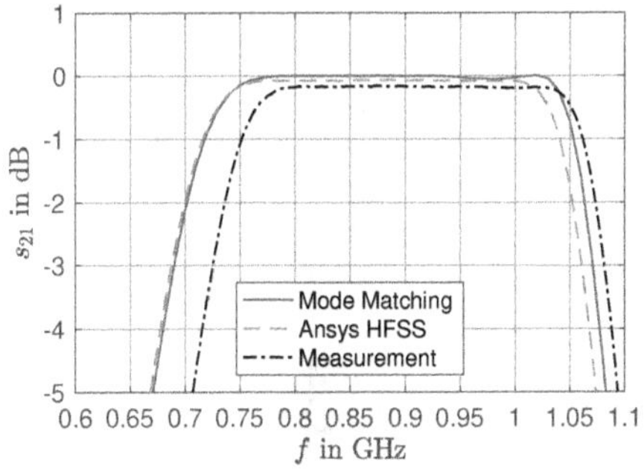

Figure 13.23.: Measured and simulated scattering parameters of the 4th order tubular direct-coupled bandpass filter. Enlarged representation of the passband.
Measurement data courtesy of SF Microwave GmbH

13.4. Rectangular Waveguide Filters: Iris-Coupled Bandpass Filter

In the previous section we discussed the realisation of a direct-coupled bandpass filter as a tubular design. We will now turn to the implementation of such filters in rectangular waveguide technology. The theory of such filters was discussed in Cohn's paper [97] and will be combined with points of view expressed in [78] and [96] in the following. In the next section we will then investigate the analysis of such filters using the Mode Matching technique.

Consider the generalised topology of direct-coupled bandpass filters depicted in figure 13.26a. In order to realise such a filter, two elements are required, namely impedance inverters and series resonators.

In section 10.2 we have seen that symmetric irises as shown in figure 10.4 on page 83 can be used to realise shunt inductances. Such irises thus can be used to implement the well-known *shunt-inductance impedance inverter* topology depicted in figure 13.24.

The shunt inductance's element value can be calculated as [96][37]

$$X_{K,i,i+1} = \frac{K_{i,i+1}}{1 - \left(\frac{K_{i,i+1}}{Z_0}\right)^2}. \tag{13.23}$$

The inductance of infinitely thin irises can be calculated using the equivalent circuit discussed in section 10.2. For irises with a non-negligible thickness, full-wave analysis of the filter is required in order to correct for the irises' thickness.

The impedance inverter's transmission lines are of negative electric length [96][38]

$$\alpha_{i,i+1} = -\mathrm{atan}\left(\frac{2X_{K,i,i+1}}{Z_0}\right) \tag{13.24}$$

and need to be absorbed in adjacent transmission lines [96].

Due to the nature of the impedance inverters used, the filters discussed in this section are referred to as *waveguide iris filters* or *shunt-inductance-coupled filters* [96].

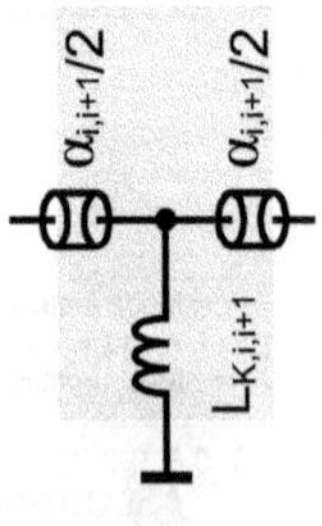

Figure 13.24.: Shunt-inductance impedance inverter

[37] see [96, fig. 8.03-1, p. 436]
[38] see ibid.

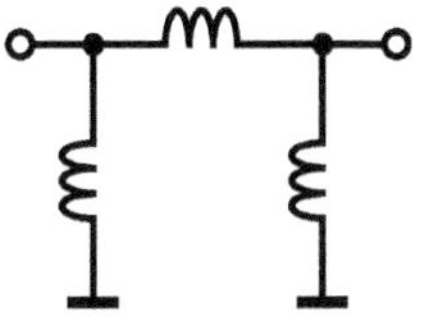

Figure 13.25.: Equivalent circuit of a half-wavelength waveguide segment

As already mentioned earlier, it is quite common to use transmission line resonators to mimic the series resonators' behaviour. Obviously such transmission line resonators can be realised as rectangular waveguide segments with a length approximately equal to half the guided wavelength at the passband's center frequency ω_0.

It can be shown that the equivalent circuit shown in figure 13.25 applies for such a waveguide segment [97].[39,40] Because even for moderately wide bandwidths the shunt susceptances B are small compared to the shunt susceptances of the adjacent coupling irises, they may be neglected in the following [97][41]. Consequently, the series arm can be used as series resonators for the direct-coupled filter design [78, 97].

Let us now determine the reactance slope of the half-wavelength waveguide segment, which is required for determining the impedance inverter's constants $K_{i,i+1}$ as we have seen in section 13.1.3. The reactance of the series arm calculates

$$X = Z_0 \sin(\theta) = Z_0 (\theta - \pi) = \pi Z_0 \left(\frac{\theta}{\pi} - 1 \right) \tag{13.25}$$

if a small angle approximation is applied [97]. $\theta \approx \pi$ denotes the electric length of the transmission line resonator and Z_0 its characteristic impedance.

The electric lengths θ and π may be rewritten using the wavenumber k_z and the guided wavelength λ_g, giving [97]

$$X = \pi Z_0 \left(\frac{\theta}{\pi} - 1 \right) = \pi Z_0 \left(\frac{k_z l}{k_{z0} l} - 1 \right) = \pi Z_0 \left(\frac{\frac{2\pi}{\lambda_g}}{\frac{2\pi}{\lambda_{g0}}} - 1 \right) = \pi Z_0 \left(\frac{\lambda_{g0}}{\lambda_g} - 1 \right)$$

for (13.25).[42]

As we have already discussed in section 13.1.3, the reactance slope (see (13.12)) is defined

$$\chi = \frac{\omega_0}{2} \frac{dX}{d\omega} \bigg|_{\omega = \omega_0}.$$

[39] see [97, p. 194]

[40] Here, the phase reversal of the half wavelength transmission line was ignored. Normally, the phase reversal would be realised by an additional 1:-1 transformer [97].

[41] see ibid.

[42] The subscript 0 indicates quantities at resonance, that is, at ω_0.

For the half-wavelength waveguide section the derivative may easily be calculated by taking the difference quotient of the reciprocal guided wavelength at the passband's edge frequencies ω_1 and ω_2. We may thus write the *reactance slope* as [96, 97][43]

$$\chi = \frac{\omega_0}{2} Z_0 \pi \frac{\frac{\lambda_{g,0}}{\lambda_{g,2}} - \frac{\lambda_{g,0}}{\lambda_{g,1}}}{\omega_2 - \omega_1} = \frac{Z_0 \pi}{2\mathfrak{w}} \left[\frac{\lambda_{g,0}}{\lambda_{g,2}} - \frac{\lambda_{g,0}}{\lambda_{g,1}} \right] \approx \frac{Z_0 \pi}{2\mathfrak{w}} \underbrace{\left[\frac{\lambda_{g,2} - \lambda_{g,1}}{\lambda_{g,0}} \right]}_{\mathfrak{w}_\lambda} = \frac{\pi \mathfrak{w}_\lambda}{2\mathfrak{w}} Z_0 \quad (13.26)$$

where $\mathfrak{w} = \omega_2 - \omega_1 / \omega_0$ denotes the relative bandwidth defined earlier and $\mathfrak{w}_\lambda$ represents the fractional wavelength.

If (13.26) is inserted into (13.17), the equations (1) - (3) for the inverter constants given in [96, fig. 8.06-1] are readily obtained.

Finally, the corrected electric lengths of the waveguide segments, which account for absorbing the impedance inverters' transmission lines, can be recalculated as [96][44]

$$\theta_i = \pi - \frac{1}{2} \left[\alpha_{i-1,i} + \alpha_{i,i+1} \right]. \quad (13.27)$$

The realisation of the filter as rectangular waveguide iris filter is depicted in figure 13.26c.

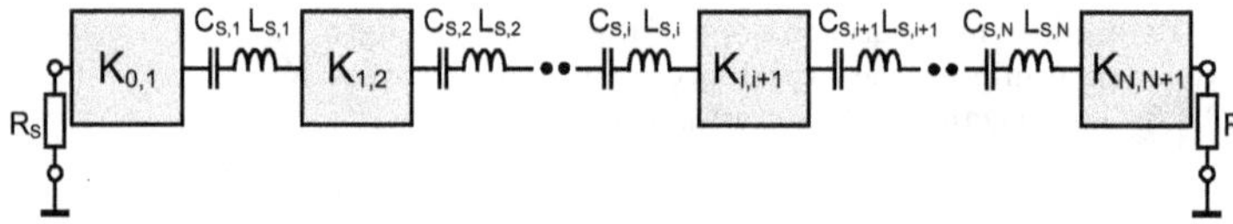

(a) Generalised direct-coupled bandpass filter

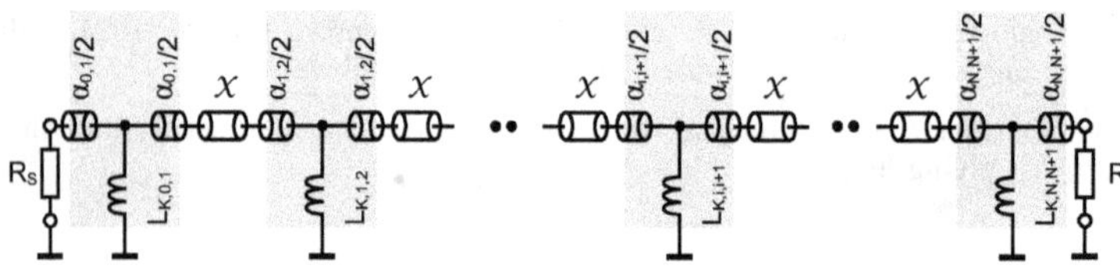

(b) Shunt-inductance-coupled filter, based on [97]

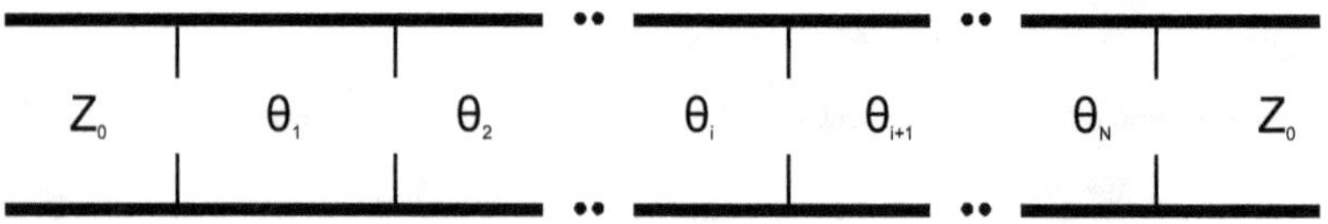

(c) Rectangular waveguide iris filter realisation of a shunt-inductance-coupled filter, based on [97]

Figure 13.26.: Rectangular waveguide iris-coupled filter

[43] see [96, fig. 8.06-1 (cont.), p. 452] and [97, p. 194]
[44] see [96, fig. 8.06-1, p. 451]

13.4.1. Results: Solving a 4th Order Waveguide Iris Filter

Having extended the theory of direct-coupled filters to rectangular waveguide realisations in the previous section, we now wish to examine the process of analysing such filters using the Mode Matching technique.

In the following, we will consider a 4th order Chebyshev bandpass filter ($\delta_{dB} = 0.1$ dB) for the K_a band (waveguide type: WR 28) with a passband center frequency of 35.45 GHz and a bandwidth of 700 MHz. The filter geometry was obtained from [101] and is depicted in figure 13.27.

While this filter is an arbitrary one used to demonstrate the capabilities of the Mode Matching solver developed during the preparation of this thesis, such rectangular waveguide filters have for example been used in *communication satellites* [10, 95] where bandpass filters are used for input and output filtering as well as for channel separation [102]. Input filtering is mainly required to remove out-of-band noise at the receiver while output filtering eliminates harmonics generated by the power amplifiers [102]. Channel separation is required because the individual channels are amplified by separate high power amplifiers in order to minimise signal degeneration due to nonlinearities of the amplifiers [10, 95].[45]

Generally, waveguide filter realisations recommend themselves for such applications due to their high power handling capability and their low loss [10]. However, waveguide filters are typically both large and heavy, which obviously contradicts with the application's demand for small, lightweight devices [10]. Consequently, considerable efforts have been made to reduce both size and weight of waveguide filters [10], which lead to more advanced designs, namely *dual-mode waveguide filters* [95]. Since in such a design each of the filter's cavities is designed to support two degenerate Eigenmodes acting as individual resonators, the number of cavities can be reduced by a factor of two. Furthermore, the cavities can be dielectrically loaded in order to further decrease their size [95]. Such filters are called *dual-mode dielectric resonator loaded cavity filters* [95]. The design and full-wave solution of such filters represents an interesting field for future work.

In contrast to the tubular filters discussed earlier, a reliable analysis of rectangular waveguide iris filters using Ansys HFSS is far from easy as the results appear to be prone to meshing issues. Unlike in the previous sections, we thus dedicate an individual section to simulating the rectangular waveguide filters using Ansys HFSS.

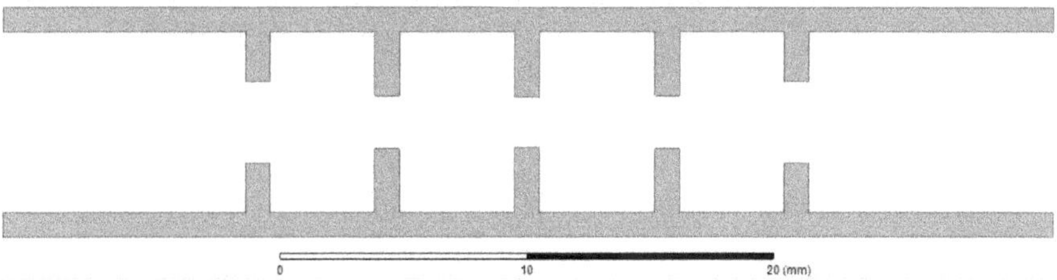

Figure 13.27.: 4th order rectangular waveguide shunt-inductance-coupled bandpass filter
Filter geometry obtained from [101]

[45] A modulated signal is impaired by a non-linear amplifier because third-order intermodulation products ($2 \cdot \omega_i - \omega_j$) appear in the bandwidth used by the signal. If the high power amplifier has to cover a larger bandwidth, it has to provide a larger total output power, which goes along with stronger intermodulation.

Solution Using the Mode Matching Technique

In chapter 7 we carried out an extensive study of a symmetric discontinuity between a narrow and a wide waveguide of equal height. The iris discontinuities can be modeled using the exact same type of discontinuity. As a consequence, our findings from chapter 7 apply and we can limit the analysis of the filter to odd-order $TE^{x}_{m,0}$ Eigenmodes in order to minimise the required computation time. However, it is emphasized that the Eigenwave solver developed in the scope of this thesis generally can handle all kinds of discontinuities using the complete TE^{x}/TM^{x} mode-set if required.

The scattering parameters of the filter shown in figure 13.27 are depicted in figure 13.28. The Mode Matching solution (blue lines) shows very good agreement with the solution obtained using Ansys HFSS (dashed red lines) over the entire recommended bandwidth of the waveguide (26.5 - 40 GHz [6]) and above.

The tubular bandpass filter discussed in chapter 13.3.2 had a relative bandwidth of about 35 %. In contrast, the waveguide filter discussed here has a relative bandwidth of about 2 %. Thus, the coupling coefficients of the filter's resonators are chosen such that the resonators exhibit a high loaded Q factor in order to realise the narrow bandwidth. As a consequence, the filter is much more sensitive to imperfect representations of the waveguide irises' shunt inductances.

Consequently, compared to the tubular filter, a considerably larger number of Eigenmodes is required. This can be seen from figure 13.29a, which depicts the convergence criterion introduced in chapter 9 for various solutions including different numbers of Eigenmodes. It can be seen that if 51 odd-order Eigenmodes are taken into account, the simulation has converged well below $\Delta s_{max} = 0.01$. If 51 odd-order Eigenmodes are included in the analysis, the highest odd-order Eigenmode is the $TE_{101,0}$ Eigenmode.

Convergence of the Mode Matching technique can also be observed in the detailed representation of the passband given in figure 13.29b. For smaller numbers of Eigenmodes included, the passband is minimally shifted to lower frequencies. As one would expect from figure 13.29a, no differences can be observed between a Mode Matching solution including either 51 or 76 Eigenmodes. The remaining difference between the latter two solutions is below 100 ppm of the filter's center frequency at the left -3 dB point.

The Mode Matching solver developed in the scope of this thesis required 330 seconds to calculate a 201 point frequency sweep taking 51 odd-order Eigenmodes into account. Details on the machine used for the presented simulations can be found in section 13.2.2.

Solution Using Ansys HFSS

Similar to the Mode Matching technique, the high Q factor of the filter's resonators represents a challenge for Ansys HFSS's FEM solver. During the preparation of the results presented here it was found that the adaptive meshing process has considerable difficulties to converge if meshing is carried out at the passband's center frequency.[46] The reason for this is that mesh modifications performed during the adaptive meshing process have a stronger impact on the scattering parameters than the improved representation of the structure due to the finer mesh.

[46] Meshing at stopband frequencies is not an option because when doing so, very little power is coupled into the filter's inner resonators. Since the fields in these resonators thus are comparably small, the mesh is not sufficiently refined in these regions.

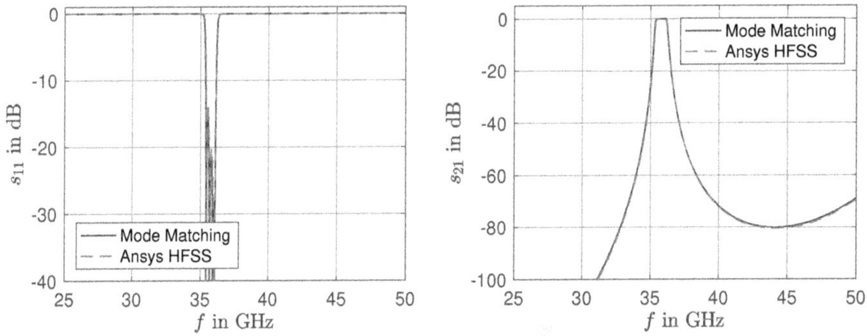

Figure 13.28.: Scattering parameters of the 4th order rectangular waveguide bandpass filter obtained using Mode Matching, 51 Eigenmodes, $\leq TE_{101,0}$

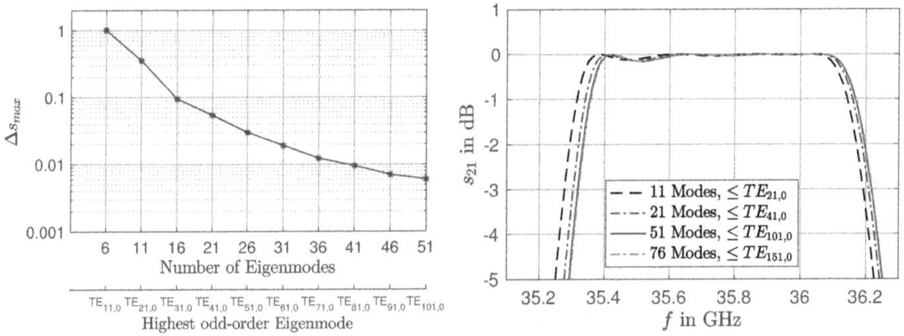

(a) Convergence Criterion Δs_{max}

(b) Enlarged passband for various numbers of Eigenmodes included

Figure 13.29.: On the convergence of the Mode Matching solution

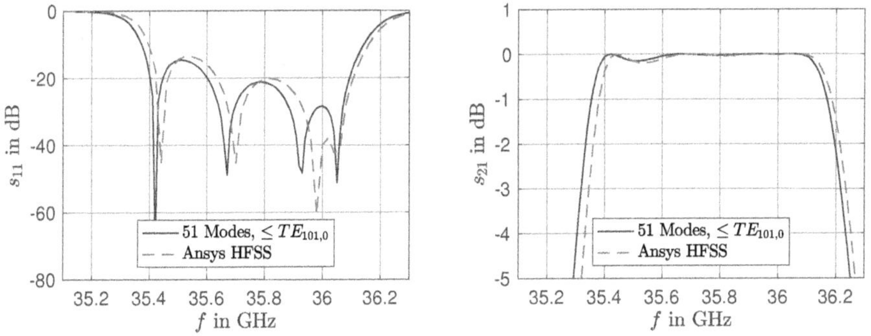

Figure 13.30.: Comparison between the Mode Matching solution and Ansys HFSS' results based on an enlarged representation of the passband

If the structure is solved in Ansys HFSS with $\Delta s_{max} = 0.1$, the passband is shifted considerably to higher frequencies. For smaller values of Δs_{max}, convergence is achieved after at least 20 mesh iterations. The size of the resulting meshes is in the region of half a million mesh cells, which is obviously undesirable in terms of computation time.

A more suitable approach for solving the filter using Ansys HFSS is to use the software's wavelength-based mesh refinement function. Using this approach, the initial mesh is optimised such that the length of each mesh element is below a given fraction of the wavelength. For the given filter, a rather small value of 0.05λ (default: 0.33λ) enabled the adaptive meshing process to converge below $\Delta s_{max} = 0.1$ after three iterations. The resulting mesh, which consists of about 160,000 elements, is depicted in figure 13.31.

The scattering parameters shown in figure 13.28 were obtained using these settings. However, for comparative purposes, the enlarged representation of the filter's passband depicted in figure 13.30 is more suitable. From figure 13.30 it can be seen that the Mode Matching solution and the solution obtained using Ansys HFSS agree to below 1 ‰ of the passband's center frequency at the left -3 dB point.

In contrast to the Mode Matching solver presented in this thesis, Ansys HFSS required about 2 hours for solving a 201 point sweep if heavy parallelisation[47] is used. Simulations were carried out on the same machine as the Mode Matching analyses.

Conclusion

Summarising the above, we can conclude that for the given waveguide filter, the Mode Matching technique is again superior to Ansys HFSS FEM approach: Two reasons support this point of view: Firstly, our Mode Matching solver presented here solves the filter in small fraction of the time required by Ansys HFSS. Secondly, the Mode Matching solution does not require any additional settings except for the number of Eigenmodes to be taken into account in order to converge properly. In contrast, Ansys HFSS' FEM solver requires additional information on how to properly mesh the structure.

Figure 13.31.: Mesh generated by Ansys HFSS at 35.75 GHz with $\Delta s_{max} = 0.1$ using an aggressive mesh refinement strategy (0.05λ)

[47] See footnote 26 on page 120.

14. Eigenmodes of Partially Filled Coaxial Waveguides

The previous chapter was concerned with the study of various tubular filters. In such filter designs, dielectric tubes as shown in figure 14.1 are often desirable as they can be used to center the inner conductor and increase the filter's power handling capability due to the increased breakdown voltage of the dielectric [103].

However, in order to still exploit the computational advantages of the Mode Matching technique, it is mandatory to know the Eigenmodes of *partially filled coaxial waveguides* as shown in figure 14.2 on the following page.

Unfortunately, to the author's best knowledge, the Eigenmodes of this type of waveguide are not readily available in standard waveguide literature.

In this chapter, we present the derivation of the partially filled coaxial waveguide's rotationally symmetric TM Eigenmodes based on the generalised theory of multi-layered cylindrical waveguides [48, 56, 57]. The results presented in this chapter were originally published by the author in [68] at the German Microwave Conference 2016.

In the following, we will firstly study the general properties of the partially filled waveguide's Eigenmodes in terms of the waveguide's wavenumbers and the separation condition. Secondly, we will solve the scalar Helmholtz equation for the waveguide's empty and filled region. Next, by enforcing the boundary conditions at the interface between the two regions, we solve the underlying boundary value problem[1] and determine the Eigenvalues ${}^{1}k_{\rho}$ and ${}^{2}k_{\rho}$ applicable for region 1 and 2. From these transverse wavenumbers we obtain k_z by using the separation condition. Eventually, the propagation constant directly follows from k_z.

Finally, we will study the properties of the Eigenmodes obtained and conclude this chapter by discussing the Mode Matching solution for a simple structure comprising a partially filled coaxial waveguide.

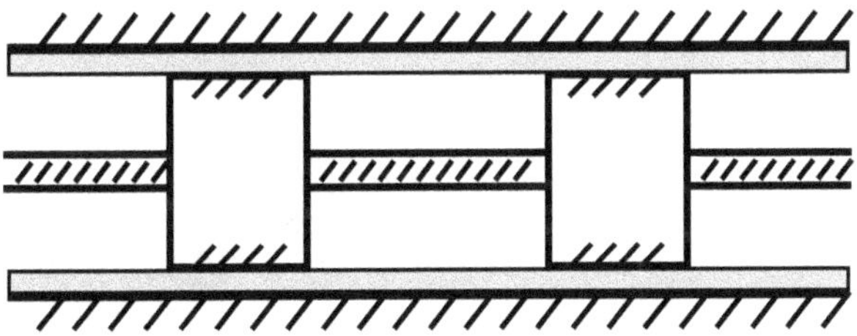

Figure 14.1.: Tubular filter with dielectric tube

[1]The boundary value problem was introduced in section 2.1.

14.1. General Discussion

14.1.1. The Fundamental Eigenmode

Consider the partially filled coaxial waveguide shown in figure 14.2. For obvious reasons, for both regions, the same wavenumber k_z has to apply [48]. It is well known that for the TEM Eigenmodes, k_z calculates

$$ {}^1k_z = {}^1k = \omega\sqrt{{}^1\varepsilon\ {}^1\mu} \qquad {}^2k_z = {}^2k = \omega\sqrt{{}^2\varepsilon\ {}^2\mu} \tag{14.1} $$

in the two regions 1 and 2. Consequently, the fundamental Eigenmode cannot be TEM because ${}^1k_z = {}^2k_z$ cannot be maintained. Thus, in contrast to an empty coaxial waveguide, the partially filled waveguide's fundamental Eigenmode must be TM or TE [68].

Now consider the expressions for the transverse electric and magnetic field components (2.13) and (2.15) for TM Eigenmodes and (2.20) and (2.22) for TE Eigenmodes. From these equations, it can be shown that the transverse field components E_ρ and H_ϕ of TM Eigenmodes calculate [48]

$$ E_\rho^{TM} = \frac{1}{j\omega\varepsilon}\,\frac{\partial^2\psi}{\partial\rho\,\partial z} \qquad H_\phi^{TM} = -\frac{\partial\psi}{\partial\rho} \tag{14.2} $$

while for TE Eigenmodes we find [48]

$$ E_\rho^{TE} = -\frac{1}{\rho}\frac{\partial\psi}{\partial\phi} \qquad H_\phi^{TE} = \frac{1}{j\omega\mu\rho}\,\frac{\partial^2\psi}{\partial\phi\,\partial z}. \tag{14.3} $$

Let us now assume that the fundamental Eigenmode of the partially filled waveguide is of comparable structure as the TEM Eigenmode of the empty coaxial waveguide. Under this assumption, we would expect that the partially filled waveguide's fundamental Eigenmode has (at least) an E_ρ and H_ϕ component and is rotationally symmetric.

By careful examination of (14.3) we find that TE Eigenmodes may not comply with this assumption: Both E_ρ^{TE} and H_ϕ^{TE} are obtained by deriving the scalar vector potential ψ for ϕ. Due to rotational symmetry, ψ does not have a ϕ dependence and thus its derivative w.r.t ϕ vanishes. Thus, the fundamental Eigenmode can only be a TM Eigenmode [68].

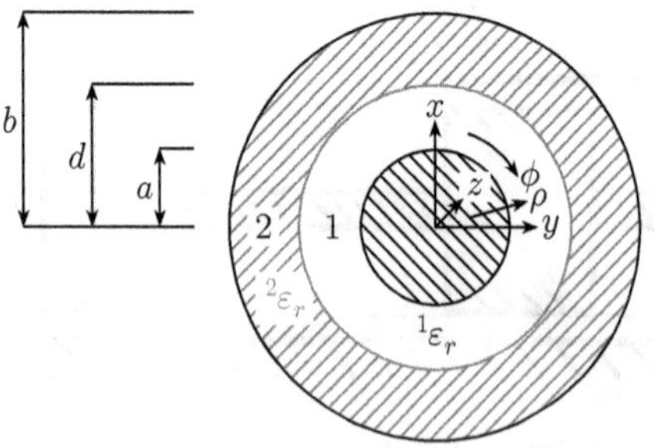

Figure 14.2.: Partially filled coaxial waveguide (with modifications from [68])

We now need to determine the proper range of values for ${}^{1}k_{\rho}$ and ${}^{2}k_{\rho}$. If a dielectric is inserted into a waveguide, the wavelength is reduced. As a consequence, for the partially filled waveguide, we have

$$\lambda_{empty} > \lambda > \lambda_{filled} \tag{14.4}$$

and thus

$$k_{empty} = k_{z,empty} < k_z < k_{z,filled} = k_{filled} \tag{14.5}$$

where variables indexed with "empty" or "filled" refer to an entirely empty or filled coaxial waveguide with TEM propagation while k_z applies for the partially filled coaxial waveguide's fundamental TM Eigenmode [68].

As ${}^{1}k = k_{empty}$ and ${}^{2}k = k_{filled}$, (14.5) yields

$${}^{1}k < k_z < {}^{2}k. \tag{14.6}$$

From the separation condition (2.42) derived in section 2.2.2, k_z can be calculated as

$$k_z = \pm\sqrt{{}^{i}k^2 - {}^{i}k_{\rho}^2} \tag{14.7}$$

in region i. As a consequence, $k_z > {}^{1}k$ can only be maintained if ${}^{1}k_{\rho} \in \mathbb{I} > 0$ [68].

We will see later (cf. figure 14.3a) that for ${}^{1}k_{\rho} \in \mathbb{I} > 0$, the waveguide's characteristic equation has a single zero only [68]. This single solution to the characteristic equation provides the waveguide's fundamental propagating $TM_{1,0}$ Eigenmode, which is equivalent to the empty waveguide's TEM Eigenmode [68].

14.1.2. Higher-Order Eigenmodes

In the scope of the discussion of tubular filters carried out in the previous chapter, we have already seen that in a rotationally symmetric tubular filter design excited by a TEM Eigenmode all higher-order Eigenmodes except for the rotationally symmetric TM Eigenmodes, which will be below cut-off, can be ignored (see section 13.2.2).

Thus, the following sections will cover rotationally symmetric TM Eigenmodes only. In order to obtain these higher-order Eigenmodes, we let ${}^{1}k_{\rho}, {}^{2}k_{\rho} \in \mathbb{R}$.

For ${}^{1}k_{\rho} \in \mathbb{R} > 0$ an infinite number of zeros of the characteristic equation exists (cf. figure 14.3b). Consequently, an infinite number of higher-order $TM_{m,0}$ Eigenmodes ($m \geq 2$) can be found [68].

14.2. Derivation of the $TM^z_{m,0}$ Eigenmodes

From the generalised theory of multi-layered cylindrical waveguides [48, 56, 57] it is well known that the Eigenmodes of a cylindrical, rotationally symmetric waveguide may be obtained by dividing the waveguide into regions for which the appropriate Eigenmodes can be obtained by solving the Helmholtz equation[2]

$$\nabla_t^2 {}^i\psi_t + {}^ik_c^2 \, {}^i\psi_t = 0$$

using a product approach ${}^i\psi_t = {}^i\psi_\rho(\rho) \cdot {}^i\psi_\phi(\phi)$.

As discussed in section 2.2.2, the functions ${}^i\psi_\rho$ must maintain Bessel's differential equation, whose solutions are Bessel functions of first or second kind as well as linear combinations of these functions.

As we wish to limit our analysis to rotationally symmetric TM Eigenmodes, we choose $n = 0$, which yields $\psi_\phi(\phi) = \cos(0) = 1$. Our analysis thus provides the $TM_{m,0}$ mode-set $(m > 0)$ of the partially filled waveguide [68].

Since both regions 1 and 2 are bounded by a PEC wall, it is reasonable to choose

$${}^1\psi^{TM} = {}^1\psi_t^{TM} e^{\mp jk_z z} \quad \text{where} \quad {}^1\psi_t^{TM} = A \left[N_0({}^1k_\rho a)\, J_0({}^1k_\rho \rho) - J_0({}^1k_\rho a)\, N_0({}^1k_\rho \rho)\right] \tag{14.8}$$

and

$${}^2\psi^{TM} = {}^2\psi_t^{TM} e^{\mp jk_z z} \quad \text{where} \quad {}^2\psi_t^{TM} = C \left[N_0({}^2k_\rho b)\, J_0({}^2k_\rho \rho) - J_0({}^2k_\rho b)\, N_0({}^2k_\rho \rho)\right] \tag{14.9}$$

as scalar vector potentials for the two regions [68]. These vector potentials inherently maintain the Dirichlet boundary conditions[3] [68]

$${}^1\psi_t\Big|_{\rho=a} = 0 \qquad {}^2\psi_t\Big|_{\rho=b} = 0 \tag{14.10}$$

applicable for TM Eigenmodes [58].

We now need to find Eigenvalues ${}^1k_c^2 = {}^1k_\rho^2$ and ${}^2k_c^2 = {}^2k_\rho^2$ so that the second boundary condition for region 1 and 2, namely the continuity of tangential fields at the air-material interface at $\rho = d$ is ensured. At the same time, ${}^1k_c^2$ and ${}^2k_c^2$ must be chosen in a fashion so that the solutions for region 1 and 2 have the same propagation constant, that is, ${}^1k_z = {}^2k_z$ must apply.

---

[2] also see (2.9)

[3] also see (2.30)

From (2.14) and (2.15) it can be seen that the $TM^z_{m,0}$ Eigenmodes' field components tangential to the air-dielectric interface can be calculated

$$E_z^{TM} = \frac{1}{j\omega\varepsilon}\left(\frac{\partial^2}{\partial z^2}+k^2\right)\psi \qquad H_\phi^{TM} = -\frac{\partial\psi}{\partial\rho}. \tag{14.11}$$

The *continuity condition* at the interface at $\rho = d$ can be put under form of a homogeneous system of equations [68]

$$\underbrace{\begin{bmatrix} {}^1E^{TM}_{z|m,0} & -{}^2E^{TM}_{z|m,0} \\ {}^1H^{TM}_{\phi|m,0} & -{}^2H^{TM}_{\phi|m,0} \end{bmatrix}}_{\mathbf{\Phi}} \underbrace{\begin{bmatrix} A \\ C \end{bmatrix}}_{\vec{x}} = \vec{0}\,\Big|_{\rho=d}, \tag{14.12}$$

which has non-trivial solutions ($\vec{x} = \Pi\,\vec{x_0}$) only if the matrix' determinant vanishes, that is,

$$\det(\mathbf{\Phi}) = 0. \tag{14.13}$$

This is the partially filled waveguide's *characteristic equation* [68]. The characteristic equation's zeros, which have to be found numerically, provide the wavenumbers k_ρ [68]. In practice this means that it is required to sweep ${}^1k_\rho$ over either ${}^1k_\rho \in \mathbb{I} > 0$ (fundamental Eigenmode) or ${}^1k_\rho \in \mathbb{R} > 0$ (higher-order Eigenmodes) while $k_{\rho,2}$ is calculated

$${}^2k_\rho = \pm\sqrt{\omega^2\varepsilon_0\mu_0({}^2\varepsilon_r - 1) + {}^1k_\rho^2}, \tag{14.14}$$

which directly follows from ${}^1k_z = {}^2k_z$ using the separation condition [68]. Note that here ${}^1\varepsilon_r = 1$ was assumed.

After suitable wavenumbers ${}^1k_\rho$, ${}^2k_\rho$ have been determined, the solution vector $\vec{x_0}$ can be found by inserting the two equations contained in (14.12) into each other. Note that while (14.12) provides the solution vector $\vec{x_0}$, that is, the ratio of the waveguide regions' amplitudes, Π and thus the actual amplitudes are determined by the excitation of the waveguide, which determines the power transported or the energy stored reactively [68].

Finally note that the explicit expressions for the Eigenmodes' fields are identical to those for the higher-order $TM^z_{m,0}$ Eigenmodes of the coaxial waveguide (B.25) given in appendix B.3.1 if the appropriate wavenumbers for region 1 or 2 are inserted.

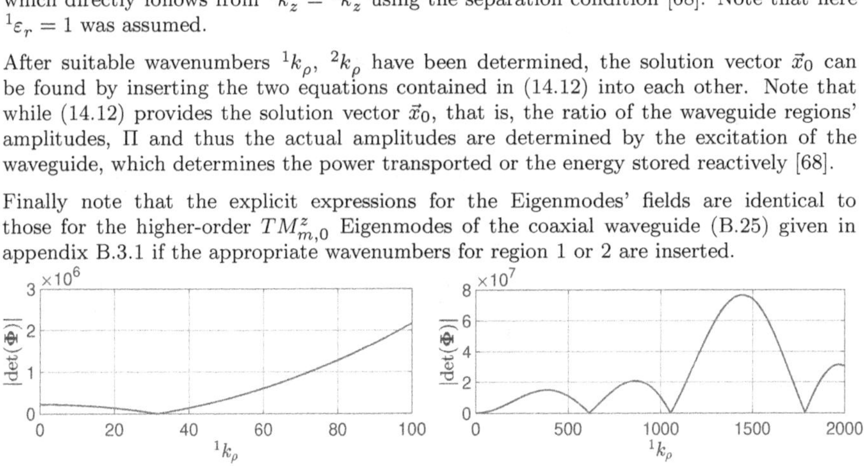

(a) Characteristic Equation for the fundamental $TM_{1,0}$ Eigenmode (${}^1k_\rho \in \mathbb{I} > 0$)

(b) Characteristic Equation for the higher-order $TM_{m,0}$ Eigenmodes (${}^1k_\rho \in \mathbb{R} > 0$)

Figure 14.3.: Characteristic equations for a partially filled coaxial waveguide $a = 4.4$ mm, $b = 10$ mm, $d = 7.5$ mm, ${}^1\varepsilon_r = 1$, ${}^2\varepsilon_r = 4$, $f = 2.5$ GHz.

14.3. Properties of the Obtained Eigenmodes

In the following section, the properties of the obtained $TM^z_{m,0}$ Eigenmodes shall be discussed. The presented results [68] apply for a partially filled waveguide with $a = 4.4$ mm, $b = 10$ mm, $d = 7.5$ mm, ${}^1\varepsilon_r = 1$, ${}^2\varepsilon_r = 4$ at $f = 2.5$ GHz unless stated otherwise.

Figures 14.4 and 14.5 depict the field distributions E_ρ, H_ϕ and E_z of the fundamental and first evanescent Eigenmode of the partially filled waveguide. For the two field components H_ϕ and E_z tangential to the air-dielectric interface, continuity clearly is maintained [68].

In contrast, the electric field component E_ρ normal to the air-dielectric interface is discontinuous. From basic electromagnetic theory we know that at such an interface the normal electric displacement field must be continuous, that is, ${}^1D_\rho = {}^2D_\rho$ [58] and thus

$$\frac{{}^1E_\rho}{{}^2E_\rho} = \frac{{}^2\varepsilon_r}{{}^1\varepsilon_r} \tag{14.15}$$

applies. Consequently, if we let ${}^1\varepsilon_r = 1$ and ${}^2\varepsilon_r = 4$ as given above, at the interface, we would expect the electric field inside the dielectric to be four times smaller than outside the dielectric. This effect can nicely be observed in figures 14.4 and 14.5.

As a matter of fact, the dielectric displaces the electric field out of the outer region of the waveguide and as a consequence, the power transported in the dielectric is decreased if ${}^2\varepsilon_r$ is increased as is shown in figure 14.6.

The wavenumbers and the characteristic impedance obtained for the fundamental propagating and the first evanescent TM Eigenmode are given in table 14.1. The results show excellent agreement with results obtained using a commercial FEM solver.

In figure 14.7, the wavenumber k_z and the characteristic impedance Z_0 is plotted for various inner diameters of the dielectric tube. As one would expect, the solutions for k_z and Z_0 approach the values applicable for a TEM Eigenmode in an empty (left) or entirely filled (right) coaxial waveguide [68].

In an empty coaxial waveguide, the fundamental TEM Eigenmode's wave impedance, which represents the ratio of E_ρ and H_ϕ, is equal to the free space impedance of 377 Ohm. In contrast, the wave impedance of the partially filled waveguide's fundamental Eigenmode varies over the cross-section as can be seen in figure 14.4 [68]. More precisely, due to the fact that at the interface H_ϕ is continuous while for E_ρ (14.15) applies, the wave impedance in the empty region is four times larger than inside the dielectrically filled region.

The same applies for the wave impedance of the first evanescent $TM_{m,0}$ Eigenmode depicted in figure 14.5. Note that Z_{wave} is negative imaginary as one would expect for TM Eigenmodes (also see section 10.2). Consequently, the higher-order Eigenmodes store energy capacitively.

	Eigenmode Solution				Ansys HFSS	
TM Eigenmode	${}^1k_\rho$	${}^2k_\rho$	k_z	Z_0	k_z	Z_0
Propagating Eigenmode ($TM^z_{1,0}$)	$j31.52$	85.1	61.15	41.4	61.14	42.3
1st evan. Eigenmode ($TM^z_{2,0}$)	613.8	620.5	$-j611.6$	-	$-j611.6$	-

Table 14.1.: Calculated wavenumbers and wave impedances (with modifications from [68]) $a = 4.4$ mm, $b = 10$ mm, $d = 7.5$ mm, ${}^1\varepsilon_r = 1$, ${}^2\varepsilon_r = 4$, $f = 2.5$ GHz
k_ρ in 1/m, k_z in rad/m or Np/m, Z_0 in Ohm

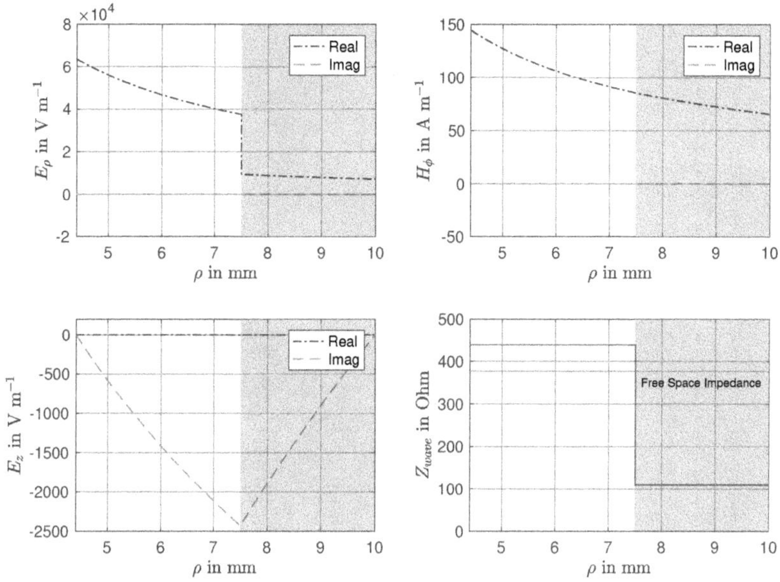

Figure 14.4.: Fields and wave impedance of the fundamental $TM_{1,0}$ Eigenmode [68]

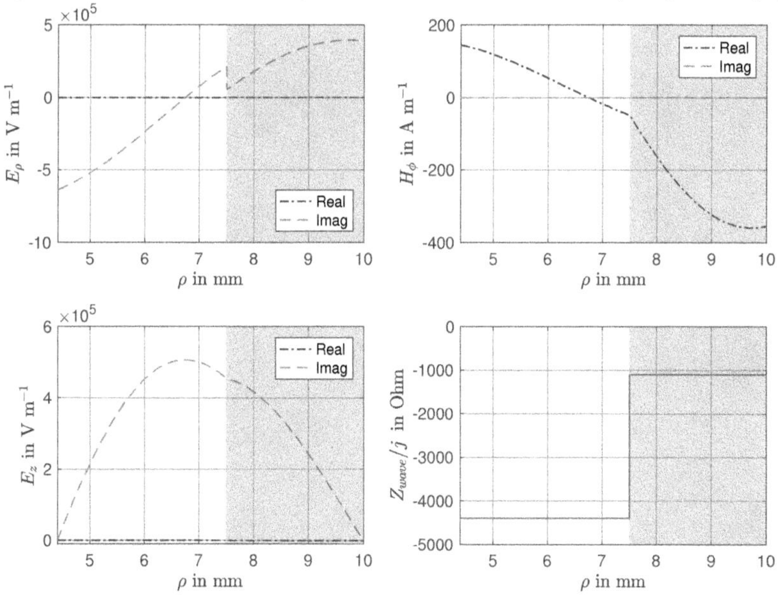

Figure 14.5.: Fields and wave impedance of the higher-order $TM_{2,0}$ Eigenmode [68]

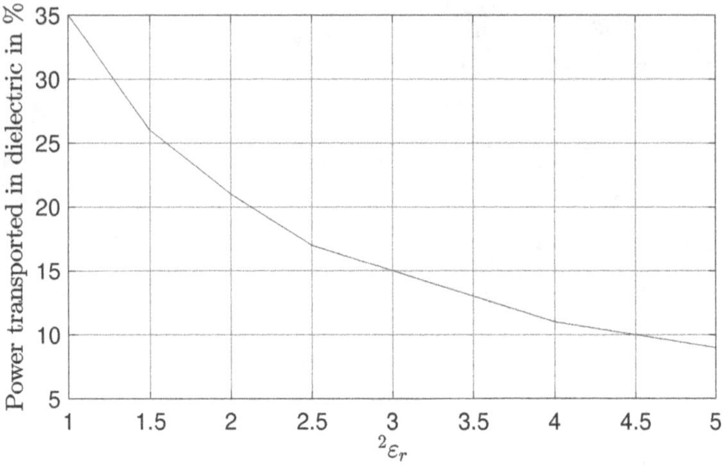

Figure 14.6.: Variation of the power transported inside the dielectric tube over the tube's permittivity, $a = 4.4$ mm, $b = 10$ mm, $d = 7.5$ mm, $^1\varepsilon_r = 1$, $f = 2.5$ GHz

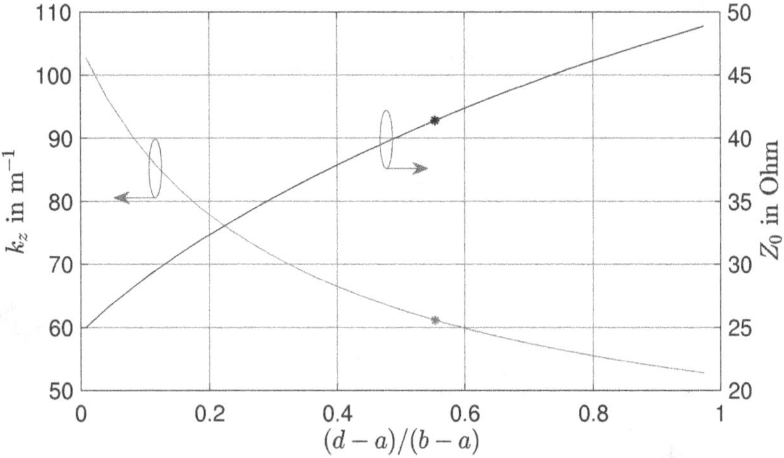

Figure 14.7.: Variation of the fundamental Eigenmode's wavenumber k_z and characteristic impedance Z_0 over the dielectric tube's inner diameter (modified from [68]) The markers (*) indicate the geometry variation ($d = 7.5$ mm) discussed in the text. $a = 4.4$ mm, $b = 10$ mm, $^1\varepsilon_r = 1$, $^2\varepsilon_r = 4$, $f = 2.5$ GHz

14.4. Solving a Structure Comprising Partially Filled Coaxial Waveguides

The final aim of our endeavours is to analyse tubular filters comprising partially filled coaxial waveguides using the Mode Matching technique in a fast and accurate fashion. When implementing a Mode Matching solver capable of dealing with partially filled waveguides, the orthogonality of the waveguide's Eigenmodes requires additional attention as already discussed in section 2.6.

The $TM^z_{m,0}$ Eigenmodes of the partially filled coaxial waveguide derived in this section only have H_ϕ as magnetic field component, that is, they do not have a magnetic field component normal to the air-dielectric interface and thus are of *Longitudinal Section Magnetic* (LSM) type (see section 2.6). An additional weighting function $1/\varepsilon_r(\rho)$ is required to maintain the Eigenmodes' orthogonality as already mentioned in section 2.6. It is absolutely mandatory to observe this weighting function when calculating the Eigenmodes' overlap integrals; failure to do so will lead to incorrect unphysical scattering parameters.

Another important point regarding the implementation of the Mode Matching formalism for partially filled coaxial waveguides is that due to the frequency-dependent nature of the solution for ${}^1k_\rho$ and ${}^2k_\rho$ (also see (14.14)) the Eigenmodes' overlap integrals become frequency-dependent as well. As a consequence, it is no longer possible to perform the computational expensive task of numerically solving overlap integrals once rather than for each individual frequency point in the simulation's sweep (also see section 11.4.2).

Let us now conclude this chapter with a brief discussion of the Mode Matching solution for an exemplary structure. While we have not yet designed filters which include partially filled coaxial waveguide segments, the required Mode Matching solver is readily available. We will thus discuss the Mode Matching solution for the structure depicted in figure 14.8.

The matching of the tangential fields at the interface between the empty and partially filled coaxial waveguide is shown in figure 14.9. It is especially interesting to observe the higher-order Eigenmodes of the empty coaxial waveguide approximating the discontinuity of E_ρ at the air-dielectric interface in the partially filled waveguide.

For completeness, the structure's scattering parameters are depicted in figure 14.10.

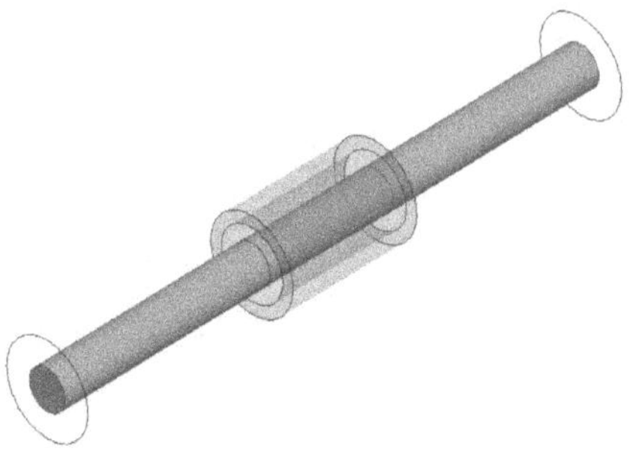

Figure 14.8.: A simple structure including a partially filled coaxial waveguide
$a = 4.4$ mm, $b = 10$ mm, $d = 7.5$ mm, ${}^1\varepsilon_r = 1$, ${}^2\varepsilon_r = 4$

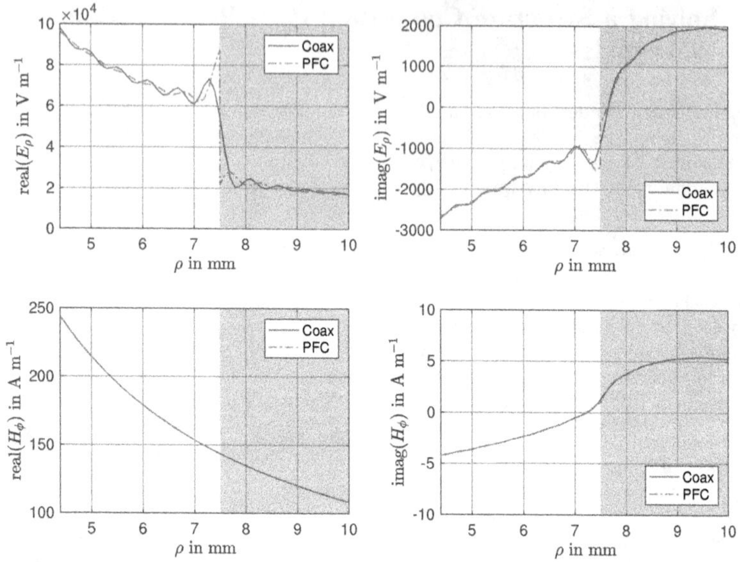

Figure 14.9.: Matching of the tangential fields at the interface between the empty and partially filled coaxial waveguide shown in figure 14.8, $P = V = 20$ Modes

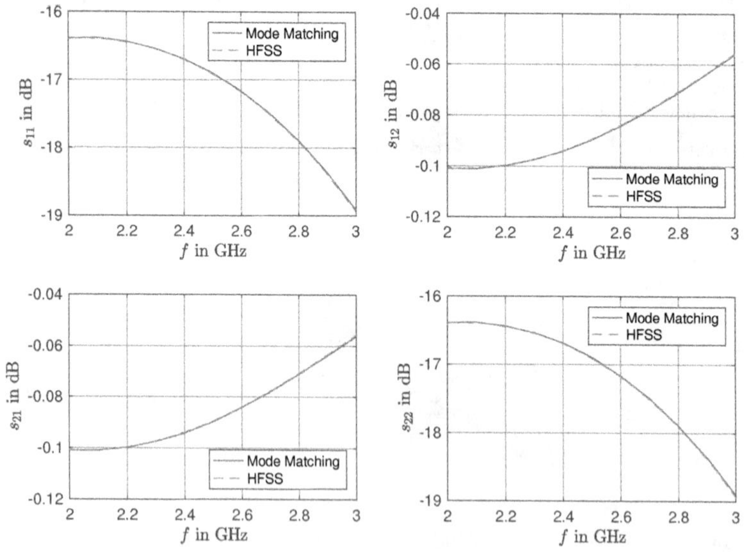

Figure 14.10.: S parameters for the structure shown in figure 14.8, $P = V = 20$ Modes

15. Cavity Resonators for Electromagnetic Material Characterisation

If an electromagnetic cavity is perturbed by a small dielectric sample as is shown in figure 15.1, both its resonant frequency and quality factor are decreased. This effect can be used for determining the complex relative permittivity ε_r of the dielectric sample perturbing the cavity.

In order to do so, expressions relating the shift of the cavity's resonant frequency and quality factor to the sample's complex relative permittivity are required. This relation is provided by *cavity perturbation theory* [48, 92] and consequently, we refer to this method as *cavity perturbation material characterisation.*

During the derivation of these expressions assumptions regarding the fields inside the sample volume, the energy stored inside the cavity and the ohmic losses in the cavity's walls have to be made. Depending on the actual measurement, these assumptions may induce "secondary effects" which skew the measurement of the sample's relative permittivity.

In this chapter we will investigate the validity of these assumptions by recalculating the approximated quantities in an exact fashion based on a Mode Matching solution for the cavity's fields. The corresponding approach for treating resonant structures using the Mode Matching technique was introduced in chapter 12.

As a lot of groundwork is required before we can begin our discussion of these approximations, the remainder of this chapter is structured as follows:

Firstly, in section 15.1 we start with a brief overview on the state of the art in electromagnetic material characterisation. Next, in section 15.2 we will review the theory of cavity perturbation material characterisation and introduce appropriate approximations where applicable. Subsequently, exemplary measurements for two materials, Polytetrafluorethylen (PTFE, Teflon) and water, are shown in section 15.3 for illustrative purposes. Finally, the analysis of the aforementioned secondary effects in section 15.4 concludes this chapter.

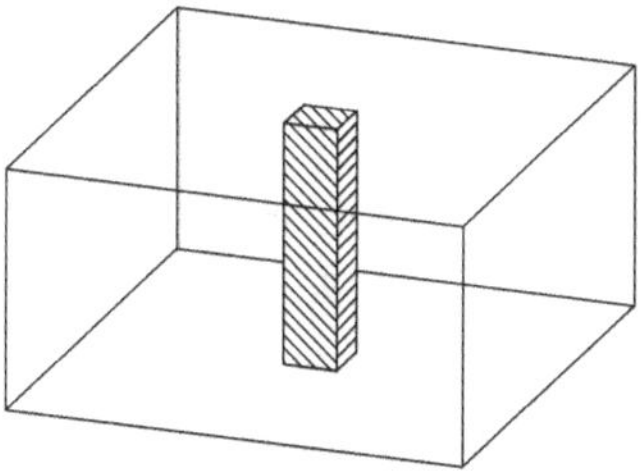

Figure 15.1.: Rectangular cavity perturbed by a square dielectric rod

15.1. State of the Art in Permittivity Measurement

Over the last decades, the accurate and reliable determination of electromagnetic material parameters has received considerable attention and several papers [104–106] summarising the state of the art in electromagnetic material characterisation have been published.

One possible approach to classify existing techniques is to divide them into resonant and transmission/reflection techniques [106]. Generally, resonant techniques provide material parameters at specific frequencies while transmission/reflection methods can be considered as broadband techniques [106]. However, resonant methods provide a higher measurement accuracy and are thus better suited for measuring low-loss dielectrics [105, 106].

15.1.1. Resonant Techniques

Resonant techniques generally can be distinguished into two groups, cavity perturbation methods and dielectric resonator techniques.

Cavity Perturbation Technique

When performing *cavity perturbation material characterisation*, the detuning of an electromagnetic cavity resonator caused by a small dielectric sample being inserted into the cavity's volume is evaluated as we have already discussed at the beginning of this chapter. We will investigate the theoretical background of this method shortly.

It is interesting to note that the history of cavity perturbation theory dates back as early as 1943 when Bethe and Schwinger published their report [92] on cavity perturbation theory at M.I.T. Radiation Laboratory. Later, in 1949, Birnbaum and Franeau [107] were one of the first to publish a paper on a method for material characterisation using cavity perturbation theory. Since then, the method has continued to receive attention [108–110].

Dielectric Resonator Techniques

In contrast to the cavity perturbation theory, *dielectric resonator techniques* use the sample itself as a resonator. In order to do so, a cylindrical sample representing a dielectric waveguide is put between two ideally conducting plates of (theoretically infinite) size. In doing so, the waveguide is shorted at both ends, thus making it a resonant structure [111].

Because the Eigenmodes of such a dielectric resonator are known, it is possible to calculate the sample's permittivity from the resonant frequencies and quality factors of the various Eigenmodes [112]. However, one of the problems involved with this method is the proper mapping of the multitude of observed resonances to the correct Eigenmodes [112].

Various approaches for coupling the dielectric resonator to a feedline can be thought of: Hakki and Coleman coupled the resonator to a rectangular waveguide using an iris in the waveguide's wall, thus making it a reaction-type measurement [112]. In contrast, a transmission-type measurement can be carried out if the structure is excited by two coaxial lines terminating in E-field probes close to the dielectric waveguide [111].

If very low loss dielectrics shall be measured, additional measures [113] are required in order to account for the finite conductivity of the plates shorting the dielectric waveguide.

15.1.2. Transmission/Reflection Techniques

In contrast to the resonant approaches discussed previously, when using transmission/reflection techniques, the complex material parameters are obtained by analysing transmission through or reflections at a dielectric sample.

Transmission-Line Methods

In order to measure a dielectric sample using *transmission-line methods*, the dielectric sample is inserted into a coaxial or rectangular waveguide test section.

When performing such measurements, accounting for multiple reflections at the air/material interfaces is absolutely mandatory. The most common approach to this issue is the well-known NWR method by Nicolson and Ross [114] and Weir [115]. Most methods applied today are variations of this method.

However, the original NWR method is afflicted with several difficulties, i.e. numerical instabilities at sample lengths equal to integer multiples of half the guided wavelength, the need for branch-picking when taking the complex logarithm of the transmission coefficient [116], air gaps [117], undesired overmoding [117] and large measurement uncertainty[1] due to the limited accuracy of Vector Network Analyser (VNA) measurements [116, 118].

Free-Space Techniques

Free-Space techniques [119, 120] are closely related to transmission-line methods as they similarly need to account for multiple reflections in the sample.

In the last years, free-space techniques have especially received attention for the characterisation of materials in the millimeter and submillimeter wave domain [121–123] where a precise fit of the sample into the test setup is difficult to achieve [121].

Open-Ended Coaxial Probes

An alternate approach to the aforementioned transmission-line methods are techniques using *open-ended coaxial probes* [124–126]. In general these techniques analyse the measured reflection coefficient of the open-ended coaxial probe using an approximate circuit model for the admittance of the open-ended probe [124].

Amongst other applications, open-ended coaxial probes are widely used for the characterisation of liquids [104].

15.1.3. Other Relevant Techniques

Besides the aforementioned approaches, especially the dielectric characterisation of microwave PCB substrates has received considerable attention. For example, in [127] a split-cylinder resonator is presented.

Finally, rather special approaches such as material characterisation using pseudonoise test signals have been reported [128, 129].

[1] See remark on terminology on p. 235.

15.2. Theory of Cavity Perturbation Material Characterisation

Let us now commence our review of the theory of cavity perturbation material characterisation. The general measurement process of the method is depicted in figure 15.2 and in the following, we will build up the method's theory along this diagram. Cavity perturbation material measurements are carried out as follows:

Firstly, the scattering parameters of the empty and perturbed cavity are measured using a *Vector Network Analyser* (VNA). In order to calibrate the network analyser, a calibration technique which is appropriate for the particular type of waveguide has to be selected. For a rectangular waveguide test setup one would typically choose to use a *Thru-Reflect-Line calibration* (TRL).

In order to assess the resonant behaviour of the cavity, either a reflection-type or transmission-type measurement can be carried out. In the following, we choose the reflection-type approach and process the reflection coefficients $\Gamma_{unpert} = s_{11,unpert}$ and $\Gamma_{pert} = s_{11,pert}$.

From Γ_{unpert} and Γ_{pert} the shift Δf of the cavity's resonant frequency is easily obtained.

In order to determine the empty and perturbed cavity's loaded quality factors $\mathfrak{Q}_{L,unpert}$ and $\mathfrak{Q}_{L,pert}$ as well as the coupling coefficients $\mathfrak{K}_{unpert}$ and $\mathfrak{K}_{pert}$, the measured reflection coefficients are fitted to a model function describing the resonator's reflection coefficient using linear fractional curve fitting. This process will be discussed in section 15.2.1.

From the fitted model function, the *loaded quality factors* $\mathfrak{Q}_{L,unpert}$ and $\mathfrak{Q}_{L,pert}$ as well as the corresponding *coupling coefficients* are readily available and it is straightforward to determine the *unloaded quality factors* using $\mathfrak{Q}_0 = (1 + \mathfrak{K})\,\mathfrak{Q}_L$ [6].

Next, the *quality factor of the dielectric sample* $\mathfrak{Q}_d$ needs to be approximated by taking the reciprocal difference between the quality factors $\mathfrak{Q}_{0,pert}$ and $\mathfrak{Q}_{0,unpert}$. This step is further discussed in section 15.2.2.

Finally, we can use analysis equations deduced from cavity perturbation theory in order to calculate the complex relative permittivity $\varepsilon_r' - j\varepsilon_r''$ from Δf and $\mathfrak{Q}d$. Cavity perturbation theory will be revised in section 15.2.3.

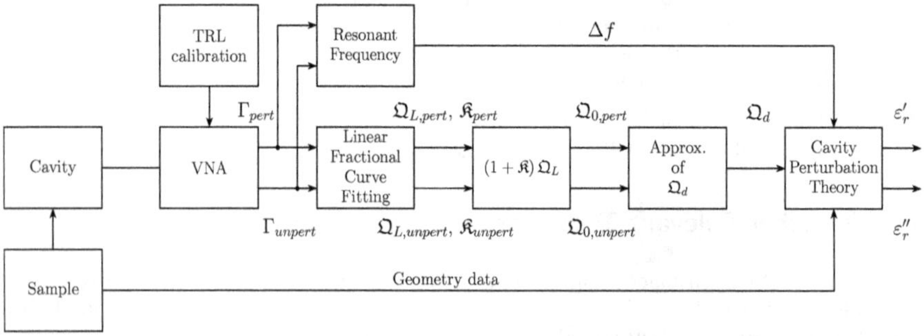

Figure 15.2.: Cavity perturbation material measurement process

15.2.1. Measurement of the Unloaded Q Factor

It is a common misassumption that $\mathfrak{Q}_L$ and $\mathfrak{Q}_0$ can be obtained by analysing the -3 dB and -7 dB points in a reflection coefficient measurement as this holds true for well-matched resonators ($k \approx 1$) only and introduces errors if applied to under- or overcritically coupled resonators.

The technique presented in the following is due to Kajfez [130–132]. Especially noteworthy is Kajfez' treatise [133] of random and systematic uncertainties in Q factor measurements. Excellent summaries on the subject are given in the two books [134, 135].

A Model Function for the Cavity's Reflection Coefficient for Linear Fractional Curve Fitting

We will now derive a model function for the cavity's reflection coefficient which can be fitted to the measured scattering parameters using *linear fractional curve fitting.*

Consider the circuit depicted in figure 15.3. On the right, the cavity resonator is shown. The box at the center of figure 15.3 represents a lossless coupling network and the resistor R_s on the left corresponds to the network analyser's source impedance.

Let us now derive an expression for the coupled resonator's reflection coefficient.

The input admittance of the unloaded resonator is [130]

$$Y_r = G_0 \left(1 + j\mathfrak{Q}_0 2 \frac{\omega - \omega_0}{\omega_0} \right). \tag{15.1}$$

Using an ABCD (chain) matrix to represent the lossless coupling network, the input impedance of the coupled resonator can be calculated [130]

$$Z_i = \frac{A + B\,Y_r}{C + D\,Y_r}. \tag{15.2}$$

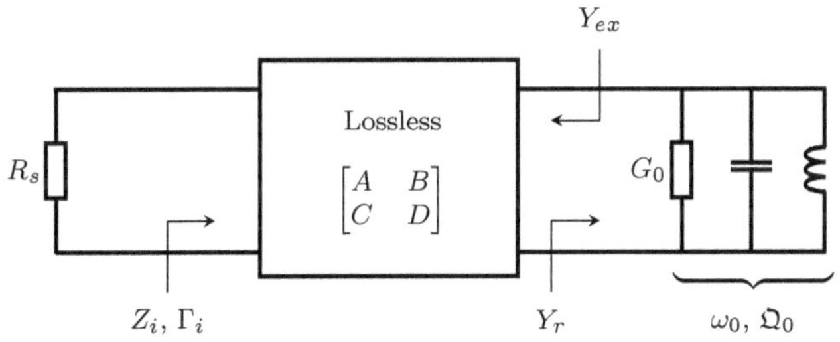

Figure 15.3.: Equivalent circuit of the coupled resonator, based on [130]

The coupled resonator's reflection coefficient Γ_i is found to be [130]

$$\Gamma_i = \frac{Z_i - R_s}{Z_i + R_s} = \frac{KY_r + L}{MY_r + 1} \tag{15.3}$$

where K, L and M denote quantities directly related to the coupling network's ABCD matrix' elements as well as R_s [135]. K, L and M can be interpreted as

$$\frac{1}{M} = Y_{ex} \qquad \frac{K}{M} = \Gamma_S \tag{15.4}$$

as shown in [130]. Γ_S denotes the detuned reflection coefficient shown in figure 15.4.

Eq. (15.3) can be rewritten as [130]

$$\Gamma_i = \frac{K}{M} + \frac{L - \frac{K}{M}}{M} \frac{1}{Y_r + \frac{1}{M}} = \frac{K}{M} + \frac{L - \frac{K}{M}}{M} \frac{1}{Y_r + Y_{ex}}. \tag{15.5}$$

The admittance $Y_L = Y_r + Y_{ex}$ of the loaded cavity can be expressed as [130]

$$Y_L = Y_r + Y_{ex} = (G_0 + G_{ex}) \left(1 + j\Omega_L 2 \frac{\omega - \omega_L}{\omega_0}\right) \tag{15.6}$$

and consequently, (15.5) is rewritten [130]

$$\Gamma_i = \Gamma_S + \underbrace{\frac{L - \frac{K}{M}}{M(G_0 + G_{ex})}}_{\mathfrak{d}e^{-j2\delta}} \frac{1}{1 + j\Omega_L 2\frac{\omega - \omega_L}{\omega_0}}. \tag{15.7}$$

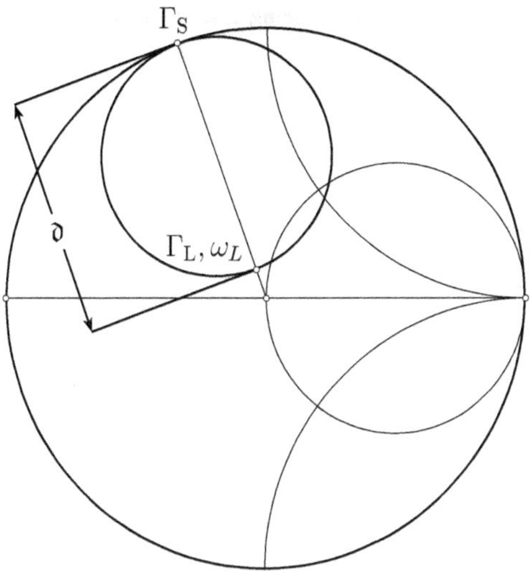

Figure 15.4.: Input reflection coefficient forming a circle in the Smith Chart, based on [130]

Since our measurement operates on a very narrow bandwidth only, we can assume the coupling network to be frequency independent [130]. Consequently, we assume the first fraction of the second addend in (15.7) to be a frequency independent complex constant $\mathfrak{d}e^{-j2\delta}$.

We may thus rewrite (15.7) as [130]

$$\Gamma_i = \Gamma_S + \frac{\mathfrak{d}e^{-j2\delta}}{1 + j\mathfrak{Q}_L 2\frac{\omega - \omega_L}{\omega_0}}. \tag{15.8}$$

Equation (15.8) forms a circle in the Smith chart as is depicted in figure 15.4 [130]. The diameter of this circle (and the distance between the detuned reflection coefficient Γ_S and the reflection coefficient Γ_L at resonance frequency ω_L) is equal to $\mathfrak{d}$.

From the diameter $\mathfrak{d}$ the coupling coefficient $\mathfrak{K}$ can be obtained using

$$\mathfrak{K} = \frac{\mathfrak{d}}{2 - \mathfrak{d}} \tag{15.9}$$

as is derived in [130].

Linear Fractional Curve Fitting for Determining $\mathfrak{Q}_0$, $\mathfrak{Q}_L$ and $\mathfrak{K}$ using the Least-Square Method

Having derived an expression for the cavity's reflection coefficient which is suitable for *linear fractional curve fitting*, we will now investigate the curve fitting process itself.

By introducing a normalised frequency [130]

$$\Upsilon = 2\,\frac{\omega - \omega_L}{\omega_0}$$

(15.8) can be rewritten as [130]

$$\Gamma_i = \frac{j\mathfrak{Q}_L \Gamma_S\,\Upsilon + \Gamma_S + \mathfrak{d} e^{-j2\delta}}{j\mathfrak{Q}_L \Upsilon + 1}. \tag{15.10}$$

By substituting [130]

$$\begin{aligned} a_1 &= j\mathfrak{Q}_L \Gamma_S \\ a_2 &= \Gamma_S + \mathfrak{d} e^{-j2\delta} \\ a_3 &= j\mathfrak{Q}_L \end{aligned} \tag{15.11}$$

we can denote (15.10) as [130]

$$\Gamma_i = \frac{a_1 \Upsilon + a_2}{a_3 \Upsilon + 1}, \tag{15.12}$$

which can be identified as fractional linear transform (Moebius transformation) [130].

After (15.12) was fitted to the measurement data, it is straightforward to obtain the desired quantities $\mathfrak{Q}_L$ and $\mathfrak{d}$ by calculating [130]

$$\begin{aligned} \mathfrak{Q}_L &= \Im\{a_3\} \\ \mathfrak{d} &= |a_2 - \tfrac{a_1}{a_3}|, \end{aligned} \tag{15.13}$$

which can easily be deduced from (15.11).

Finally, the *coupling coefficient* $\mathfrak{K}$ can be calculated from $\mathfrak{d}$ using (15.9) and $\mathfrak{Q}_0$ is found from $\mathfrak{Q}_L$ and $\mathfrak{K}$ using

$$\mathfrak{Q}_0 = (1 + \mathfrak{K})\,\mathfrak{Q}_L. \tag{15.14}$$

For the fitting process, (15.12) is again rearranged as [130]

$$\Gamma_{i,n} = a_1 \Upsilon_n + a_2 - a_3 \Gamma_{i,n} \Upsilon_n \qquad n = 1 \dots N \tag{15.15}$$

where $\Gamma_{i,n}$ represents the nth measurement at the normalised frequency Υ_n. [130].

For N measurements, (15.15) provides us with a linear system of equations [130]

$$\mathbf{A} \cdot \vec{a} = \vec{\Gamma}_i \tag{15.16}$$

with N equations but only three unknowns, which denotes as [130]

$$\begin{bmatrix} \Upsilon_1 & 1 & -\Upsilon_1 \Gamma_{i,1} \\ \vdots & \vdots & \vdots \\ \Upsilon_N & 1 & -\Upsilon_n \Gamma_{i,N} \end{bmatrix} \begin{bmatrix} a_1 \\ a_2 \\ a_3 \end{bmatrix} = \begin{bmatrix} \Gamma_{i,1} \\ \vdots \\ \Gamma_{i,N} \end{bmatrix} \tag{15.17}$$

under vector-matrix form.

If this system shall be solved in a least-square sense, $\vec{a}$ has to satisfy the normal equation

$$(\mathbf{A}^{\mathbf{H}} \mathbf{A}) \vec{a} = (\mathbf{A}^{\mathbf{H}}) \vec{\Gamma} \tag{15.18}$$

where $\mathbf{A}^{\mathbf{H}}$ denotes the hermitian (conjugate transpose) of $\mathbf{A}$ [136].

This can be achieved using the *pseudoinverse matrix* $\mathbf{A}^{+}$ of $\mathbf{A}$ defined as [136]

$$\mathbf{A}^{+} = (\mathbf{A}^{\mathbf{H}} \mathbf{A})^{-1} \mathbf{A}^{\mathbf{H}}, \tag{15.19}$$

which allows solving for the solution vector $\vec{a}$ [136]

$$\vec{a} = \mathbf{A}^{+} \vec{\Gamma}_i. \tag{15.20}$$

15.2.2. Approximation of $\mathfrak{Q}_d$

Having determined the quality factor of the empty and perturbed cavity by means discussed in the previous section, we now wish to determine the dielectric sample's quality factor $\mathfrak{Q}_d$.

As we have already seen in chapter 12 of this thesis the *unloaded quality factor* of a cavity can be calculated as [78]

$$\mathfrak{Q}_0 = \frac{\omega_0 \, \mathfrak{W}}{\mathfrak{P}_l} \tag{15.21}$$

where $\mathfrak{W} = 2\,\mathfrak{W}_e = 2\,\mathfrak{W}_m$ denotes the energy stored inside the cavity and $\mathfrak{P}_l$ represents the total power dissipated in the cavity.

By splitting the total loss $\mathfrak{P}_l$ in conductive and dielectric losses, we can define individual quality factors for both types of loss, that is,

$$\mathfrak{Q}_c = \frac{\omega_0 \, \mathfrak{W}}{\mathfrak{P}_c} \tag{15.22}$$

for conductive loss and

$$\mathfrak{Q}_d = \frac{\omega_0 \, \mathfrak{W}}{\mathfrak{P}_d} \tag{15.23}$$

for dielectric loss .

Thus, the quality factor of the perturbed cavity can be denoted as

$$\mathfrak{Q}_{0,pert} = \left(\frac{1}{\mathfrak{Q}_{c,pert}} + \frac{1}{\mathfrak{Q}_d} \right)^{-1} \tag{15.24}$$

and rearranging for $\mathfrak{Q}_d$ gives

$$\mathfrak{Q}_d = \left(\frac{1}{\mathfrak{Q}_{0,pert}} - \frac{1}{\mathfrak{Q}_{c,pert}} \right)^{-1} . \tag{15.25}$$

If we assume the quality factor $\mathfrak{Q}_c$ to be unaffected from the perturbation, we can approximate

$$\mathfrak{Q}_{c,pert} = \mathfrak{Q}_{c,unpert} = \mathfrak{Q}_{0,unpert} \tag{15.26}$$

because the unperturbed cavity does not incorporate any dielectric losses. The validity of this approximation will be discussed in section 15.4.

Using the approximation (15.26), we find the desired expression

$$\mathfrak{Q}_d = \left(\frac{1}{\mathfrak{Q}_{0,pert}} - \frac{1}{\mathfrak{Q}_{0,unpert}} \right)^{-1} \tag{15.27}$$

for the sample's quality factor (see e.g. [108]).

15.2.3. Cavity Perturbation Theory

Having determined the shift of the cavity's resonant frequency Δf and the dielectric sample's quality factor $\mathfrak{Q}_d$, we now wish to calculate the sample's complex permittivity. In the following, we will briefly revise *cavity perturbation theory* and derive the required analysis equations from cavity perturbation theory.

Lossless Case

Consider the unperturbed lossless cavity shown in figure 15.5a. The cavity's fields denote $\vec{E}_0$, $\vec{H}_0$ and ω_0 represents the cavity's resonant frequency. For obvious reasons, the cavity's fields must maintain the field equations [48]

$$\begin{aligned} -\quad \nabla \times \vec{E}_0 &= j\omega_0 \mu \vec{H}_0 \\ \nabla \times \vec{H}_0 &= j\omega_0 \varepsilon \vec{E}_0. \end{aligned} \tag{15.28}$$

Now consider figure 15.5b depicting a perturbed loss-free cavity. The fields of the perturbed cavity denote $\vec{E}$, $\vec{H}$ and the cavity's resonant frequency is denoted ω. Similarly to the unperturbed case, the fields must maintain the field equations [48]

$$\begin{aligned} -\quad \nabla \times \vec{E} &= j\omega\left(\mu + \Delta\mu\right)\vec{H} \\ \nabla \times \vec{H} &= j\omega\left(\varepsilon + \Delta\varepsilon\right)\vec{E}. \end{aligned} \tag{15.29}$$

After several manipulations and by making use of the divergence theorem as shown in [48], an exact expression (without any approximations)

$$\frac{\omega - \omega_0}{\omega} = -\frac{\iiint_{\mathcal{V}_0} \Delta\varepsilon \vec{E}\vec{E}_0^* + \Delta\mu \vec{H}\vec{H}_0^* \, dV}{\iiint_{\mathcal{V}_0} \varepsilon \vec{E}\vec{E}_0^* + \mu \vec{H}\vec{H}_0^* \, dV} \tag{15.30}$$

for the change of the cavity's resonant frequency is found [48].

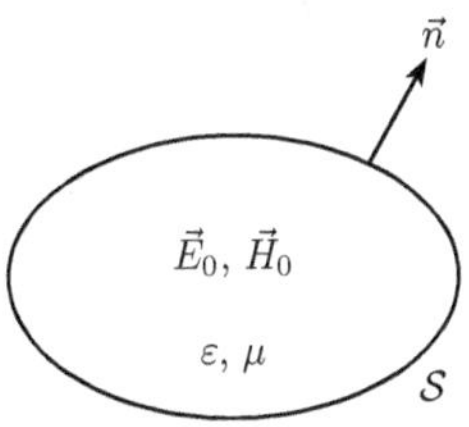

(a) Unperturbed cavity (without sample)

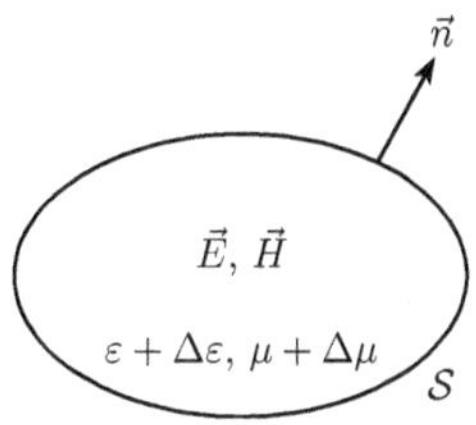

(b) Perturbed cavity (with sample)

Figure 15.5.: On the derivation of the cavity perturbation theory, based on [48]

For the nominator integral, the integration range can be limited to the sample volume $\mathcal{V}_S$ because we have $\Delta\varepsilon = 0$, $\Delta\mu = 0$ outside the sample.

Since the overall field distribution and the resonant frequency are minimally changed by the perturbation only, we may approximate $\vec{E} = \vec{E}_0$, $\vec{H} = \vec{H}_0$ in the denominator integrals and $\omega = \omega_0$, which leads to [48]

$$\frac{\omega - \omega_0}{\omega_0} = -\frac{\iiint_{\mathcal{V}_S} \Delta\varepsilon \vec{E}\vec{E}_0^* + \Delta\mu \vec{H}\vec{H}_0^* \, dV}{\iiint_{\mathcal{V}_0} \varepsilon|\vec{E}_0|^2 + \mu|\vec{H}_0|^2 \, dV.} \tag{15.31}$$

The denominator on the right side of (15.31) can be split into two integrals, one integrating the electric flux density, thus calculating the energy $\mathfrak{W}_e$ stored inside the cavity electrically and another one integrating the magnetic flux density, thus calculating the energy $\mathfrak{W}_m$ stored magnetically.

At resonance, we have $\mathfrak{W}_e = \mathfrak{W}_m$ and consequently (15.31) may be rewritten [48]

$$\frac{\omega - \omega_0}{\omega_0} = -\frac{\iiint_{\mathcal{V}_S} \Delta\varepsilon \vec{E}\vec{E}_0^* + \Delta\mu \vec{H}\vec{H}_0^* \, dV}{2\iiint_{\mathcal{V}_0} \varepsilon|\vec{E}_0|^2 \, dV}. \tag{15.32}$$

Lossy Case

We now extend cavity perturbation theory to lossy materials by introducing complex quantities

$$\begin{array}{lllllllll} \omega_0 & \rightarrow & \Omega_0 & = & \omega_0(1 + j\frac{1}{2\mathfrak{Q}_0}) & \varepsilon_r & \rightarrow & \varepsilon_r & = & \varepsilon_r' - j\varepsilon_r'' \\ \omega & \rightarrow & \Omega & = & (\omega_0 + \Delta\omega)(1 + j\frac{1}{2\mathfrak{Q}}) & \mu_r & \rightarrow & \mu_r & = & \mu_r' - j\mu_r'' \end{array} \tag{15.33}$$

for ω, ω_0 and ε_r, μ_r [137]. For simplicity's sake, we again assume non-magnetic materials, and thus $\mu_r = 1 \rightarrow \Delta\mu = 0$ applies.

After some manipulations, (15.32) may be rewritten as

$$\frac{\Delta f}{f_0} + j\frac{1}{2\mathfrak{Q}_d} = \left[(1 - \varepsilon_r') + j\varepsilon_r''\right] \frac{\iiint_{\mathcal{V}_S} \vec{E}\vec{E}_0^* \, dV}{2\iiint_{\mathcal{V}_0} |\vec{E}_0|^2 \, dV} \tag{15.34}$$

where Δf denotes the cavity's frequency shift and $\mathfrak{Q}_d$ represents the dielectric sample's quality factor. These quantities have been determined in the previous sections.

By identifying the real and imaginary parts in (15.34), the real and imaginary part of the permittivity may be calculated as

$$\varepsilon_r' = 1 - \frac{2\Delta f}{f_0} \underbrace{\frac{\iiint_{\mathcal{V}_0} |\vec{E}_0|^2 \, dV}{\iiint_{\mathcal{V}_S} \vec{E}\vec{E}_0^* \, dV}}_{\mathfrak{G}} \qquad \varepsilon_r'' = \frac{1}{\mathfrak{Q}_d} \underbrace{\frac{\iiint_{\mathcal{V}_0} |\vec{E}_0|^2 \, dV}{\iiint_{\mathcal{V}_S} \vec{E}\vec{E}_0^* \, dV}}_{\mathfrak{G}} \tag{15.35}$$

where $\mathfrak{G}$ denotes the cavity's *geometry factor*, which will be determined in the following.

Exemplary Calculation of the Geometry Factor for a Thin Dielectric Rod in a Cavity

Let us now determine the *geometry factor* $\mathfrak{G}$ for the sample configuration depicted in figure 15.6. In order to do so, approximations regarding the electric field in the sample volume $\mathcal{V}_s$ have to be made.

Since at the air-dielectric material interface continuity of tangential fields has to be maintained, we can assume the electric field inside the sample to be equal to the electric field outside the sample. Obviously, this approximation is only valid for samples with a sufficiently small cross-section. We will discuss the validity of this approximation later in section 15.4. It should be noted that this approximation is valid independently of the sample's actual cross-sectional shape [48].

In order to determine the nominator integral, we have to integrate the square of the cavity's unperturbed electric field distribution depicted in figure 15.6 (red lines) over the entire volume of the cavity. This integration yields

$$\iiint_{\mathcal{V}_0} |\vec{E}_0|^2 \, dV = |\hat{E}_0|^2 \frac{\mathcal{V}_0}{4}$$

where $\hat{E}_0$ denotes the Eigenmode's amplitude. In order to determine the denominator integral, we assume a constant field strength over the sample's cross-section and obtain

$$\iiint_{\mathcal{V}_S} \vec{E}\vec{E}_0^* \, dV = \iiint_{\mathcal{V}_s} |\vec{E}_0|^2 \, dV \approx |\hat{E}_0|^2 \mathcal{V}_s.$$

The geometry factor $\mathfrak{G}$ thus denotes

$$\mathfrak{G} = \frac{\mathcal{V}_0}{4\mathcal{V}_s} \tag{15.36}$$

and we can put (15.35) under the final form

$$\varepsilon_r' = 1 - \frac{2\Delta f}{f_0}\mathfrak{G} = 1 - \frac{\Delta f}{f_0}\frac{\mathcal{V}_0}{2\mathcal{V}_s} \qquad \varepsilon_r'' = \frac{1}{\mathfrak{Q}_d}\mathfrak{G} = \frac{1}{\mathfrak{Q}_d}\frac{\mathcal{V}_0}{4\mathcal{V}_s}. \tag{15.37}$$

This concludes our review of cavity perturbation material characterisation. In the last section of this chapter, we will validate the approximations used in the scope of the preceding derivations.

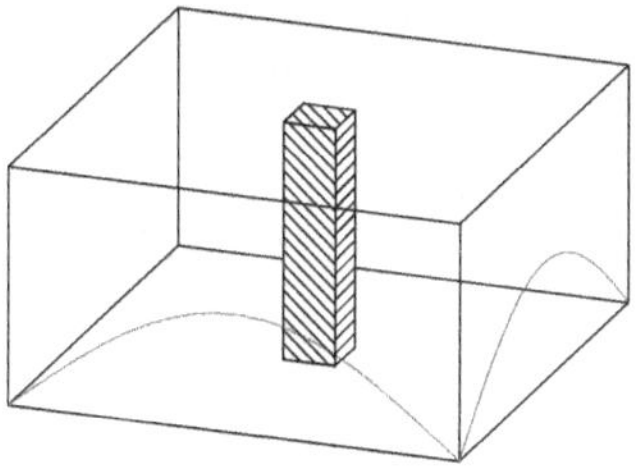

Figure 15.6.: Cavity perturbed by a dielectric rod. Red lines indicate the electric field distribution.

15.3. Exemplary Measurement Data

In order to illustrate the concept of cavity perturbation material characterisation outlined in the preceding sections, we will now study exemplary measurement data for Polytetrafluorethylen (PTFE, Teflon) and water.

All measurements were carried out using the WR-340 cavity depicted in figure 15.7a. Depending on the size and permittivity the cavity's resonant frequency is in the region of 2400 MHz. Measurements were carried out using an Agilent E8361C vector network analyser. A TRL calibration was performed using the waveguide-to-coaxial transition's flange as reference plane.

In order to process the obtained measurement data, an extended version of the fitting procedure described in section 15.2.1 was used. In contrast to the linear fractional curve fitting technique described earlier, this extended version minimises the effect of outlier values on the result of the fitting process by weighting the individual measurements Γ_i depending on how well they comply with (15.15) [135]. Moreover, this technique provides uncertainties for the coefficients a_1, a_2 and a_3, from which by means of standard error propagation[2] the *uncertainty*[3] of the cavity's unloaded Q factor can be obtained [133].

While a complete measurement uncertainty budget still remains future work, an approximate uncertainty for the loss tangent measurement was calculated. Apart from the uncertainty of the fitting process, this analysis also accounts for the VNA's residual uncertainty and the uncertainty due to source (mis-)match by means discussed in [133].

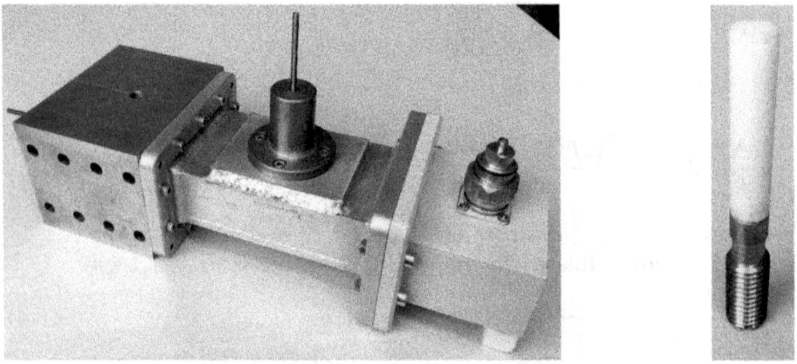

(a) Cavity (b) Sample

Figure 15.7.: Cavity perturbation material characterisation test setup

[2] Uncertainty propagation by means similar to standard error propagation, that is, by using a linear approximation of the model function, is advised by the Guide to the Expression of Uncertainty in Measurement (GUM) [138]. Here, we ignored the non-linearity of the model function which may (or may not) require either the inclusion of higher order derivatives [138] or the application of Monte Carlo methods as discussed in Supplement 1 [139] to [138]. Input quantities were assumed to be uncorrelated.

[3] See remark on terminology on p. 235.

15.3.1. Polytetrafluorethylen (PTFE, Teflon)

PTFE is among the most commonly studied materials in literature on electromagnetic material characterisation.

Figure 15.7b depicts a PTFE sample to be used with the test setup shown in figure 15.7a.

In figures 15.8a and 15.8b the measured reflection coefficient of the empty and perturbed cavity (blue line) is shown. By comparing the measured data with exemplary data points (red diamonds) calculated from the model function (15.12) for various Υ, we find that the fit shows very good agreement with the measured data.

The empty and perturbed cavity's resonant frequency and quality factor are given in table 15.1. The same table also provides the *uncertainty*[4] of the unloaded Q factors due to the imperfections discussed earlier. It can be seen that the residual measurement uncertainty of the network analyser has the largest impact on the measurement uncertainty of the unloaded quality factors.

The shift of the cavity's resonant frequency and the dielectric sample's quality factor as well as its uncertainty are given in table 15.2.

Using (15.37), from the data given in table 15.2 we obtain the desired material parameters $\varepsilon_r' = 2.03$ and $\tan(\delta) = 1.6 \cdot 10^{-4}$. As can be seen from table 15.3, these results show good agreement with values available from literature.

By applying standard error propagation[5] to (15.37), we find that the loss tangent's uncertainty is $\sigma(\tan(\delta)) = 1.2 \cdot 10^{-4}$. We thus have to conclude that at least for the given cavity, measuring loss tangents in the region of 10^{-4} represents the method's lower limit in terms of measurement uncertainty.

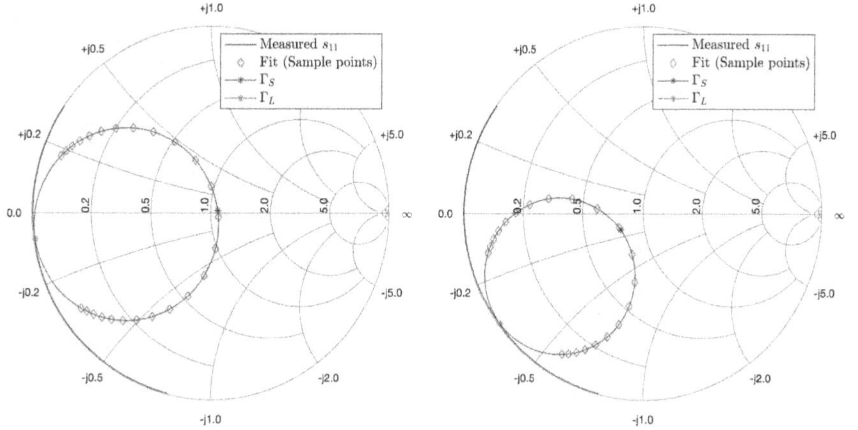

(a) Unperturbed cavity (without sample)

(b) Perturbed cavity (with sample)

Figure 15.8.: Scattering parameters of an empty and perturbed cavity for a measurement of PTFE

[4]See remark on terminology on p. 235.

[5]See footnote 2 on page 158.

Unperturbed f_0 in MHz	$\mathfrak{Q}_0$	Total $\sigma(\mathfrak{Q}_0)$	Fitting $\sigma(\mathfrak{Q}_{0,F})$	Source match $\sigma(\mathfrak{Q}_{0,M})$	Residual $\sigma(\mathfrak{Q}_{0,R})$
2430	4240	133	26	63	114
Perturbed f_0 in MHz	$\mathfrak{Q}_0$	Total $\sigma(\mathfrak{Q}_0)$	Fitting $\sigma(\mathfrak{Q}_{0,F})$	Source match $\sigma(\mathfrak{Q}_{0,M})$	Residual $\sigma(\mathfrak{Q}_{0,R})$
2378	4012	105	20	48	91

Table 15.1.: Resonant frequency and quality factor of the empty and perturbed cavity

Δf in MHz	$\mathfrak{Q}_d$	$\sigma(Q_d)$
52	74609	54850

Table 15.2.: Effect of the perturbation on the cavity's resonant frequency and quality factor

Reference	ε_r'	$\varepsilon_r'' \cdot 10^4$	$\tan(\delta) \cdot 10^4$	f in MHz
This thesis	2.03	3.2	1.6	2400
[109]	2.05	5.9	2.9	2200
[140]	2.10	3.2	1.5	3000

Table 15.3.: Complex permittivity of Polytetrafluorethylen (PTFE, Teflon)

15.3.2. Water

The need to investigate the temperature dependence of the permittivity of water occurred during the development of a water load for a calorimetric power measurement setup to characterise mass-produced magnetrons used in industrial microwave heating.

During test runs of this setup it was found that the water load's match is considerably deteriorated if the water's temperature increases. This is obviously a highly undesirable effect in a calorimetric power measurement system. The reason for this effect is the strong temperature dependence of the permittivity of water.

Since we were initially unaware of this temperature dependence, cavity perturbation measurements were carried out using the test setup depicted in figure 15.7a. Although these measurements were realised in a rather simplistic fashion, the obtained results showed surprisingly good agreement with the widely accepted single Debye relaxation model [141–143] for the temperature-dependent behaviour of water. This model will be briefly summarised in the following.

The measurements were carried out as follows: A glass capillary tube with an approximate inner diameter of 400 μm was inserted into the cavity shown in figure 15.7a. The resonant frequency and the unloaded quality factor of this cavity were measured and considered the "empty" cavity's measurement.

Next, hot tap water was pumped circularly from a bucket through the cavity's capillary tube back into the bucket. During the cooling down of the water contained in the system, measurements were taken at various temperatures. These measurements represent the perturbed cavity's measurements. The temperature was monitored at the outlet of the capillary tube using a digital multimeter.

Single Debye Relaxation Model

It is well known [141–143] that the permittivity of (pure) water can be modelled using a *single Debye relaxation process* [141]

$$\varepsilon_r(f, \mathfrak{T}) = \varepsilon_\infty(\mathfrak{T}) + \frac{\Delta(\mathfrak{T})}{1 - j2\pi f \tau(\mathfrak{T})} \tag{15.38}$$

in the lower GHz range.[6]

Here, ε_∞ denotes the permittivity at "infinite" frequency that should correspond to the optical refractive index $\sqrt{\varepsilon_\infty}$ [141]. Moreover, the spectral amplitude of the relaxation process is denoted $\Delta(\mathfrak{T})$ while $\tau(\mathfrak{T})$ denotes the characteristic relaxation time. Both quantities are temperature-dependent [141]. For clarity, note that f is the frequency.

For the DC case, we obtain $\varepsilon_s(\mathfrak{T}) = \varepsilon_\infty(\mathfrak{T}) + \Delta(\mathfrak{T})$ where ε_s shall be defined as the static dielectric constant [141]. Using this expression, we can rewrite (15.38) as

$$\varepsilon_r(f, \mathfrak{T}) = \varepsilon_s(\mathfrak{T}) - \Delta(\mathfrak{T}) + \frac{\Delta(\mathfrak{T})}{1 - j2\pi f \tau(\mathfrak{T})},$$

which can be expanded as

$$\varepsilon_r(f, \mathfrak{T}) = \varepsilon_s(\mathfrak{T}) - \frac{[1 - j2\pi f \tau(\mathfrak{T})]\,\Delta(\mathfrak{T})}{1 - j2\pi f \tau(\mathfrak{T})} + \frac{\Delta(\mathfrak{T})}{1 - j2\pi f \tau(\mathfrak{T})}$$

giving

$$\varepsilon_r(f, \mathfrak{T}) = \varepsilon_s(\mathfrak{T}) + j2\pi f \frac{\tau(\mathfrak{T})\,\Delta(\mathfrak{T})}{1 - j2\pi f \tau(\mathfrak{T})}. \tag{15.39}$$

For the temperature dependence of the static dielectric constant, [141] proposes to use the polynomial function

$$\varepsilon_s(\mathfrak{T}) = 87.9144 - 0.404399\,\mathfrak{T} + 9.58726 \cdot 10^{-4}\,\mathfrak{T}^2 - 1.32892 \cdot 10^{-6}\,\mathfrak{T}^3 \tag{15.40}$$

originally presented in [144].

For the temperature dependence of the characteristic properties of the relaxation process [141] suggests to use simple exponential models, that is

$$\tau(\mathfrak{T}) = \mathfrak{p}_1 \cdot e^{\mathfrak{p}_2/\mathfrak{T} + \mathfrak{T}_c} \qquad \Delta(\mathfrak{T}) = \mathfrak{p}_3 \cdot e^{-\mathfrak{p}_4 \cdot \mathfrak{T}} \tag{15.41}$$

where $\mathfrak{T}_c$ denotes a "critical" temperature [141].

For the parameters $\mathfrak{p}_1$, $\mathfrak{p}_2$, $\mathfrak{p}_3$, $\mathfrak{p}_4$ and $\mathfrak{T}_c$ [141] proposes fitted values

$$\begin{aligned} \mathfrak{p}_1 &= 80.69715 & \mathfrak{p}_2 &= 0.004415996 \\ \mathfrak{p}_3 &= 1.367283 \cdot 10^{-13} & \mathfrak{p}_4 &= 651.4728 \\ \mathfrak{T}_c &= 133.0699 \end{aligned} \tag{15.42}$$

based on 983 experimental values over the frequency range from 0 to 50 GHz.

[6] [141] states that the single Debye function model can be used up to 30 GHz.

Illustrative Results

Before discussing the results obtained using the test setup discussed above, it is again stressed that due to the rather simplistic measurement setup, the presented measurements have a proof-of-concept character only.

As can be seen from figure 15.9 the measured real part of the permittivity agrees to the single Debye relaxation model to about 1 %. Similarly the imaginary part of the permittivity corresponds to this model to about 5 %. However, regarding the imaginary part, it should be noted that the slope of the measured data somewhat differs from the Debye model.

Table 15.4 compares the measurement data to reference values for pure water at 25 °C available from the body of literature. Considering the spread of the reference values, the quality of the obtained data again appears to be surprisingly good.

According to the author's today knowledge, the largest sources of uncertainty are the inner diameter of the capillary and the accurate determination of the water temperature inside the cavity. After the measurement was carried out, the capillary tube was cut into small segments[7] and the inner diameter of these segments was measured using a microscope.

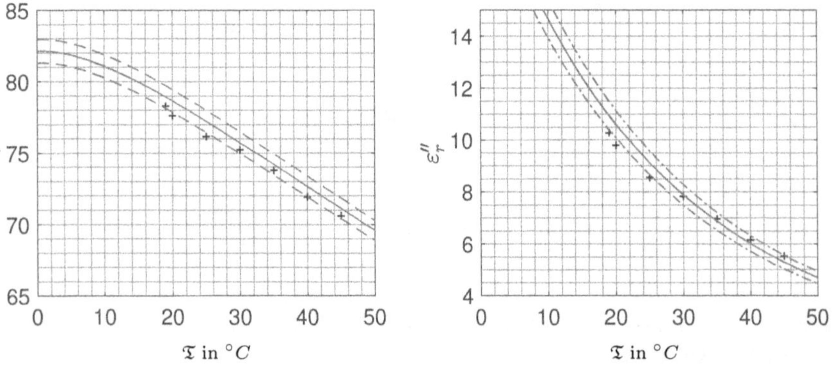

Figure 15.9.: Temperature dependence of the complex permittivity of water. Blue crosses indicate measurement data while the red solid line depicts the single Debye relaxation model (15.31). Dashed limits indicate 1% error limits while dash-dotted lines represent 5% error limits.

Reference	ε_r'	ε_r''	$\mathfrak{T}$ in °C	f in MHz
This thesis (tap water)	76.2	8.6	25	2400
Single Debye relaxation model, based on [141]	77.2	9.1	25	2400
[142]	77.5	8.22	25	2200
[142]	76.9	9.77	25	2610
[143]	76.7	9.4	25	2500

Table 15.4.: Complex permittivity of Water at 25 °C

[7] Due to the residual moisture in the capillary the glass tubes were not reused anyway.

15.4. Results: Investigation of Secondary Effects in Cavity Perturbation Material Measurements

Having studied the theory of cavity perturbation material characterisation and two application examples, we can now examine the validity of the assumptions made in the previous sections. Two major approximations have been made:

Firstly, due to the continuity of tangential fields, the electric field inside the sample volume was assumed to be equal to the field outside the sample (see section 15.2.3).

Secondly, in order to approximate the quality factor of the dielectric $\mathfrak{Q}_d$, we assumed the cavity's quality factor due to ohmic loss to be equal for the empty and perturbed case (see section 15.2.2).

As can be seen from $\mathfrak{Q}_c = {}^{\omega_0\,\mathfrak{W}}/_{\mathfrak{P}_c}$ this assumption is true to some extent only. This is because of three reasons:

- Due to the perturbation, the resonant frequency ω_0 is decreased.
- The energy stored inside the cavity $\mathfrak{W} = 2 \cdot \mathfrak{W}_e$ is determined by calculating

$$\mathfrak{W}_e = \frac{\varepsilon'}{2} \iiint\limits_{\mathcal{V}} |\vec{E}|^2 dV, \tag{15.43}$$

  which is an integration of the dielectric flux $\vec{D} = \varepsilon\vec{E}$ over the cavity's volume $\mathcal{V}$. As a consequence, the energy stored inside the cavity increases on insertion of the dielectric sample.

- Since at resonance $\mathfrak{W}_m = \mathfrak{W}_e$ applies, the energy stored inside the cavity magnetically increases and thus the magnetic field at the cavity's walls becomes larger as well.

  This would lead to an increase of the ohmic losses inside the cavity's walls because ohmic loss is proportional to the square of the magnetic field tangential to the cavity's walls as can be seen from

$$\mathfrak{P}_c = R_s \iint\limits_{Walls} |H_{tan}|^2 dA. \tag{15.44}$$

  At the same time, due to the downshift of the cavity's resonant frequency, the cavity's walls exhibit a decreased surface resistance $R_s = \sqrt{{}^{\omega_0\mu}/_{2\sigma}}$, which would lead to a reduction of the ohmic losses.

Unfortunately, to the author's best knowledge, this matter is not discussed in today's literature in a complete, satisfying manner:

For example, in [109], it is erroneously proposed to use a correction term which is valid only for entirely filled cavities. Another example is [145,146] where a correction is proposed which is based on several misassumptions: While the author correctly points out the increase of the stored energy inside the cavity, he oversees the lowering of the resonant frequency as well as the change of the conductive loss as a consequence of the reduced surface resistance and of the increased amplitude of the magnetic field.

In order to assess the impact of these "secondary effects" on cavity perturbation material measurements, in the following, we recalculate the empty and perturbed cavity's resonant frequencies and quality factors based on a Mode Matching solution for the cavity's fields.

15.4.1. Simulation Methodology

Full-wave simulation of the empty and perturbed cavity was carried out using the extension of the Mode Matching technique for electromagnetic cavities presented in chapter 12.

The original intention to use a mesh-free method for determining the cavity's fields was to avoid incorrect results due to meshing issues. When solving the cavity for various permittivities of the dielectric slab a new mesh is generated for each point in the sweep. Thus, it is not clear whether the observed variations are due to the change of the permittivity or the variation of the mesh. However, it turned out that Ansys HFSS' Eigenmode solver provided results which show very good agreement with the presented Mode Matching solution.

Let us assume that the perturbed cavity's Eigenmode is of comparable form as the empty cavity's fundamental odd-order Eigenmode. By carrying out an analysis of the interface between the empty and partially filled waveguide similar to the analysis in chapter 7 it can be shown that for the perturbed field configuration only odd-order TE Eigenmodes have to be taken into account.

It is thus possible to introduce symmetry planes as is shown in figure 15.10a and as consequence, the cavity can be simplified as depicted in figure 15.10b. In order to describe this structure in terms of the Mode Matching technique, the segment matrix representation shown in figure 12.5 can be used (also see section 12.2).

For the empty waveguide $\mathcal{I}$, the odd-order $TE^x_{m,0}$ Eigenmodes given in appendix B.1.1 have to be taken into account. For the partially filled waveguide $\mathcal{III}$, the odd-order $TE^x_{m,0}$ Eigenmodes derived in appendix B.2 have to be included in our analysis.

The solution process and post-processing was carried out as described in chapter 12.

For the calculation of ohmic losses as per (15.44) some remarks have to be made on properly determining the *surface resistance* R_s in the presence of *surface roughness*.

In order to calculate an *effective conductivity* for the cavity's walls, which takes the surface roughness into account, the *surface roughness model* by Groiss et. al. [147] was used.

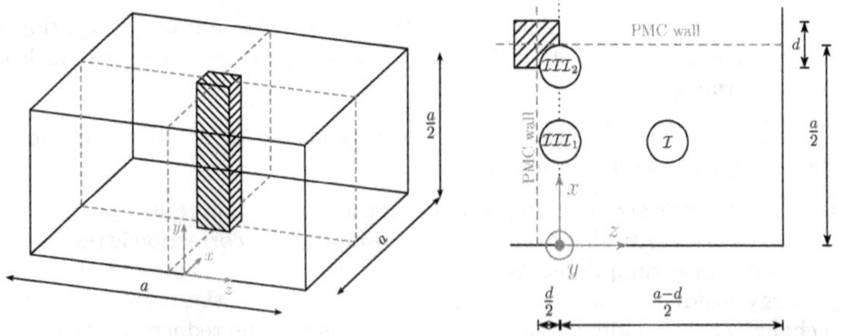

(a) Symmetry planes (b) Simplified representation

Figure 15.10.: Rectangular cavity perturbed by a small dielectric slab

We assumed the surface roughness to be equal to the surface's *root mean square roughness*. However, in mechanical engineering, it is much more common to use the *arithmetic average roughness*. It is common practice to estimate $R_q = 1.1 \ldots 1.4 \cdot R_a$, although this estimation depends on the exact surface topology [148, 149]. Here we assumed a value of 1.4.

For milled surfaces, $R_a = 1.6\ \mu\text{m}$ is the best surface quality to be expected in an average application [149].

15.4.2. Assessment of Secondary Effects

Very Low Loss Dielectrics Perturbing a Milled Cavity

Let us now assess the severity of the aforementioned "secondary effects" on a measurement carried out using a brass-made milled cavity. While brass has a *conductivity* of $2.564 \cdot 10^7$ S/m [6], due to the surface roughness, the effective conductivity obtained from the roughness model discussed earlier is $8.29 \cdot 10^6$ S/m.

Although this decrease of the conductivity appears large at first sight, the quality factor calculated for this cavity is in the region of 6000, which is still considerably larger than the measured quality factor (4200) of the cavity discussed in section 15.3 (see table 15.1).

For the sample's dimensions we assumed $d = 3.5$ mm (see figure 15.10b).

Figure 15.11 depicts the resonant frequency and quality factors calculated for various permittivities of the dielectric slab. As one would expect, both the cavity's resonant frequency as well as its unloaded Q factor (stars) decrease if the sample's permittivity increases.

However, at the same time, the quality factor $\mathfrak{Q}_c$, which accounts for ohmic loss is increased as can be seen in figure 15.11b (triangles).

This fact poses a problem for material measurements on very low loss dielectrics (here $\tan(\delta) = 2 \cdot 10^{-4}$) because the dielectric's quality factor $\mathfrak{Q}_d$ is estimated to be

$$\mathfrak{Q}_d = \left(\frac{1}{\mathfrak{Q}_{0,pert}} - \frac{1}{\mathfrak{Q}_{0,unpert}} \right)^{-1}, \tag{15.45}$$

which assumes $\mathfrak{Q}_{c,pert} = \mathfrak{Q}_{c,unpert} = \mathfrak{Q}_{0,unpert}$ as discussed in section 15.2.2.

Because $\mathfrak{Q}_c$ is not constant but increases slightly, approximated values for $\mathfrak{Q}_d$ (fig. 15.11b, diamonds) are slightly larger than the sample's actual quality factor (fig. 15.11b, squares) calculated from

$$\mathfrak{Q}_d = \frac{\omega_0\, \mathfrak{W}}{\mathfrak{P}_d} \tag{15.46}$$

where the energy stored inside the cavity is obtained by means of (15.43) and the power dissipated due to dielectric loss is

$$\mathfrak{P}_d = \omega_0 \varepsilon'' \iiint\limits_{\mathcal{V}_S} |\vec{E}|^2\, dV.$$

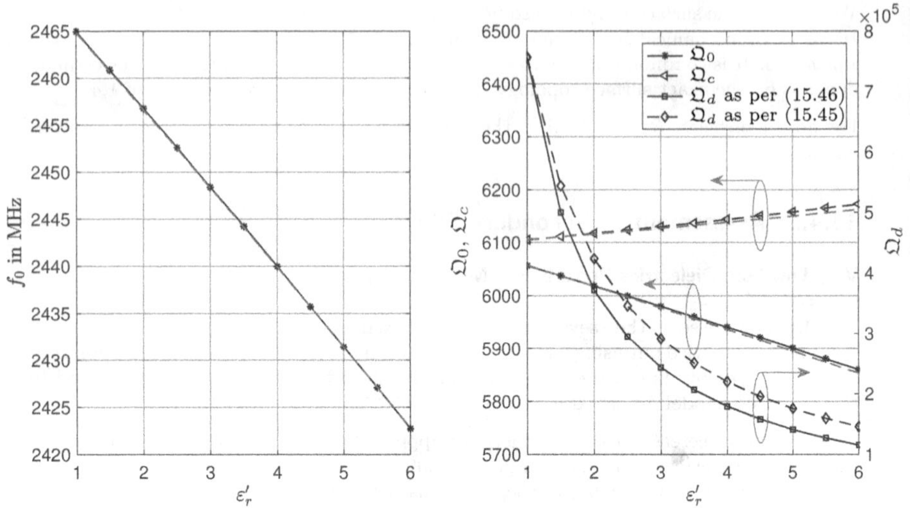

(a) (b)

Figure 15.11.: Variation of the resonant frequency and quality factors of a milled cavity depending on the permittivity of a small lossy slab perturbing the cavity. Red lines show results obtained using Ansys HFSS.

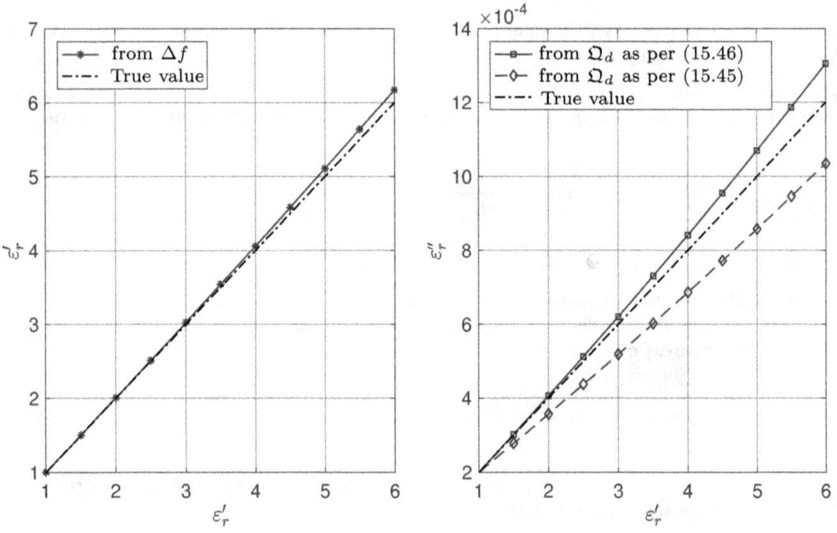

(a) (b)

Figure 15.12.: Material parameters calculated from the cavity parameters given in figure 15.11

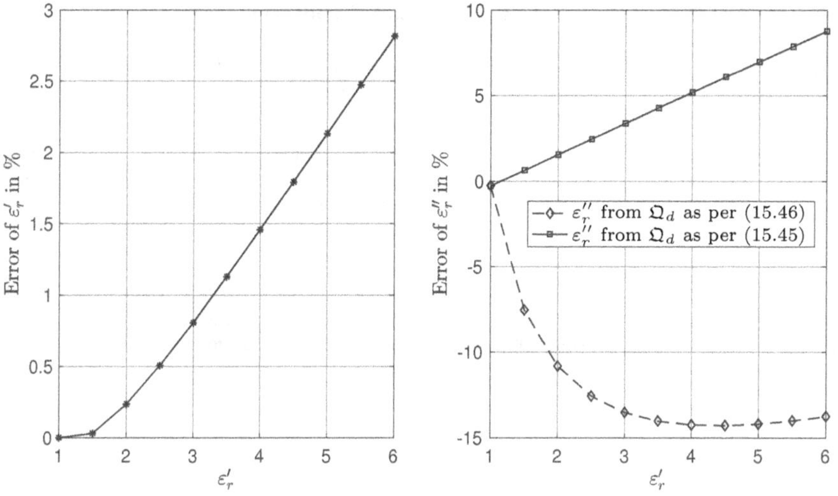

(a) (b)

Figure 15.13.: Error of the material parameters shown in figure 15.12 relative to the true material parameters used for the simulation

By using

$$\varepsilon_r' = 1 - \frac{\Delta f}{f_0}\frac{\mathcal{V}_0}{2\mathcal{V}_s}$$

$$\varepsilon_r'' = \frac{1}{\mathfrak{Q}_d}\frac{\mathcal{V}_0}{4\mathcal{V}_s}$$

derived in section 15.2.3 the dielectric sample's material parameters can be calculated from the shift Δf of the cavity's resonant frequency and the sample's quality factor $\mathfrak{Q}_d$. The obtained material parameters are given in figure 15.12 and the error[8] of these values with respect to the true material parameters used for the simulation are shown in figure 15.13.

From figures 15.12a and 15.13a it can be seen that ε_r' is accurately determined with an error below 3 % for $\varepsilon_r' < 6$. This error is due to the field approximation for the sample volume. In the following subsection, an approach using permittivity-dependent geometry factors, which allows to correct this deviation, will be introduced.

In contrast, if ε_r'' is calculated from the approximated value of $\mathfrak{Q}_d$ as per (15.45) the error of the measurement is in the region of 10 % even for materials with a comparably small permittivity ($\varepsilon_r' \approx 2$) as can be seen from figure 15.13b (diamonds).

On the contrary, the error when calculating ε_r'' from the exact value of $\mathfrak{Q}_d$ obtained by integration as per (15.46) is below 5% for small permittivities ($\varepsilon_r' < 3.5$) as is shown in figure 15.13b (squares). This remaining error is again due to the field approximation and can be compensated using the permittivity-dependent geometry factors to be introduced in the following section.

[8] See remark on terminology on p. 235.

Very Low Loss Dielectrics Perturbing a Cavity with Additional Plating

For the previously discussed milled cavity we observed that the influence of $\mathfrak{Q}_c$ on the total quality factor $\mathfrak{Q}_0$ is too large as the "parasitic" increase of $\mathfrak{Q}_c$ impairs the accuracy of $\mathfrak{Q}_d$ and thus skews loss tangent measurements.

If $\mathfrak{Q}_c$ is increased by improving the effective *conductivity* of the cavity's walls (here $\sigma = 4.098 \cdot 10^7$ S/m, ideal gold [6]), the influence of $\mathfrak{Q}_d$ on $\mathfrak{Q}_0$ becomes more dominant compared to $\mathfrak{Q}_c$. As a consequence, the approximated values for $\mathfrak{Q}_d$ obtained from (15.45) show better agreement with the values calculated from (15.46) as is shown in figure 15.14a.

Thus, the recovered values for ε_r'' show better agreement with the true values used for the simulations as can be seen from figure 15.14b (diamonds). However, this is to some extent due to the fact that the errors introduced by the two approximations are superimposed in a favourable - but arbitrary - fashion.

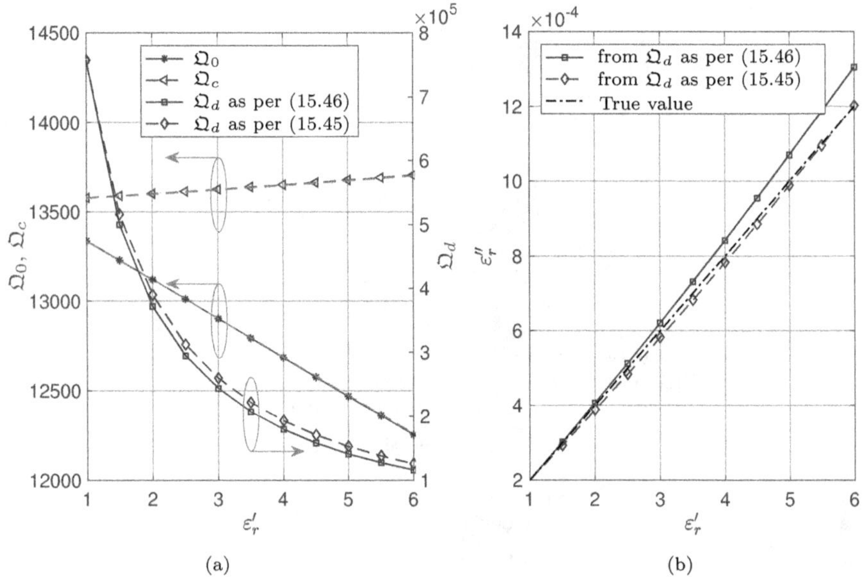

Figure 15.14.: Variation of the quality factors of a gold-plated cavity depending on the permittivity of a small lossy slab perturbing the cavity (a) and imaginary part of the permittivity obtained from these quality factors (b). Red lines show results obtained using Ansys HFSS.

15.4.3. Compensation of the Observed Secondary Effects

In the previous section 15.4.2 we saw that the discussed secondary effects may have a considerable impact on the quality of the material parameters obtained using the cavity perturbation technique. While the problem of the undesired variation of $\mathfrak{Q}_c$ can only be circumvented by using a cavity with a sufficiently high quality factor, the deviations caused by the field approximation can be eliminated by correcting the geometry factor $\mathfrak{G}$ in order to account for the imperfections of the field approximation within the sample.

For the development of this novel approach, we firstly reformulate the analysis of the simulated data obtained from the Mode Matching solver and Ansys HFSS, which is based on (15.37), to provide the "correct" geometry factor $\mathfrak{G}(\varepsilon_r')$ for a given permittivity ε_r' supplied to the solver as an input quantity.

Figure 15.15a depicts the "correct" geometry factors obtained from Mode Matching data for various sample geometries. It is interesting to note that the geometry factor $\mathfrak{G}$, which is used in (15.37), splits up into two different geometry factors $\mathfrak{G}(\varepsilon_r')$ and $\mathfrak{G}''(\varepsilon_\mathrm{t}')$ valid for either the real or imaginary part of the complex permittivity.

In a similar fashion, the geometry factor $\mathfrak{G}(\varepsilon_r')$ obtained from data provided by Ansys HFSS is depicted in figure 15.16a. At first sight, the geometry factors $\mathfrak{G}(\varepsilon_r')$ depicted in figure 15.15a and 15.16a show good agreement. However, due to the large range of values for $\mathfrak{G}(\varepsilon_r')$, an exact comparison of the geometry factor's variation is not possible from the two plots 15.15a and 15.16a.

A much better comparison of the geometry factors obtained from Mode Matching and FEM data is possible if we compare the term $\mathfrak{G}'(\varepsilon_r') - \mathfrak{G}'(\varepsilon_r' = 1)$ for both datasets shown in figure 15.15b and 15.16b. While the two plots show good agreement for relatively large samples, for smaller samples, unphysical variations of the geometry factor obtained from the FEM dataset are observed as can be seen from figure 15.16b. It appears to be impossible to remove these artifacts from the geometry factor obtained from the FEM data by choosing tight meshing restrictions and as a consequence, it can be concluded that the Mode Matching solver presented in this thesis is the tool of choice for this analysis.

Having determined "correct" geometry factors for the given setup from a Mode Matching solution, the remaining question is how to apply these geometry factors to measurement data to be processed. At this point, the reader may make the objection that in order to apply the "correct" geometry factors shown in figure 15.15a, the permittivity of the sample, which we wish to measure, has to be known beforehand.

Fortunately, there are three approaches which allow us to apply the corrected geometry factor without knowing the permittivity in advance:

As can be seen from figure 15.15b, for large sample sizes (e.g. $d = 10$ mm), the geometry factor hardly shows any variation over the permittivity. Thus, the first approach simply assumes the geometry factors $\mathfrak{G}'$ and $\mathfrak{G}''$ to be constant.

The second approach is based on the assumption that for comparably small sample sizes (e.g. $d = 3$ mm), the geometry factors may be linearised w.r.t. their dependence on ε_r'. If this linear model is inserted into (15.37), the two equations denote

$$\varepsilon_r' = 1 - \frac{2\Delta f}{f_0}\,\mathfrak{G}' = 1 - \frac{2\Delta f}{f_0}\left[\alpha_1\varepsilon_r' + \alpha_2\right] \qquad \varepsilon_r'' = \frac{1}{\mathfrak{Q}_d}\,\mathfrak{G}'' = \frac{1}{\mathfrak{Q}_d}\left[\alpha_3\varepsilon_r' + \alpha_4\right], \tag{15.47}$$

where α_1 to α_4 represent the model parameters obtained from the Mode Matching data.

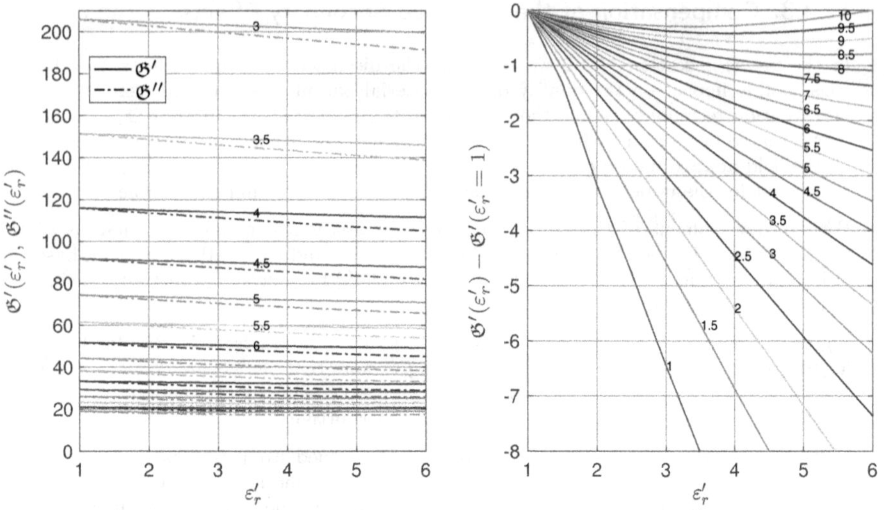

(a) (b)

Figure 15.15.: Variation of the geometry factors $\mathfrak{G}'(\varepsilon_r')$ and $\mathfrak{G}''(\varepsilon_r')$ for a small lossy slab perturbing the cavity. The width d of the slab is indicated in the plots. Results obtained using the Mode Matching solver presented in this thesis.

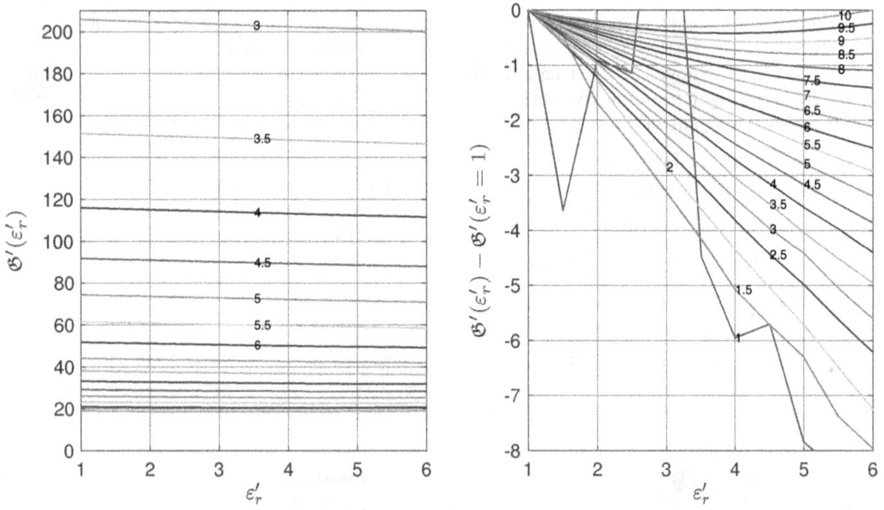

(a) (b)

Figure 15.16.: Variation of the geometry factor $\mathfrak{G}'(\varepsilon_r')$ for a small lossy slab perturbing the cavity. The width d of the slab is indicated in the plots. Results obtained using Ansys HFSS.

In order to make the expression for ε_r' usable for measurement data processing, it has to be rearranged as

$$\varepsilon_r' = \frac{\dfrac{f_0}{2\Delta f} - \alpha_2}{\dfrac{f_0}{2\Delta f} + a_1}. \tag{15.48}$$

This expression allows us to immediately calculate the desired real part of the complex permittivity from the measured data. In a second step, the expression for ε_r'' in (15.47) can readily be used to obtain the imaginary part of the permittivity.

Figure 15.17 depicts the complex material parameters calculated for simulated data shown in figure 15.11, which was simulated for the milled cavity discussed in the previous section. Note that the imaginary part of the permittivity was calculated from the exact value of $\mathfrak{Q}_d$ obtained from (15.46). The recovered complex permittivity shows excellent agreement with the material parameters initially provided to the solver.

The third approach to embed permittivity-dependent geometry factors into measurement data processing is required to carry out measurements on medium-sized samples (e.g. $d = 8$ mm). For such samples, neither the linear model nor a constant geometry factor can be used because the geometry factor shows a relatively large non-linear variation with ε_r'. For such samples it is possible to obtain a self-consistent solution for the geometry factor and the permittivity by means of an iterative process.

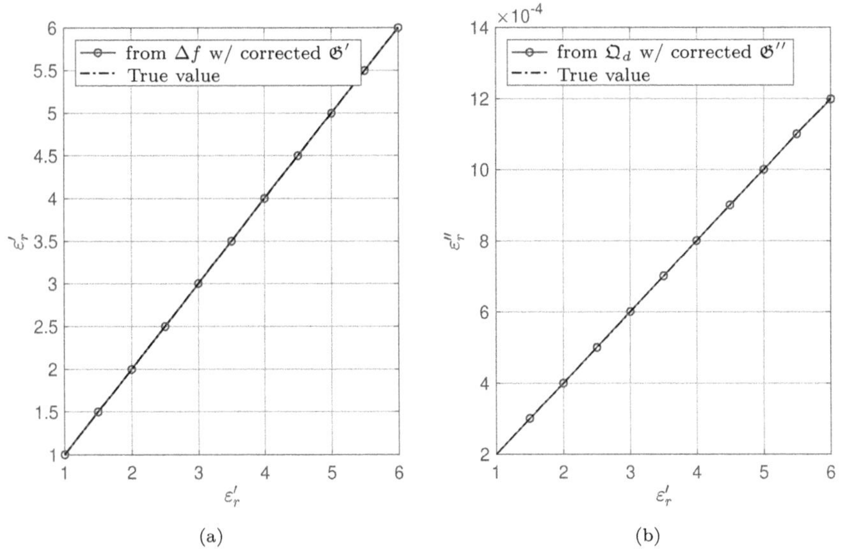

Figure 15.17.: Material parameters calculated from the cavity parameters given in figure 15.11 using the permittivity-dependent geometry factors shown in figure 15.15a

15.4.4. Conclusion

From the simulated measurements of low-loss dielectrics in cavities with a high[9] or moderate[10] quality factor we can conclude that the secondary effects introduced at the beginning of this section indeed skew accurate measurements of the complex permittivity.

The permittivity ε_r' recovered from the simulated data exhibited an acceptable error of below 3 % for a permittivity $\varepsilon_r' < 6$. This error is only due to the field approximation and can be accounted for by introducing the concept of permittivity-dependent geometry factors presented in the previous subsection.

For the imaginary part ε_r'' of the permittivity the situation is considerably more complex:

When using a cavity with a moderate quality factor, the error of ε_r'' is dominated by the error introduced by the approximation of $\mathfrak{Q}_d$ as per (15.45), introducing an error in the region of 10 % even for comparably small permittivities ($\varepsilon_r' \approx 2$). In contrast, when a cavity with a large quality factor is used, the error due to the field approximation and the approximation of $\mathfrak{Q}_d$ cancel out in a favourable fashion. Obviously, this is a somewhat arbitrary result, which should not be relied on.

As we have seen in the previous subsection, deviations of the imaginary part of the complex permittivity, which are caused by the imperfect field approximation, can also be corrected using permittivity-dependent geometry factors.

In contrast, the problem to correctly estimate $\mathfrak{Q}_d$ in the presence of an undesired variation of $\mathfrak{Q}_c$ is not solved by this concept and as a consequence, a cavity with the highest quality factor possible should be used to minimise the impact of $\mathfrak{Q}_c$ in real-world measurements.

A rigorous model for the variation of $\mathfrak{Q}_c$ would allow to correct the errors introduced by the approximation of $\mathfrak{Q}_d$. Such a model is considered future work.

As a final remark, it should be stressed that for lossy dielectrics, the error introduced due to the approximation of $\mathfrak{Q}_d$ can be ignored as can be shown by carrying out simulations similar to the ones presented here. The reason for this is that if the sample's loss tangent is increased, the sample's quality factor $\mathfrak{Q}_d$ is decreased and thus the impact of $\mathfrak{Q}_c$ on $\mathfrak{Q}_0$ becomes small compared to the impact of $\mathfrak{Q}_d$. Consequently, the error introduced by the increase of $\mathfrak{Q}_c$ is reduced.

[9] ≈ 12000, see figure 15.14b
[10] ≈ 6000, see figure 15.11b

16. Conclusion and Future Work

16.1. Conclusion

In the current thesis we investigated both the concept and theoretical foundations of the Mode Matching technique in great detail.

We demonstrated the superiority of the Mode Matching technique for solving closed structures over Partial Differential Equation methods, namely Ansys HFSS' commercial FEM implementation.

Table 16.1 compares the computation time required for solving the three waveguide filters considered in this thesis. While the Mode Matching codes presented here required a few minutes computation time at maximum, solving the structures in Ansys HFSS takes between half an hour and more than two hours depending on the complexity of the structure and on whether parallelisation is used.

It was also found that the Mode Matching technique is more robust in terms of convergence because the FEM code requires initial mesh restrictions in order to converge properly.

The results on simulating the filter structures were discussed in the context of a concise review of filter theory in order to illustrate the close interrelation between filter design and electromagnetic simulation. We discussed both the manual optimisation of filter designs using full-wave simulation of the filter's geometry as well as the implications of the various filter's topologies on the application of the Mode Matching technique.

Filter	**Mode Matching Solver**	**Ansys HFSS (FEM)** Single Core	Parallelised
11th order stepped-impedance tubular filter, 300 point sweep, see sec. 13.2.2	< 1 minute	30 minutes	5 minutes
4th order direct-coupled tubular filter, 166 point sweep, see sec. 13.3.2	< 1 minute	150 minutes	23 minutes
4th order direct-coupled shunt-iris waveguide filter 201 point sweep, see sec. 13.4.1	5.5 minutes	n.a.	120 minutes

Table 16.1.: Approximate computation time required for solving the filters considered in this thesis. All calculations were carried on a machine equipped with two Intel Xeon E5-2630 (Sandy Bridge) CPUs (each providing 6 cores and 12 threads) and 64 GByte RAM.

Several implementation aspects of the Mode Matching technique were addressed. We found the structure's system matrix to be a sparse band-limited matrix, investigated the efficient population of this matrix and assessed using LU decomposition for such matrices. An outlook on possible approaches for future parallelisation was given.

In tubular filter designs, the use of dielectric tubes is often desirable to both center the inner conductor as well as to increase the filter's power handling capability. In order to analyse such filters using the Mode Matching technique, the Eigenmodes of partially filled coaxial waveguides must be known.

We derived the rotationally symmetric $TM^z_{m,0}$ Eigenmodes of such waveguides, which are relevant in a tubular filter design, studied their properties in great detail and demonstrated Mode Matching for a structure comprising such waveguides.

Finally, we investigated "secondary effects" in cavity perturbation material measurements based on a Mode Matching solution for the perturbed cavity's fields.

For the given test setup, the error of ε_r' due to the field approximation in the sample region was found to be below 3 % for a permittivity $\varepsilon_r' < 6$.

In contrast, we showed that the error of the imaginary part of the permittivity may be in the region of 10 % when measuring low-loss dielectrics in a cavity with a moderate quality factor. This deviation already occurs if $\varepsilon_r' \approx 2$.

It was shown that the deviation of the real and imaginary part of the complex permittivity caused by the field approximation can be corrected by employing the concept of permittivity-dependent geometry factors introduced in this thesis.

16.2. Future Work

The Mode Matching framework presented in the scope of this thesis makes an excellent starting point for future work on the subject.

Exciting future work could be carried out on extending the Mode Matching solver's capability towards treating the more advanced filter designs used in today's satellite communication systems, which were discussed in chapter 13.4.1.

Besides that, the realisation of a tubular filter design making use of the partially filled waveguides discussed in chapter 14 is still future work.

Additional work could be carried out on the analysis of e.g. corrugated horn antennas [18] or Pickett-Potter horns [150, 151] using the Mode Matching technique. This work is especially interesting because a suitable method for calculating the antenna's far field would have to be implemented. This could either be done using an aperture field method using Huygens equivalent sources or by Spherical Wave expansion.

While the current Mode Matching framework implemented in Matlab already performs quite well, porting the code to C++ most likely would enable an additional speed-up. The Armadillo library [152, 153] for linear algebra and scientific computing appears to be a well-suited foundation for this undertaking.

In the current implementation, the geometry to be analysed is described to the Mode Matching solver by means of simple text files. Since this method is both cumbersome and error-prone, a graphical modelling interface would be of the greatest use in order to simplify the usage of the software. Alternatively, one could imagine a pre-processsor which checks the validity of the geometry described in the text file.

When carrying out any modifications on the definition of the geometry description file, it should be kept in mind that the structure of this file is closely related to the algorithm used by the Mode Matching solver to populate the system matrix.

Apart from continuing work on the existing Mode Matching solver, it would be interesting to implement the Boundary Contour Mode Matching technique, which alleviates the geometric limitations imposed by the Mode Matching technique. This would for example enable the analysis of waveguide iris filters with milling radii.

The cavity perturbation material measurements discussed in chapter 15 make another large field of future work. Apart from the obvious need for a better model to treat the "secondary effects" discussed in this thesis, a complete measurement uncertainty budget is required.

The measurements of the relative permittivity of water, which were presented in chapter 15.3.2, had a proof-of-concept character only. However, since the first results discussed here look promising, the method should be pursued further. As a first step, temperature measurements should be improved. Moreover, simulations of the temperature distribution inside the glass capillary would provide valuable insights.

The question of dielectric material characterisation was originally presented to the author in the context of an industrial microwave heating application. For such processes, the temperature dependence of the relative permittivity is of great interest. A research project on this problem is currently carried out at the institute, which is based on the insights presented in this thesis.

Part IV.

Appendices

A. Brief Review of Variational Methods

Variational methods represent the underlying fundamental concept of several important computational electromagnetics techniques. In order to complete the discussion of these methods in chapter 1, this appendix provides a concise overview of variational methods.

Variational methods can be grouped into two categories: The *direct variational method*, namely the *Rayleigh-Ritz method* and *indirect variational methods*, i.e. *methods of weighted residuals* [20].

In order to apply variational methods to the electromagnetic problem under consideration, the problem has to be put under form of an *operator equation* [20]

$$\mathcal{L}\,\mathfrak{u} = \mathfrak{f} \tag{A.1}$$

where $\mathcal{L}$ denotes a linear operator, which is either based on an integral representation of the fields or on a suitable partial differential equation such as the wave equation or Laplace's equation [154]. $\mathfrak{f}$ is called the *source function* or *forcing function* while $\mathfrak{u}$ is the *unknown function* to be determined [20, 154].

A.1. Direct Method

It can be shown that the solution to (A.1) extremises the *functional*[1]

$$I\,(\mathfrak{u}) = \langle \mathcal{L}\,\mathfrak{u}, \mathfrak{u}\rangle - 2\langle \mathfrak{u}, \mathfrak{f}\rangle, \tag{A.2}$$

that is, renders the functional *stationary* [20, 154]. $\langle\cdot,\cdot\rangle$ denotes an inner product.

By applying (A.2) to the partial differential equation under consideration, the partial differential equation's functional can be obtained [20]. An overview of the functionals of various partial differential equations can be found in [20, tab. 4.1, p. 244].

The functionals usually take a form[2] [20]

$$I\,(\mathfrak{u}) = \iiint \mathfrak{F}\,(x, y, z, \mathfrak{u}, \mathfrak{u}_x, \mathfrak{u}_y, \mathfrak{u}_z)\, dV \tag{A.3}$$

and depending on the underlying partial differential equation sometimes can be interpreted in a physical meaningful sense [155]. For example, the functional of *Laplace's equation* $\nabla^2\mathfrak{u} = 0$, which in electrostatics governs the potential in charge-free regions [156], can be interpreted as stored energy [155]. However, in electromagnetics "no physical significance can be attached to the *stationary point* of the functional" [154, p. 27].

[1] [20] uses the terms functional and variational principle in an exchangeable fashion.

[2] $\mathfrak{u}_x$, $\mathfrak{u}_y$ and $\mathfrak{u}_z$ denote partial derivatives [20]

We now wish to establish an approximate solution

$$\tilde{\mathfrak{u}} = \sum_{i=1}^{N} a_i \mathfrak{u}_i \tag{A.4}$$

to (A.3) in terms of *basis functions*[3] $\mathfrak{u}_i$ [20, 21]

By inserting (A.4) in (A.3), (A.3) becomes a function of N coefficients [20]

$$I(\mathfrak{u}) = I(a_1, ..., a_N). \tag{A.5}$$

If we demand the partial derivatives of (A.5) to be zero, that is,

$$\frac{\partial I}{\partial a_i} = 0, \ i = 1, ..., N \tag{A.6}$$

the functional is extremised and we obtain a set of N linear equations, which can be solved to obtain the desired coefficients a_i [20].

An alternative approach to obtain a system of linear equations is to insert (A.4) into (A.2): After some manipulations[4], taking the partial derivatives of the obtained expression yields [20]

$$\sum_{i=1}^{N} \langle \mathfrak{u}_i, \mathcal{L}\, \mathfrak{u}_j \rangle a_i = \langle \mathfrak{u}_j, \mathfrak{f} \rangle, \ j = 1, ..., N, \tag{A.7}$$

which may be put under form of a matrix equation $\boldsymbol{A} \cdot \vec{x} = \vec{b}$ [20].

As we will see later, the form (A.7) is especially important because it highlights the variational nature of indirect methods if the Galerkin approach is used [154].

[3] also: expansion functions [20]

[4] see [20, p. 246f]

A.2. Indirect Methods - Methods of Weighted Residuals

Indirect variational methods represent an alternative approach to solve an electromagnetic problem under form of (A.1).

Let us again approximate the solution to (A.1) by means of (A.4) [20]. The linear operator equation (A.1) now becomes [20]

$$\mathcal{L}\,\tilde{\mathfrak{u}} \approx \mathfrak{f} \tag{A.8}$$

and we may introduce a *residual* [20]

$$R = \mathcal{L}\,\tilde{\mathfrak{u}} - \mathfrak{f} \tag{A.9}$$

We now strive to minimise the average weighted value of R over the domain of interest [21]. In order to so, we introduce *weighting functions* (also: *testing functions*) [21]

$$\sigma_j,\ j = 1, ..., M \tag{A.10}$$

and let the inner products of (A.9) and σ_j vanish, that is,

$$\langle \sigma_j, R \rangle = \sum_{i=1}^{N} a_i \langle \sigma_j, \mathcal{L}\,\mathfrak{u}_i \rangle - \langle \sigma_j, \mathfrak{f} \rangle = 0,\ j = 1, ..., M, \tag{A.11}$$

which may be rewritten as [20]

$$\sum_{i=1}^{N} \langle \sigma_j, \mathcal{L}\,\mathfrak{u}_i \rangle a_i = \langle \sigma_j, \mathfrak{f} \rangle,\ j = 1, ..., M. \tag{A.12}$$

This expression again may be put under form of a matrix equation $\boldsymbol{A} \cdot \vec{x} = \vec{b}$ [20, 21], that is, we have converted the linear operator equation into a matrix equation. Note that usually the number of basis functions and weighting functions is chosen equal, that is, $N = M$ [21].

Various approaches for selecting proper weighting functions can be thought of and depending on the choice of the weighting functions, different names for the method have been coined. [20].

The simplest choice is to use the Dirac delta function as weighting functions [20]. This method is called *collocation method* or *point-matching method* [20].

Another important approach is to choose the weighting functions identical to the basis functions, that is, $\mathfrak{u}_i = \sigma_i$ [20]. This approach is called *Galerkin's method* [20]. If we apply $\mathfrak{u}_i = \sigma_i$ to (A.12), we obtain

$$\sum_{i=1}^{N} \langle \mathfrak{u}_j, \mathcal{L}\,\mathfrak{u}_i \rangle a_i = \langle \mathfrak{u}_j, \mathfrak{f} \rangle,\ j = 1, ..., M, \tag{A.13}$$

which is (A.7). This is an important result as it points towards the fact that methods of weighted residuals are indeed variational methods [19]. For a more complete discussion of this topic refer to [19].

B. Eigenmodes of Standard Waveguides

B.1. Rectangular Waveguides

Consider the *rectangular waveguide* depicted in figure B.1. All problems discussed in this thesis which comprise such rectangular waveguides were solved based on the set of the TE^x / TM^x Eigenmodes. Note that the superscript x indicates that the Eigenmodes included are transverse to x, that is, the TE^x Eigenmodes exhibit no electric field in x direction while the TM^x Eigenmodes do not have a magnetic field in the direction of x.

As already discussed in section 2.1.2, due to the completeness of the TE^z / TM^z mode-set, the set of TE^x / TM^x Eigenmodes can be constructed using linear combinations of the transverse-to-z mode-sets' Eigenmodes.

However, an easier approach to obtain the desired sets of Eigenmodes is to directly derive the Eigenmodes from suitable vector potentials

$$\psi(x, y, z) = \psi(x)\,\psi(y)\,\psi(z) \tag{B.1}$$

as we will see in the following. Separation in rectangular coordinates has already been discussed in section 2.2.1.

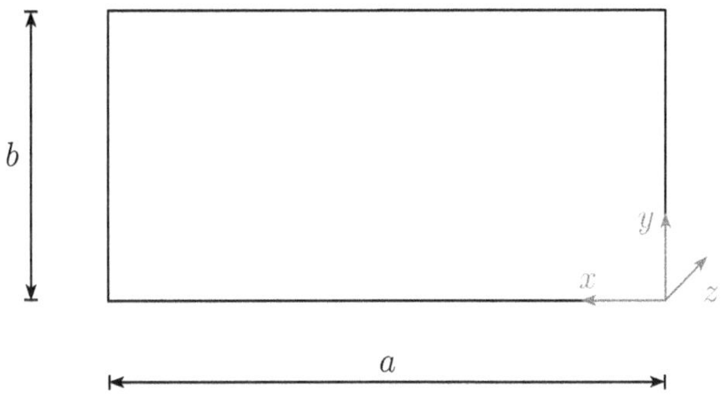

Figure B.1.: Rectangular waveguide

B.1.1. $TE^x_{m,n}$ Eigenmodes

In order to obtain the set of $TE^x_{m,n}$ Eigenmodes, an electric vector potential $\vec{\mathfrak{F}} = \psi\ \vec{e}_x$ where [48]

$$\psi^{TE} = \sin(k_x x)\cos(k_y y)\, e^{\mp j k_z z} \tag{B.2}$$

is chosen. The TE Eigenmodes' fields can be calculated using [48][1]

$$\begin{aligned}
E_x^{TE} &= 0 & H_x^{TE} &= \frac{1}{j\omega\mu}\left(\frac{\partial^2}{\partial x^2}+k^2\right)\psi^{TE}\\
E_y^{TE} &= -\frac{\partial}{\partial z}\psi^{TE} & H_y^{TE} &= \frac{1}{j\omega\mu}\frac{\partial^2}{\partial x\partial y}\psi^{TE}\\
E_z^{TE} &= \frac{\partial}{\partial y}\psi^{TE} & H_z^{TE} &= \frac{1}{j\omega\mu}\frac{\partial^2}{\partial x\partial z}\psi^{TE},
\end{aligned} \tag{B.3}$$

which directly follows from

$$\vec{E} = -\nabla\times\vec{\mathfrak{F}} \qquad \text{and} \qquad \vec{H} = \frac{1}{j\omega\mu}\nabla\times\nabla\times\vec{\mathfrak{F}}. \tag{B.4}$$

The $TE^x_{m,n}$ Eigenmodes' fields denote

$$\begin{aligned}
E^{TE}_{x|m,n} &= 0\\
E^{TE}_{y|m,n} &= \pm\, jk_{z|m,n}\,\sin\left(k_{x|m}x\right)\cos\left(k_{y|n}y\right)e^{\mp jk_{z|m,n}z}\\
E^{TE}_{z|m,n} &= -k_{y|n}\,\sin\left(k_{x|m}x\right)\sin\left(k_{y|n}y\right)e^{\mp jk_{z|m,n}z}
\end{aligned} \tag{B.5}$$

and

$$\begin{aligned}
H^{TE}_{x|m,n} &= \frac{k^2_{y|n}+k^2_{z|m,n}}{j\omega\mu}\,\sin\left(k_{x|m}x\right)\cos\left(k_{y|n}y\right)e^{\mp jk_{z|m,n}z}\\
H^{TE}_{y|m,n} &= -\frac{k_{x,m}k_{y,n}}{j\omega\mu}\,\cos\left(k_{x|m}x\right)\sin\left(k_{y|n}y\right)e^{\mp jk_{z|m,n}z}\\
H^{TE}_{z|m,n} &= \mp\,\frac{jk_{x|m}k_{z|m,n}}{j\omega\mu}\,\cos\left(k_{x|m}x\right)\cos\left(k_{y|n}y\right)e^{\mp jk_{z|m,n}z}
\end{aligned} \tag{B.6}$$

where we have introduced additional subscripts $m \geq 1$, $n \geq 0$ [48] to enable indexing of the individual Eigenmodes.

[1] See [48, (4-32), p. 153]

B.1.2. $TM^x_{v,w}$ Eigenmodes

In contrast, the set of $TM^x_{v,w}$ Eigenmodes follows from a magnetic vector potential $\vec{\mathfrak{A}} = \psi\, \vec{e}_z$ with a scalar vector potential [48]

$$\psi = \cos(k_x x)\sin(k_y y)\, e^{\mp j k_z z} \tag{B.7}$$

and the Eigenmodes' fields are calculated using [48][2]

$$\begin{aligned} E_x^{TM} &= \frac{1}{j\omega\varepsilon} \left(\frac{\partial^2}{\partial x^2} + k^2\right) \psi & H_x^{TM} &= 0 \\ E_y^{TM} &= \frac{1}{j\omega\varepsilon} \frac{\partial^2}{\partial x \partial y} \psi & H_y^{TM} &= \frac{\partial}{\partial z} \psi \\ E_z^{TM} &= \frac{1}{j\omega\varepsilon} \frac{\partial^2}{\partial x \partial z} \psi & H_z^{TM} &= -\frac{\partial}{\partial y} \psi, \end{aligned} \tag{B.8}$$

which follows from

$$\vec{E} = \frac{1}{j\omega\varepsilon} \nabla \times \nabla \times \vec{\mathfrak{A}} \qquad \text{and} \qquad \vec{H} = \nabla \times \vec{\mathfrak{A}}. \tag{B.9}$$

The TM-Eigenmodes' fields denote

$$\begin{aligned} E^{TM}_{x|v,w} &= \frac{k^2_{y|w} + k^2_{z|v,w}}{j\omega\varepsilon} \cos\left(k_{x|v}x\right)\sin\left(k_{y|w}y\right) e^{\mp j k_{z|v,w} z} \\ E^{TM}_{y|v,w} &= -\frac{k_{x|v}k_{y|w}}{j\omega\varepsilon} \sin\left(k_{x|v}x\right)\cos\left(k_{y|w}y\right) e^{\mp j k_{z|v,w} z} \\ E^{TM}_{z|v,w} &= \mp \frac{j k_{x|v} k_{z|v,w}}{j\omega\varepsilon} \sin\left(k_{x|v}x\right)\sin\left(k_{y|w}y\right) e^{\mp j k_{z|v,w} z} \end{aligned} \tag{B.10}$$

and

$$\begin{aligned} H^{TM}_{x|v,w} &= 0 \\ H^{TM}_{y|v,w} &= \mp j k_{z|v,w} \cos\left(k_{x|v}x\right)\sin\left(k_{y|w}y\right) e^{\mp j k_{z|v,w} z} \\ H^{TM}_{z|v,w} &= -k_{y|w} \cos\left(k_{x|v}x\right)\cos\left(k_{y|w}y\right) e^{\mp j k_{z|v,w} z} \end{aligned} \tag{B.11}$$

where we have introduced additional subscripts $v \geq 0$, $w \geq 1$ [48] to enable indexing of the individual Eigenmodes.

[2] See [48, (4-30), p. 153]

B.2. Odd-Order $TE^x_{m,0}$ Eigenmodes of a Waveguide Partially Filled with a Symmetric Dielectric Slab

For our analysis of the perturbed fundamental Eigenmode of a rectangular cavity carried out in section 15.4.1 the odd-order $TE^x_{m,0}$ Eigenmodes of a waveguide *partially filled* with a symmetric dielectric slab (cf. figure B.2) are required.

In the scope of this analysis, interfaces between empty and partially filled waveguides have to be treated using the Mode Matching technique. As the cavity's fundamental TE Eigenmode is an odd-order mode, at these interfaces, only odd-order $TE^x_{m,0}$ Eigenmodes of the partially filled waveguide can be excited because overlap integrals of even- and odd-order Eigenmodes vanish.

Consider the magnetic field distributions of an even- and odd-order $TE^x_{m,0}$ Eigenmode of an empty waveguide depicted in figures B.3 and B.4. The magnetic field distributions of even-order Eigenmodes are tangential to the waveguide's centerline at $x = a/2$. In contrast, the magnetic fields of the odd-order Eigenmodes are normal to the waveguide's centerline and consequently, for odd-order Eigenmodes, the waveguide's centerline can be interpreted as a PMC wall.[3]

If we assume the partially filled waveguide's Eigenmodes to be of comparable form, in order to limit our analysis to odd-order Eigenmodes as well as to avoid the need to ensure continuity of tangential fields at two air-material interfaces, we introduce a PMC wall at $x = a/2$ as depicted in figure B.5.

For completeness sake, it should be mentioned that overlap integrals between the waveguides' $TE^x_{m,0}$ Eigenmodes and Eigenmodes with $n > 0$ vanish because the air-material interface lies parallel to the y-z plane. Thus $TE^x_{m,n>0}$ Eigenmodes can be neglected.

Under the assumptions given above, the partially filled waveguide's odd-order $TE^x_{m,0}$ Eigenmodes can be derived as discussed in the following. Additional information on determining the Eigenmodes of partially filled rectangular waveguides can be found in [48].

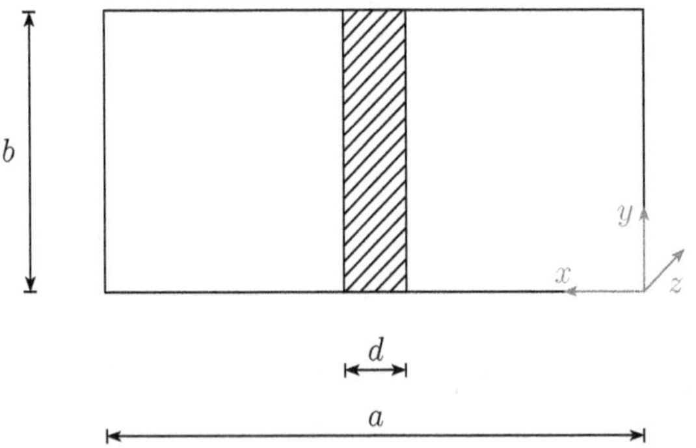

Figure B.2.: Rectangular waveguide partially filled with a symmetric dielectric slab

[3] An additional requirement for a PMC wall is that there is no normal electric field. This is true because $TE^x_{m,0}$ Eigenmodes have an E_y component only.

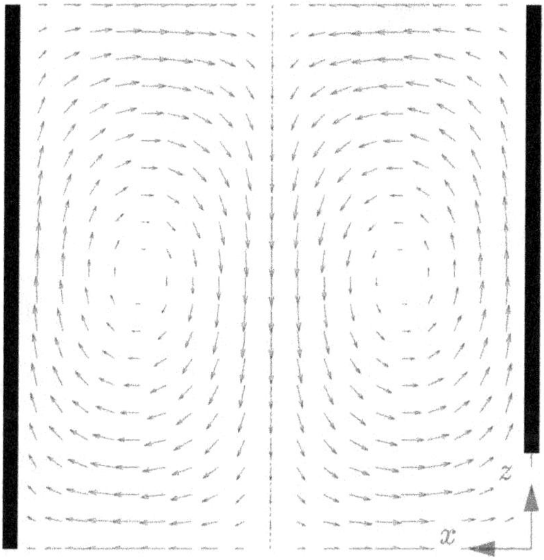

Figure B.3.: Magnetic field distribution of the $TE^x_{2,0}$ Eigenmode (even-order) of an empty rectangular waveguide

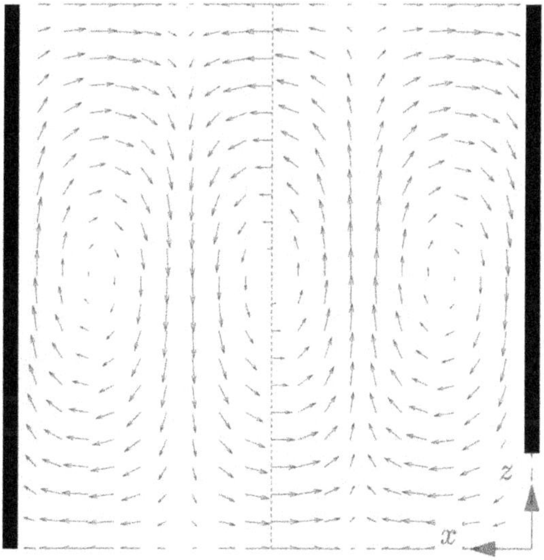

Figure B.4.: Magnetic field distribution of the $TE^x_{3,0}$ Eigenmode (odd-order) of an empty rectangular waveguide

B.2.1. Derivation

At the PEC wall at $x = 0$ the tangential electric component E_y has to be zero. For region 1, we thus choose a scalar vector potential

$$ {}^{1}\psi^{TE} = A \sin\left({}^{1}k_{x|m}x\right) e^{\mp j\,{}^{1}k_{z|m}z}. \tag{B.12}$$

Similarly, at the PMC wall at $x = a/2$, the tangential magnetic component H_z has to vanish. We thus choose

$$ {}^{2}\psi^{TE} = C \cos\left({}^{2}k_{x|m}\left[x - \frac{a}{2}\right]\right) e^{\mp j\,{}^{2}k_{z|m}z}. \tag{B.13}$$

For obvious reasons, $k_z = {}^{1}k_{z|m} = {}^{2}k_{z|m}$ must apply.

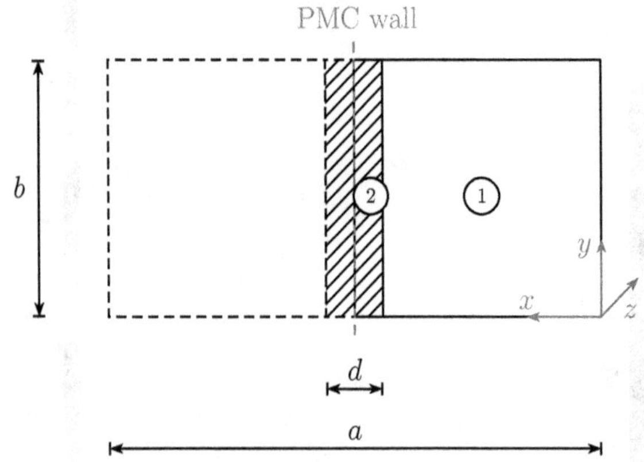

Figure B.5.: Partially filled rectangular waveguide with PMC symmetry plane

From $\vec{\mathfrak{F}} = \psi\,\vec{e}_x$, using (B.3) provided in the previous section of this appendix, the fields

$$\begin{aligned}
{}^1E^{TE}_{y|m} &= \pm \; jk_{z|m} \; A \; \sin\left({}^1k_{x|m}x\right) e^{\mp jk_{z|m}z} \\
{}^1H^{TE}_{x|m} &= \frac{{}^1k^2 - {}^1k^2_{x|m}}{j\omega\mu} \; A \; \sin\left({}^1k_{x|m}x\right) e^{\mp jk_{z|m}z} \\
{}^1H^{TE}_{z|m} &= \mp \; \frac{j\,{}^1k_{x|m}k_{z|m}}{j\omega\mu} \; A \; \cos\left({}^1k_{x|m}x\right) e^{\mp jk_{z|m}z}
\end{aligned} \tag{B.14}$$

and ${}^1E^{TE}_{x|m} = {}^1E^{TE}_{z|m} = 0$, ${}^1H^{TE}_{y|m} = 0$ are calculated for region 1 and similarly

$$\begin{aligned}
{}^2E^{TE}_{y|m} &= \pm \; jk_{z|m} \; C \; \cos\left({}^2k_{x|m}\left[x - \tfrac{a}{2}\right]\right) e^{\mp jk_{z|m}z} \\
{}^2H^{TE}_{x|m} &= \frac{{}^2k^2 - {}^2k^2_{x|m}}{j\omega\mu} \; C \; \cos\left({}^2k_{x|m}\left[x - \tfrac{a}{2}\right]\right) e^{\mp jk_{z|m}z} \\
{}^2H^{TE}_{z|m} &= \pm \; \frac{j\,{}^2k_{x|m}k_{z|m}}{j\omega\mu} \; C \; \sin\left({}^2k_{x|m}\left[x - \tfrac{a}{2}\right]\right) e^{\mp jk_{z|m}z}
\end{aligned} \tag{B.15}$$

and ${}^2E^{TE}_{x|m} = {}^2E^{TE}_{z|m} = 0$, ${}^2H^{TE}_{y|m} = 0$ for region 2.

Similar to section 14.2, we now need to find the two regions' Eigenvalues ${}^1k_c = {}^1k_x$ and ${}^2k_c = {}^2k_x$ which maintain continuity of tangential fields at $x = (a-d)/2$. In order to do so, we again put the *continuity condition* under form of a homogeneous system of equations

$$\underbrace{\begin{bmatrix} {}^1E^{TE}_{y|m} & -{}^2E^{TE}_{y|m} \\ {}^1H^{TE}_{z|m} & -{}^2H^{TE}_{z|m} \end{bmatrix}}_{\mathbf{\Phi}} \underbrace{\begin{bmatrix} A \\ C \end{bmatrix}}_{\vec{x}} = \vec{0}\,\Big|_{x=\frac{a-d}{2}}, \tag{B.16}$$

which has non-trivial solutions ($\vec{x} = \Pi\,\vec{x}_0$) only if the matrix' determinant vanishes.

While (14.12) contains Bessel functions, (B.16) contains harmonic functions (sin, cos) and thus the expression for the matrix' determinant can be further simplified.

The determinant of $\mathbf{\Phi}$ denotes

$$\begin{aligned}
& \left[\left(jk_{z|m}\sin\left({}^1k_{x|m}\tfrac{a-d}{2}\right)\right)\cdot - \left(\frac{j\,{}^2k_{x|m}k_{z|m}}{j\omega\mu}\sin\left({}^2k_{x|m}\left[\tfrac{a-d}{2} - \tfrac{a}{2}\right]\right)\right)\right] \\
- & \left[\left(jk_{z|m}\cos\left({}^2k_{x|m}\left[\tfrac{a-d}{2} - \tfrac{a}{2}\right]\right)\right)\cdot - \left(-\frac{j\,{}^1k_{x|m}k_{z|m}}{j\omega\mu}\cos\left({}^1k_{x|m}\tfrac{a-d}{2}\right)\right)\right] \\
= & \;\; 0,
\end{aligned}$$

which can be rearranged as

$$\begin{aligned}
&- \quad jk_{z|m} \sin\left({}^1k_{x|m}\tfrac{a-d}{2}\right) \cdot \frac{j\,{}^2k_{x|m}k_{z|m}}{j\omega\mu} \sin\left({}^2k_{x|m}\left[\tfrac{a-d}{2} - \tfrac{a}{2}\right]\right) \\
&- \quad jk_{z|m} \cos\left({}^2k_{x|m}\left[\tfrac{a-d}{2} - \tfrac{a}{2}\right]\right) \cdot \frac{j\,{}^1k_{x|m}k_{z|m}}{j\omega\mu} \cos\left({}^1k_{x|m}\tfrac{a-d}{2}\right) \\
&= \quad 0
\end{aligned}$$

and

$$\begin{aligned}
&- \quad \sin\left({}^1k_{x|m}\tfrac{a-d}{2}\right) \cdot {}^2k_{x|m} \sin\left({}^2k_{x|m}\left[\tfrac{-d}{2}\right]\right) \\
&- \quad \cos\left({}^2k_{x|m}\left[\tfrac{-d}{2}\right]\right) \cdot {}^1k_{x|m} \cos\left({}^1k_{x|m}\tfrac{a-d}{2}\right) \\
&= \quad 0.
\end{aligned}$$

By exploiting $\cos(-x) = \cos(x)$ and $\sin(-x) = -\sin(x)$ we find

$$\begin{aligned}
&+ \quad \sin\left({}^1k_{x|m}\tfrac{a-d}{2}\right) \cdot {}^2k_{x|m} \sin\left({}^2k_{x|m}\tfrac{d}{2}\right) \\
&- \quad \cos\left({}^2k_{x|m}\tfrac{d}{2}\right) \cdot {}^1k_{x|m} \cos\left({}^1k_{x|m}\tfrac{a-d}{2}\right) \\
&= \quad 0,
\end{aligned}$$

which can be rewritten as

$$\sin\left({}^1k_{x|m}\frac{a-d}{2}\right) \cdot {}^2k_{x|m} \sin\left({}^2k_{x|m}\frac{d}{2}\right) = \cos\left({}^2k_{x|m}\frac{d}{2}\right) \cdot {}^1k_{x|m} \cos\left({}^1k_{x|m}\frac{a-d}{2}\right)$$

and finally may be put under form of an transcendental equation

$${}^1k_{x|m} \cot\left({}^1k_{x|m}\frac{a-d}{2}\right) = {}^2k_{x|m} \tan\left({}^2k_{x|m}\frac{d}{2}\right). \tag{B.17}$$

This is the waveguide's *characteristic equation* (also see section 14.2).

The characteristic equation's zeros provide the wavenumbers $k_{x|m}$, which have to be found numerically. In order to do so, we sweep ${}^1k_x$ over ${}^1k_x \in \mathbb{R} > 0$ while ${}^2k_x$ is calculated

$${}^2k_{x|m} = \sqrt{\omega^2\varepsilon_0\mu_0({}^2\varepsilon_r - 1) + {}^1k_{x|m}^2}. \tag{B.18}$$

For consistency with the numbering scheme of the empty waveguide's Eigenmodes, the characteristic equation's zeros are numbered $m = 1, 3, 5, ...$ and so on.

A problem afflicted with solving the waveguide's characteristic equation is that for sufficiently large widths d of the dielectric slab the wavenumber ${}^1k_x$ becomes positive imaginary as is shown in figure B.6 for the waveguide's fundamental Eigenmode. In contrast, ${}^2k_x$ remains real-valued and k_z continues to increase with the width of the dielectric slab. This is a physically meaningful result because the guided wavelength $\lambda_g = 2\pi/k_z$ of a waveguide will decrease if a larger portion of the waveguide's cross-section contains a dielectric.

An interesting point to note is that the field distribution of an Eigenmode with ${}^1k_x \in \mathbb{I} > 0$ exhibits an opposite curvate in region 1 compared to an "ordinary" sine-shaped field distribution, because the sine function effectively becomes a hyperbolic function [69]. Figure B.7 depicts the field component E_y of such an Eigenmode.

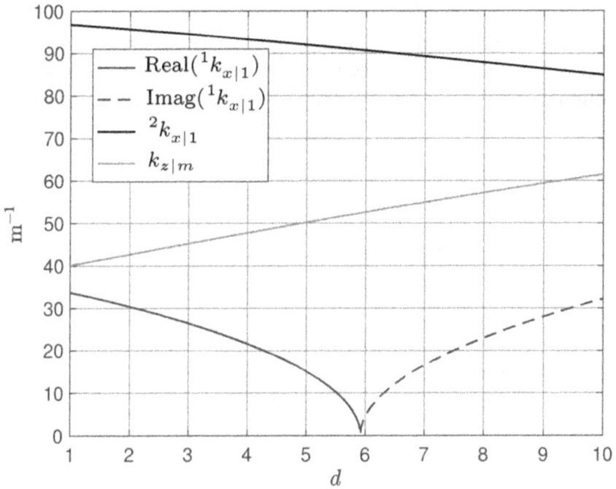

Figure B.6.: Variation of the wavenumbers ${}^1k_x$, ${}^2k_x$, k_z of the fundamental Eigenmode ($m = 1$) over the width d of a dielectric slab with a permittivity of $\varepsilon_r' = 4$. $a = 86$ mm, $b = 43$ mm, $f = 2.5$ GHz.

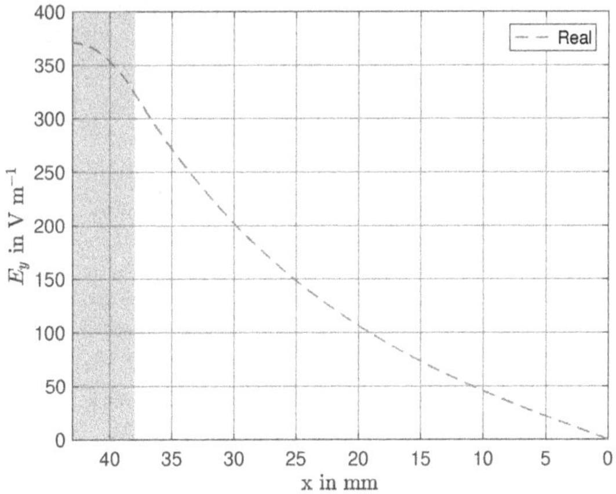

Figure B.7.: Field component E_y for an Eigenmode with an imaginary wavenumber ${}^1k_x$

B.2.2. Properties of the Obtained Eigenmodes

The field distributions of the fundamental $TE^x_{1,0}$ Eigenmode with ${}^1k_x, {}^2k_x \in \mathbb{R} > 0$ as well as the corresponding wave impedance are shown in figure B.8:

At the air-dielectric interface continuity of the tangential fields E_y, H_z can be observed.

For both regions 1 and 2 the same wave impedance $Z^{TE}_{wave} = {}^{\omega\mu}/_{k_z}$ (see appendix C) applies because the regions' solutions obviously must have an identical propagation constant. Thus Z_{wave} is constant over the waveguide's cross-section.

As the tangential electric field component E_y has to be continuous across the air-dielectric interface and identical wave impedances apply for region 1 and 2, the normal magnetic field component H_x has to be continuous across the interface as well.

Tables B.1 and B.2 summarise the characteristic properties of the partially filled waveguide's Eigenmodes. As can be seen from table B.1, the wavenumbers obtained from the analytical solution show excellent agreement with results obtained using a commercial FEM solver, namely Ansys HFSS.

The same applies for the fundamental Eigenmode's wave impedance as well as the characteristic impedances given in table B.2. As one would expect from the fact that the characteristic impedance of a transmission line is proportional to $\sqrt{L'/C'}$ where L', C' denote the inductance and capacitance per unit length [6], the characteristic impedance is decreased (empty waveguide: $Z_0 = 525.5$ Ohm) because the dielectric slab increases C'.

Several remarks have to be made on the obtained characteristic impedances Z_0:

As a matter of fact, calculating the "correct" characteristic impedance of a rectangular waveguide is a difficult task because neither voltage nor current are uniquely defined. Therefore, in appendix C we will introduce definitions for voltage waves U^+ and current waves I^+, which ensure that

$$\mathcal{P} = \iint_{\mathfrak{S}} \vec{E} \times \vec{H}^* \, d\vec{A} \qquad \text{and} \qquad \mathcal{P} = U^+ \cdot {I^+}^*$$

provide consistent results. In brief, a voltage wave is calculated by integrating the electric field along the waveguide's centerline at $x = {}^a/_2$ (see C.21) while a current wave is obtained by integrating the magnetic field along the centerline at $y = {}^b/_2$ with an additional prefactor ${}^\pi/_4$ (see C.22).

Using this definition, if an empty waveguide is considered, calculating the characteristic impedance $Z_{0,UI}$ from U^+ and I^+ yields a result identical to the characteristic impedance $Z_{0,PU}$ calculated from P and U^+.

As can be seen from table B.2, for the partially filled waveguide, the situation is different because we have $Z_{0,UI} \neq Z_{0,PU}$. The reason for this is that based on our definition of voltage and current waves, consistency between the two expressions for the complex power transported is only ensured in empty waveguides as can be seen from (C.22). For a partially filled waveguide, the integral over x does not yield exactly 2 because the integral has to be decomposed into two regions for which different k_x apply.

Another remark has to be made on comparing characteristic impedances calculated analytically with values obtained from Ansys HFSS. As Ansys HFSS does not include the prefactor ${}^\pi/_4$ when calculating the current wave at a port, the characteristic impedance calculated from U^+ and I^+ has to be multiplied by ${}^4/_\pi$ in order to obtain results consistent to our analytical treatise.

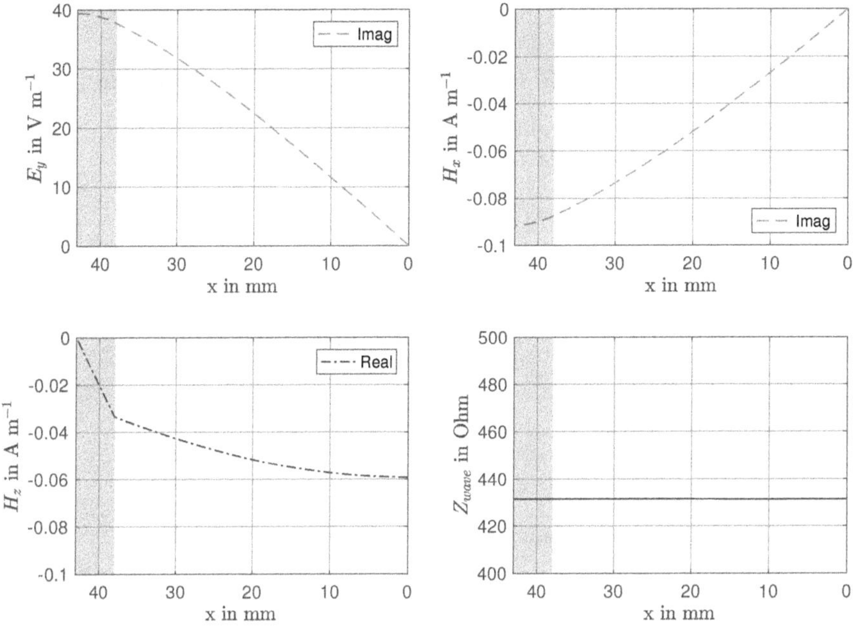

Figure B.8.: Fields and wave impedance of the fundamental $TE^x_{1,0}$ Eigenmode of the partially filled waveguide shown in figure B.2

TM Eigenmode	Eigenmode Solution ^1k_x	^2k_x	k_z	Ansys HFSS k_z
Prop. Eigenmode ($TE_{1,0}$)	25.54	58.29	45.75	45.75
2nd evan. Eigenmode ($TE_{3,0}$)	107.00	119.14	$-j93.3$	$-j93.3$

Table B.1.: Calculated wavenumbers
$a = 86$ mm, $b = 43$ mm, $d = 10$ mm, $^1\varepsilon_r = 1$, $^2\varepsilon_r = 2$, $f = 2.5$ GHz
k_x in 1/m, k_z in rad/m or Np/m, Z_0 in Ohm

$\mathrm{TM}^x_{1,0}$ Eigenmode	Eigenmode Solution	Ansys HFSS
Z_{wave}	431.4	431.4
$Z_{0,UI}$	478.1	$375.5 \cdot 4/\pi = 478.1$
$Z_{0,PU}$	510.7	510.6

Table B.2.: Calculated characteristic and wave impedances
$a = 86$ mm, $b = 43$ mm, $d = 10$ mm, $^1\varepsilon_r = 1$, $^2\varepsilon_r = 2$, $f = 2.5$ GHz
Z_{wave}, Z_0 in Ohm

B.3. Coaxial Waveguides

B.3.1. The Mode-Set of Rotationally Symmetric $TM^z_{m,0}$ Eigenmodes

Consider *coaxial waveguide* depicted in figure B.9. The Eigenmodes of such a waveguide may be derived from a vector potential

$$\psi(\rho, \phi, z) = \psi(\rho)\,\psi(\phi)\,\psi(z)\,. \tag{B.19}$$

$\psi(\rho)$ is governed by Bessel's differential equation while $\psi(\phi), \psi(z)$ are governed by an ODE of similar form as the differential equation of the harmonic oscillator.

Keeping in mind that linear combinations of solutions to differential equations are again solutions to the same differential equation, we make the ansatz

$$\psi^{TM} = [A\ J_n(k_\rho \rho) + B\ N_n(k_\rho \rho)] \begin{Bmatrix} \sin(n\phi) \\ \cos(n\phi) \end{Bmatrix} e^{\mp j k_z z} \tag{B.20}$$

in order to obtain the required additional degree of freedom to maintain the boundary conditions at both the inner and outer conductor. J_n denotes Bessel functions of first kind while N_n denotes Bessel functions of second kind.

In the following, we wish to limit our analysis to rotationally symmetric Eigenmodes, that is, to the $TM^z_{m,0}$ mode-set. Thus, we let $n = 0$, which yields $\sin(0) = 0$, $\cos(0) = 1$ and find

$$\psi^{TM} = \psi_t^{TM} e^{\mp j k_z z} \quad \text{where} \quad \psi_t^{TM} = [A\ J_0(k_\rho \rho) + B\ N_0(k_\rho \rho)]\,. \tag{B.21}$$

For TM Eigenmodes, the Dirichlet boundary condition (cf. (2.25))

$$\left.\psi_t^{TM}\right|_{\rho=a} = 0 \qquad \left.\psi_t^{TM}\right|_{\rho=b} = 0 \tag{B.22}$$

must be maintained.

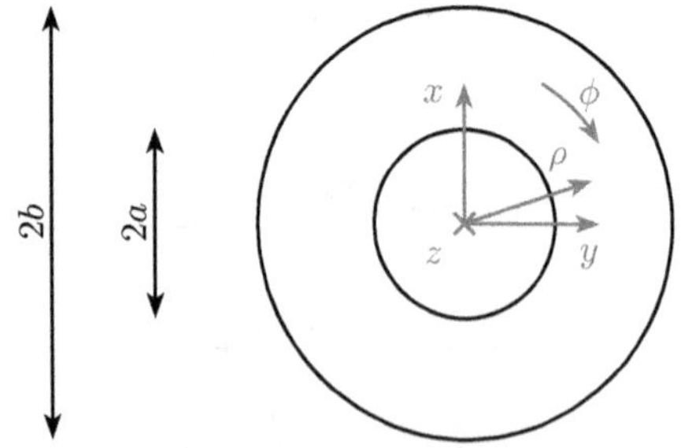

Figure B.9.: Coaxial waveguide

In order to fulfil the boundary condition at the inner conductor ($\rho = a$), we choose $A = N_0(k_\rho a)$ and $B = -J_0(k_\rho a)$ and rewrite (B.21) as [17]

$$\psi_t^{TM} = \left[N_0(k_\rho a)\; J_0(k_\rho \rho) - J_0(k_\rho a)\; N_0(k_\rho \rho)\right]. \tag{B.23}$$

In order to also fulfil the boundary condition at the outer conductor ($\rho = b$) the waveguide's characteristic equation

$$0 = N_0(k_\rho a)\; J_0(k_\rho b) - J_0(k_\rho a)\; N_0(k_\rho b) \tag{B.24}$$

has to be solved numerically for k_ρ. This is equivalent to determining the waveguide's Eigenvalues $k_c^2 = k_\rho^2$. Figure B.10 exemplary depicts the characteristic equation for $2a = 7$ mm and $2b = 16$ mm.

Finally, the Eigenmodes' fields can be calculated from $\vec{\mathfrak{A}} = \psi\; \vec{e}_z$ by means discussed in section 2.1.2 as

$$\begin{aligned}
E_{\rho|m}^{TM} &= \mp \frac{j k_{\rho|m} k_{z|m}}{j\omega\varepsilon} \left[N_0(k_{\rho|m} a)\; J_0'(k_{\rho|m}\rho) - J_0(k_{\rho|m} a)\; N_0'(k_{\rho|m}\rho)\right] e^{\mp j k_{z|m} z} \\
H_{\phi|m}^{TM} &= - k_{\rho|m} \left[N_0(k_{\rho|m} a)\; J_0'(k_{\rho|m}\rho) - J_0(k_{\rho|m} a)\; N_0'(k_{\rho|m}\rho)\right] e^{\mp j k_{z|m} z} \\
E_{z|m}^{TM} &= \frac{k^2 - k_{z|m}^2}{j\omega\varepsilon} \left[N_0(k_{\rho|m} a)\; J_0(k_{\rho|m}\rho) - J_0(k_{\rho|m} a)\; N_0(k_{\rho|m}\rho)\right] e^{\mp j k_{z|m} z}
\end{aligned} \tag{B.25}$$

where we have introduced an additional subscript m to enable indexing of the individual Eigenmodes.

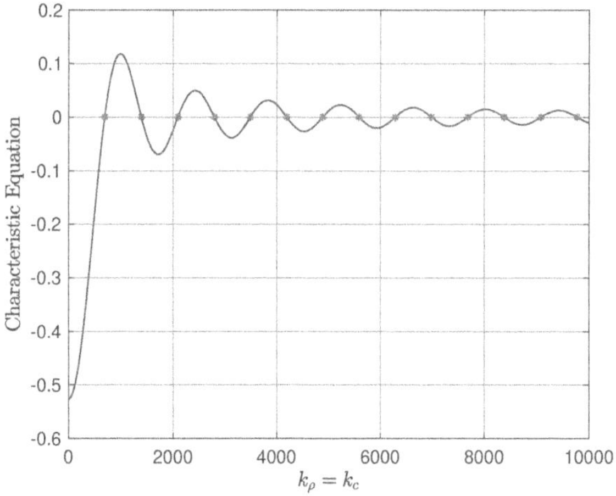

Figure B.10.: First zeros of the coaxial waveguide's characteristic equation for rotationally symmetric TM Eigenmodes, $2a = 7$ mm and $2b = 16$ mm

B.3.2. The TEM Eigenmode as Special Case of TM Eigenmodes

The TEM Eigenmode is a special case of the TM Eigenmodes where $k_\rho = 0$. For this special case we have [48]

$$\begin{aligned} J_{n=0}(0\cdot\rho) &\rightarrow 1 \\ J_{n>0}(0\cdot\rho) &\rightarrow \rho^n \\ N_{n=0}(0\cdot\rho) &\rightarrow \ln(\rho) \\ N_{n>0}(0\cdot\rho) &\rightarrow \rho^{-n}. \end{aligned} \tag{B.26}$$

Inserting the above expressions for $n > 0$ into the characteristic equation (B.24), we find that no solution exists since $a^{-n}b^n - a^n b^{-n} \neq 0$ [48].

In contrast, for $n = 0$ we find $\ln(a) - 1 \cdot \ln(\rho)$, which can be replaced by $\ln(\rho)$ and similarly maintains Bessel's differential equation.

The scalar vector potential for the TEM Eigenmode thus denotes

$$\psi^{TEM} = \ln(\rho)\, e^{\mp jk_z z} \tag{B.27}$$

and by similar means as in the previous section, from $\vec{\mathfrak{A}} = \psi\,\vec{e}_z$ we obtain the Eigenmode's fields

$$\begin{aligned} E_\rho^{TEM} &= \mp \tfrac{jk_z}{j\omega\varepsilon}\tfrac{1}{\rho} e^{\mp jk_z z} \\ H_\phi^{TEM} &= -\tfrac{1}{\rho} e^{\mp jk_z z}, \end{aligned} \tag{B.28}$$

which is obviously a TEM Eigenmode field.

B.4. Circular Waveguides

B.4.1. The Mode-Set of Rotationally Symmetric $TM^z_{m,0}$ Eigenmodes

Similarly to the Eigenmodes of the coaxial waveguide derived in the previous section, the Eigenmodes of a *circular waveguide* are derived from a scalar vector potential

$$\psi(\rho, \phi, z) = \psi(\rho)\, \psi(\phi)\, \psi(z)\,. \tag{B.29}$$

As discussed in section 2.2.2, $\psi(\rho)$ is governed by Bessel's differential equation while $\psi(\phi), \psi(z)$ follow from ODEs of similar form as the differential equation of the harmonic oscillator.

We thus choose the ansatz [48]

$$\psi^{TM} = J_n(k_\rho \rho) \begin{Bmatrix} \sin(n\phi) \\ \cos(n\phi) \end{Bmatrix} e^{\mp j k_z z} \tag{B.30}$$

where in contrast to coaxial waveguides only Bessel functions of first kind are included. This is because the point $\rho = 0$ is included in the Eigenmodes' domain of definition. Since Bessel functions of second kind are singular for $\rho \to 0$ [48], they do not represent physically meaningful solutions to the waveguide's boundary value problem [48]. Bessel functions of first and second kind are depicted in figure B.12.

In chapter 13.2 and 14 as well as in section B.3.1 we have seen that in a tubular filter design only rotationally symmetric TM Eigenmodes have to be considered. We thus choose $n = 0$ and (B.30) simplifies to

$$\psi^{TM}_m = \psi^{TM}_{t|m}\, e^{\mp j k_{z,m} z} \quad \text{where} \quad \psi^{TM}_{t|m} = J_0(k_{\rho|m} \rho) \tag{B.31}$$

where an additional subscript m was introduced in order to index the individual Eigenmode's scalar vector potentials.

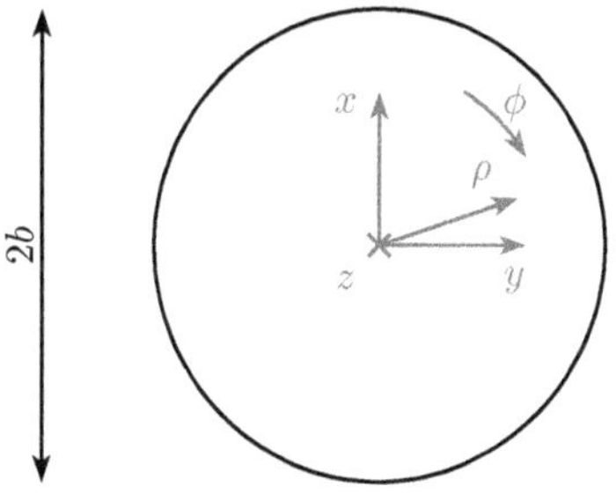

Figure B.11.: Circular waveguide

The Dirichlet boundary condition (cf. (2.25))

$$\psi_t^{TM}\Big|_{\rho=b} = 0 \tag{B.32}$$

applies for TM Eigenmodes and thus we find the Eigenmode's Eigenvalue $k_c^2 = k_{\rho,m}^2$ from

$$k_{\rho,m} b = \mathfrak{z}_m \leftrightarrow k_{\rho,m} = \frac{\mathfrak{z}_m}{b} \tag{B.33}$$

where $\mathfrak{z}_m$ denotes the mth zero of the Bessel function J_0 of first kind [48]. The first few zeros of this Bessel function are given in table B.3.

Finally, the Eigenmodes' fields can be calculated from $\vec{\mathfrak{A}} = \psi\, \vec{e}_z$ using the expressions given in section 2.1.2 as

$$\begin{aligned}
E_{\rho|m}^{TM} &= \mp \frac{jk_{\rho|m}k_{z|m}}{j\omega\varepsilon} J_0'(k_{\rho|m}\rho)e^{\mp jk_{z|m}z} \\
H_{\phi|m}^{TM} &= - k_{\rho|m} J_0'(k_{\rho|m}\rho)e^{\mp jk_{z|m}z} \\
E_{z|m}^{TM} &= \frac{k^2-k_{z|m}^2}{j\omega\varepsilon} J_0(k_{\rho|m}\rho)e^{\mp jk_{z|m}z}.
\end{aligned} \tag{B.34}$$

As usual, the wavenumber k_z follows from the separation condition (2.42).

J_n	$\mathfrak{z}_1$	$\mathfrak{z}_2$	$\mathfrak{z}_3$	$\mathfrak{z}_4$
$n=0$	2.405	5.520	8.654	11.792

Table B.3.: First four zeros of the Bessel function J_0 of first kind [48]

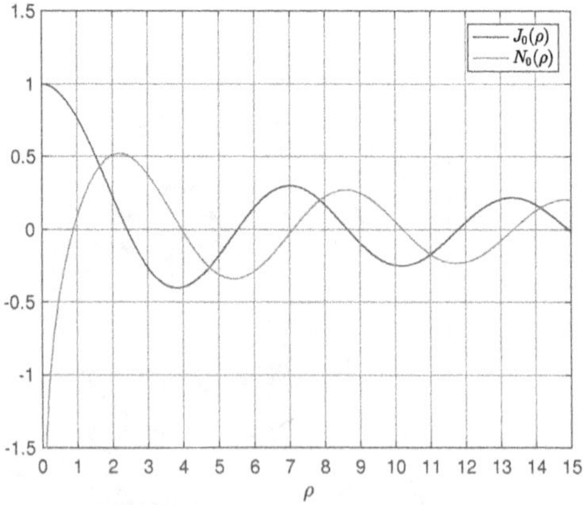

Figure B.12.: Bessel functions of first and second kind, $n = 0$

C. Voltage and Current in TEM and Non-TEM Waveguides

In order to calculate scattering parameters from the electromagnetic solution provided by the Mode Matching technique, a sound understanding of voltage and current on the waveguide as well as of the waveguide's characteristic impedance is crucial.

It is well known that for two-conductor TEM waveguides voltage and current are clearly defined. An exemplary field distribution of an Eigenmode of such a waveguide is shown in figure C.1.

The voltage on the waveguide may be calculated by solving the line integral

$$U = \int_{L_E} \vec{E}\, d\vec{l} \tag{C.1}$$

for any arbitrary *integration line* L_E between the positive and negative conductor. Because of "the electrostatic nature of the transverse electric field" [6] the value of (C.1) does not depend on the actual integration path, that is, the electric field is a conservative field.

Following a similar reasoning for the magnetic field, the current on the waveguide is calculated by solving the line integral

$$I = \oint_{L_H} \vec{H}\, d\vec{l} \tag{C.2}$$

for any arbitrary loop (integration line) L_H around a single conductor.

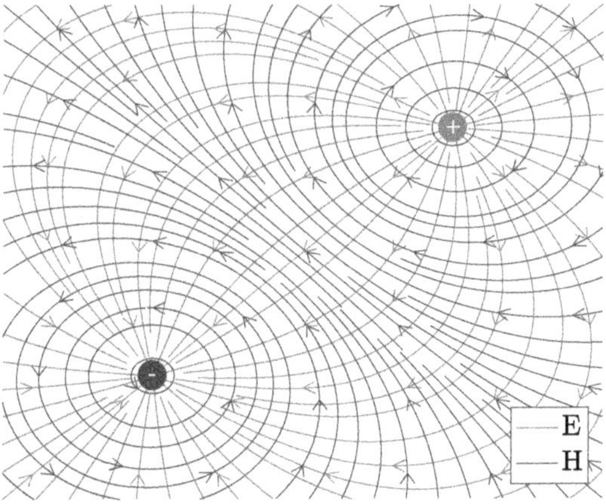

Figure C.1.: Electric and magnetic field of a two-conductor TEM waveguide

For a coaxial waveguide operating in its fundamental TEM Eigenmode, the above definitions obviously apply. We will briefly review this matter in the following section.

In contrast, when dealing with single-conductor non-TEM waveguides such as e.g. rectangular waveguides, the situation gets more complex because voltage and current are no longer uniquely defined. Of course, the same problem applies for the waveguide's characteristic impedance. In fact, the definition of voltage and current may be chosen arbitrarily [6]. However, in order to obtain most meaningful quantities, the following points should be considered when defining voltage and current [6]:

- Voltage and current should be proportional to the applicable electric and magnetic field component perpendicular to the direction of propagation.
- $\mathcal{P} = \iint_{\mathfrak{S}} \vec{E} \times \vec{H}^* \, d\vec{A}$ and $\mathcal{P} = U^+ \cdot I^{+^*}$ should be consistent.

In section C.2, we will introduce a possible definition for voltage and current in a rectangular waveguide. In the scope of this thesis, this definition is applied for all scattering parameter calculations.

C.1. Coaxial Waveguide

Consider the coaxial waveguide shown in figure C.2. The fields of the waveguide's fundamental TEM Eigenmode propagating in positive z direction[1] can be calculated as (see section B.3.2)

$$\begin{aligned} E_\rho^{TEM} &= -\frac{k_z}{\omega\varepsilon}\frac{1}{\rho}\, A^{TEM}\, e^{-jk_z z} \\ H_\phi^{TEM} &= -\frac{1}{\rho}\, A^{TEM}\, e^{-jk_z z}. \end{aligned} \tag{C.3}$$

The field distributions are depicted in figure C.2 on the right.

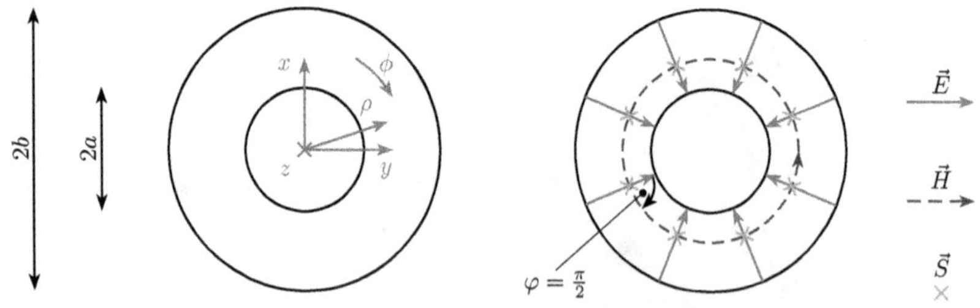

Figure C.2.: Coaxial waveguide

[1]Here we will limit our analysis to Eigenmodes travelling in positive z direction. For Eigenmodes travelling in opposite direction, similar results may be obtained.

We now define a *wave impedance* as the ratio

$$Z_{wave} = \frac{E_\rho}{H_\phi} = \frac{k_z}{\omega\varepsilon} \tag{C.4}$$

of the electric and magnetic field components E_ρ and H_ϕ, which are perpendicular to the direction of propagation.

Noting that the TEM Eigenmode is a special case of TM Eigenmodes for which $k_\rho = 0$ applies, from the separation condition $k^2 = k_\rho^2 + k_z^2$ we find $k = k_z$. Thus, the wave impedance may be rewritten as

$$Z_{wave} = \frac{k_z}{\omega\epsilon} = \frac{k}{\omega\epsilon} = \frac{\omega\sqrt{\varepsilon\mu}}{\omega\varepsilon} = \sqrt{\frac{\mu}{\varepsilon}}. \tag{C.5}$$

As a consequence of (C.5), in an air-filled coaxial waveguide, the wave impedance, i.e. the ratio of the field components E_ρ and H_ϕ is 377 Ω, which is the free-space impedance.

C.1.1. Power Transport

We now calculate the power transported in a coaxial waveguide from the Eigenmode's field components in order to compare this result with the transported power calculated from voltage and current waves later. The power transported in the waveguide may be obtained by solving the integral

$$\mathcal{P} = \iint_{\mathfrak{S}} \vec{S}\, d\vec{A} = \iint_{\mathfrak{S}} E_\rho H_\phi^*\, dA \tag{C.6}$$

over the Poynting vector

$$\vec{S} = \vec{E} \times \vec{H}^* = |\vec{E}| \cdot |\vec{H}^*| \cdot \sin(\varphi = \pi/2) \cdot \vec{e}_n = E_\rho \cdot H_\phi^* \cdot \vec{e}_z, \tag{C.7}$$

which is always real valued because the TEM Eigenmode has no cut-off frequency. $\mathfrak{S}$ denotes the waveguide's cross-section.

Inserting (C.3) into (C.6) gives

$$\begin{aligned}
\mathcal{P} &= \iint_{\mathfrak{S}} \frac{k_z}{\omega\varepsilon}\, A\, A^*\, \frac{1}{\rho^2}\, dA \\
&= \frac{k_z}{\omega\varepsilon}|A|^2 \int_0^{2\pi} d\phi \int_a^b \frac{1}{\rho^2}\, \rho\, d\rho \\
&= \frac{2\pi k_z}{\omega\varepsilon}|A|^2 \int_a^b \frac{1}{\rho}\, d\rho \\
&= \frac{2\pi\omega\sqrt{\varepsilon\mu}}{\omega\varepsilon}|A|^2 \left[\ln(\rho)\right]_a^b \\
&= 2\pi\sqrt{\mu/\varepsilon}\, |A|^2 \left(\ln(b) - \ln(a)\right) \\
&= 2\pi\sqrt{\mu/\varepsilon}\, |A|^2 \ln(b/a).
\end{aligned} \tag{C.8}$$

C.1.2. Voltage and Current

As already discussed previously, the voltage on a TEM waveguide is calculated by integrating the electric field along an integration line from the inner to the outer conductor, giving a voltage wave

$$\begin{aligned} U^{+} &= \int_{L_E} \vec{E}\, d\vec{l} \\ &= - \frac{k_z}{\omega\varepsilon} A\, e^{-jk_z z} \int_a^b \frac{1}{\rho}\, d\rho = - \frac{k_z}{\omega\varepsilon} A\, e^{-jk_z z} \left[\ln(\rho)\right]_a^b \\ &= - \frac{k_z}{\omega\varepsilon} A \ln\left({}^{b}/_{a}\right) e^{-jk_z z}. \end{aligned} \tag{C.9}$$

Similarly, the current wave is calculated by integrating the magnetic field along an integration line around the inner conductor, which gives

$$\begin{aligned} I^{+} &= \oint_{L_H} \vec{H}\, d\vec{l} \\ &= - A\, e^{-jk_z z} \int_0^{2\pi} \frac{1}{\rho}\, \rho\, d\phi \\ &= - 2\pi A\, e^{-jk_z z}. \end{aligned} \tag{C.10}$$

The *characteristic impedance* is defined as the ratio of voltage and current wave, which is

$$Z_0 = \frac{U^{+}}{I^{+}} = \frac{-\frac{k_z}{\omega\varepsilon} A \ln\left({}^{b}/_{a}\right)}{-2\pi A} = \frac{k}{2\pi\omega\varepsilon} \ln\left({}^{b}/_{a}\right) = \frac{\sqrt{{}^{\mu}/_{\varepsilon}}}{2\pi} \ln\left({}^{b}/_{a}\right). \tag{C.11}$$

The characteristic impedance is of high importance for calculating scattering parameters.

Finally, using (C.9) and (C.10), we validate that

$$\mathcal{P} = U^{+} \cdot I^{+^{*}} = \frac{2\pi k_z}{\omega\varepsilon} A\, A^{*} \ln\left({}^{b}/_{a}\right) = 2\pi \sqrt{{}^{\mu}/_{\varepsilon}}\, |A|^2 \ln({}^{b}/_{a}) \tag{C.12}$$

indeed provides the very same result as (C.8).

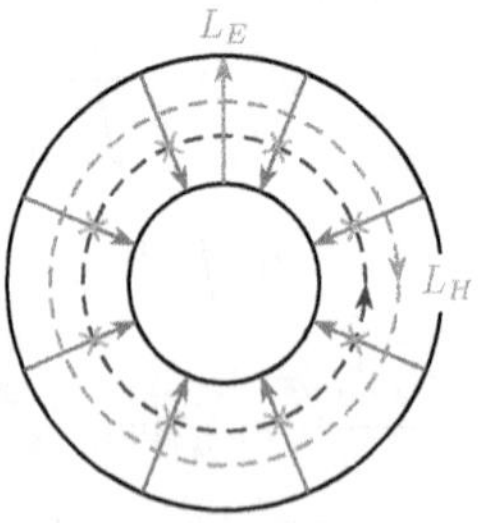

Figure C.3.: Integration lines in a coaxial waveguide

C.2. Rectangular Waveguide

Consider the rectangular waveguide shown in figure C.4. As we have seen in section B.1, the field components of the waveguide's TE_{10} Eigenmode propagating in positive z direction[2] denote

$$\begin{aligned} E^{TE}_{y|1,0} &= + \; j\, k_{z|1,0}\, A^{TE}_{1,0} \sin\left(k_{x|1} x\right) e^{-jk_{z|1,0} z} \\ H^{TE}_{x|1,0} &= - \; j \frac{k^2_{z|1,0}}{\omega\mu}\, A^{TE}_{1,0} \sin\left(k_{x|1} x\right) e^{-jk_{z|1,0} z}. \end{aligned} \tag{C.13}$$

For brevity's sake, the indices of the electric and magnetic field components shall be omitted, that is, $E_y = E^{TE}_{y|1,0}$, $H_x = H^{TE}_{x|1,0}$ and so forth.

Let us firstly define the *wave impedance* as the ratio

$$Z_{wave} = -\frac{E_y}{H_x} \tag{C.14}$$

of the electric and magnetic field components perpendicular to the direction of propagation.

Using (C.13), the wave impedance calculates

$$Z_{wave} = -\frac{E_y}{H_x} = \frac{jk_z}{jk_z^2/\omega\mu} = \frac{\omega\mu}{k_z}. \tag{C.15}$$

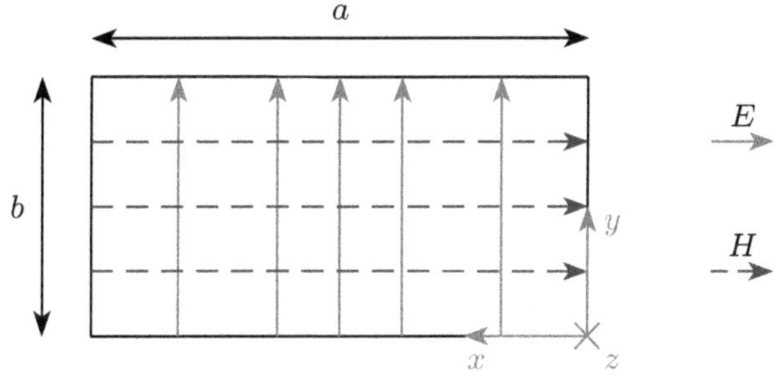

Figure C.4.: Rectangular waveguide

[2]Here we will limit our analysis to Eigenmodes travelling in positive z direction. For Eigenmodes travelling in opposite direction, similar results may be obtained.

C.2.1. Power Transport

Before discussing the definitions of voltage and current in a rectangular waveguide, it is quite instructive to review power transport. The complex power transported by the waveguide's fundamental Eigenmode in positive z direction can be calculated using the surface integral

$$\mathcal{P} = \iint_{\mathfrak{S}} \vec{S}\, d\vec{A} = -\iint_{\mathfrak{S}} E_y H_x^*\, dA \tag{C.16}$$

over the complex Poynting vector

$$\vec{S} = \vec{E} \times \vec{H}^* = |\vec{E}| \cdot |\vec{H}^*| \cdot \sin(\varphi = -\pi/2) \cdot \vec{e}_n = -E_y \cdot H_x^* \cdot \vec{e}_z \tag{C.17}$$

as is illustrated in figure C.5. $\mathfrak{S}$ denotes the waveguide's cross-section.

It is important to carefully interpret this complex quantity: If the Eigenmode is propagating, that is, $k_z \in \mathbb{R} > 0$, $\mathcal{P}$ is real-valued and thus represents a time-averaged power flow in positive z direction [48]. If the Eigenmode is below cut-off, we have $k_z \in \mathbb{I} < 0$. No power is transported and $\mathcal{P}$ expresses reactive power [48].

The power transported by the TE_{10} Eigenmode is calculated[3] from (C.16) using (C.13)

$$\begin{aligned}
\mathcal{P} &= -\iint_{\mathfrak{S}} \underbrace{j \cdot (-j)^*}_{-1} \frac{k_z \left(k_z^2\right)^*}{\omega\mu} A\, A^* \sin\left(k_x x\right)\, dA \\
&= \frac{k_z^3}{\omega\mu} |A|^2 \iint_{\mathfrak{S}} \sin\left(k_x x\right)^2 dA = \frac{k_z^3}{\omega\mu} |A|^2 \int_0^a \sin\left(k_x x\right)^2\, dx \int_0^b dy \\
&= \frac{k_z^3}{\omega\mu} \frac{ab}{2} |A|^2.
\end{aligned} \tag{C.18}$$

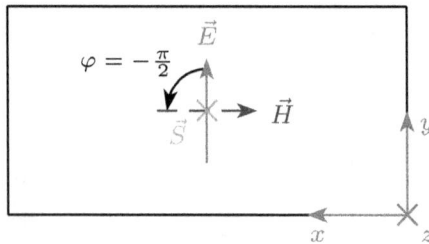

Figure C.5.: Field components and Poynting vector in a rectangular waveguide. Wave propagation in positive z direction.

[3] For $k_z \in \mathbb{R} > 0$, this is an obvious result. For $k_z \in \mathbb{I} < 0$, this result applies similarly because $k_z \cdot \left(k_z^2\right)^* = |k_z| \cdot (-j) \cdot \left(|k_z|^2 (-j)^2\right)^* = |k_z|^3 \cdot (-j) \cdot (-1) = |k_z|^3 \cdot j = |k_z|^3 \cdot (-j)^3 = k_z^3$

It is interesting to note that (C.16) and (C.17) can be rewritten as

$$\mathcal{P} = -\iint_{\mathfrak{S}} E_y H_x^* \, dA = -\iint_{\mathfrak{S}} E_y \left(-\frac{E_y}{Z_{wave}}\right)^* dA = \iint_{\mathfrak{S}} \frac{|E_y|^2}{Z_{wave}^*} \, dA, \tag{C.19}$$

which provides the same result as (C.18)[4]

$$\begin{aligned} \mathcal{P} &= \frac{1}{Z_{wave}^*} \iint_{\mathfrak{S}} |j\, k_z\, A\, \sin(k_x x)\,|^2 \, dA \\ &= \frac{|k_z|^2}{Z_{wave}^*} |A|^2 \iint_{\mathfrak{S}} \sin(k_x x)^2 \, dA \\ &= \frac{k_z^* |k_z|^2}{\omega\mu} |A|^2 \iint_{\mathfrak{S}} \sin(k_x x)^2 \, dA \\ &= \frac{k_z^3}{\omega\mu} \frac{ab}{2} |A|^2. \end{aligned} \tag{C.20}$$

C.2.2. Voltage and Current

Following the reasoning at the beginning of this chapter, we choose to define the voltage as line integral over the electric field along the *integration line* L_E shown in figure C.6, giving a voltage wave

$$\begin{aligned} U^+ &= \int_{L_E} \vec{E}\, d\vec{l} = \int_0^b E_y \, dy \\ &= j\, k_z\, A\, \underbrace{\sin\left(\frac{\pi}{a}\frac{a}{2}\right)}_{1} e^{-jk_z z} \int_0^b dy \\ &= j\, k_z\, A\, b\, e^{-jk_z z}. \end{aligned} \tag{C.21}$$

Thus, we have $U^+ \propto E_y$.

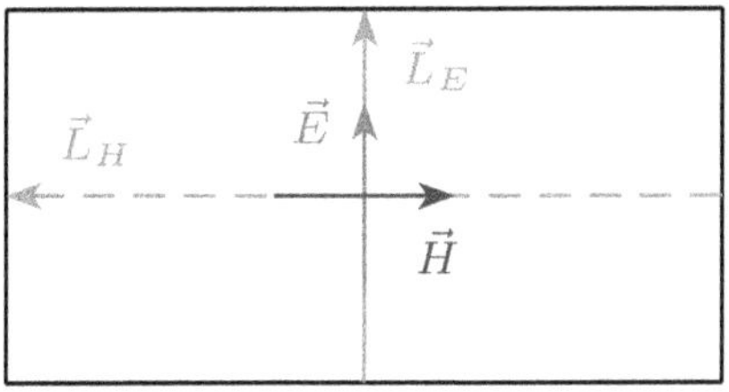

Figure C.6.: Integration lines in a rectangular waveguide

[4] For $k_z \in \mathbb{R} > 0$, this is an obvious result. For $k_z \in \mathbb{I} < 0$, this result applies similarly because $k_z^* |k_z|^2 = |k_z| \cdot j \cdot ||k_z|\,(-j)\,|^2 = |k_z|^3 \cdot j \cdot |\,(-j)\,|^2 = |k_z|^3 \cdot j = |k_z|^3 \cdot (-j)^3 = k_z^3$.

Because the electric and magnetic field distributions are of opposite sign (cf. (C.13)), it is necessary to define I in a way so that $I \propto -H_x$ in order to obtain a positive characteristic impedance. Moreover, because consistency between $\mathcal{P} = \iint_A \vec{E} \times \vec{H}^* \, d\vec{A}$ and $\mathcal{P} = U^+ \cdot I^{+^*}$ shall be ensured, an additional factor $\pi/4$ is required.

Consequently, the current is calculated as line integral over the magnetic field along the integration line L_H shown in figure C.6 with an additional pre-factor[5] $-\pi/4$, giving a current wave

$$\begin{aligned}
I^+ &= -\frac{\pi}{4} \oint_{L_H} \vec{H} \, d\vec{l} = -\frac{\pi}{4} \int_0^a H_x \, dx \\
&= j \cdot \frac{k_z^2}{\omega\mu} \, A \, \frac{\pi}{4} \, e^{-jk_z z} \int_0^a \sin(k_x x) \, dx \\
&= j \cdot \frac{k_z^2}{\omega\mu} \, A \, \frac{\pi a}{4\pi} \, e^{-jk_z z} \, [-\cos(k_x x)]_0^a \\
&= j \cdot \frac{k_z^2}{\omega\mu} \, A \, \frac{a}{4} \, 2 \, e^{-jk_z z} \\
&= j \cdot \frac{k_z^2}{\omega\mu} \, A \, \frac{a}{2} \, e^{-jk_z z}.
\end{aligned} \tag{C.22}$$

Consistency between $\mathcal{P} = \iint_A \vec{S} \, d\vec{A}$ and $\mathcal{P} = U^+ \cdot I^{+^*}$ may be validated by calculating[6]

$$\mathcal{P} = U^+ \cdot I^{+^*} = \underbrace{j \, j^*}_{1} \, \frac{k_z \, (k_z^2)^*}{\omega\mu} \, \frac{ab}{2} \, A \, A^* = \frac{k_z^3}{\omega\mu} \, \frac{ab}{2} \, |A|^2, \tag{C.23}$$

which is in compliance with (C.18).

The waveguide's *characteristic impedance* is calculated as the ratio of a voltage and current wave, that is,

$$Z_0 = \frac{U^+}{I^+} = \frac{jk_z \, Ab}{j\frac{k_z^2}{\omega\mu} \, \frac{a}{2} A} = \frac{2b}{a} \frac{\omega\mu}{k_z} = \frac{2b}{a} Z_{wave}, \tag{C.24}$$

which is an important result for scattering parameter calculation. Note that for a waveguide with $a = 2b$ we have $Z_0 = Z_{wave}$.

[5] Ansys HFSS does not include this prefactor. Thus, the characteristic impedance calculated from U and I in Ansys HFSS differs from the value calculated analytically by a factor $4/\pi$. Also see section B.2.2.

[6] also see footnote 3.

D. Analytical Model for an Inner Conductor Step in a Coaxial Waveguide

In this section, the analytical model of a step of the inner conductor of a coaxial waveguide, which is used in section 10.2 is reviewed. The structure is shown in figure D.1.

This model was originally published in Marcuvitz' Waveguide Handbook [17, sec. 5-27, pp. 310ff]. The model reuses results obtained in [17, sec. 5.3, pp. 229ff] where a disk on the inner conductor is treated. In the following, all required formulas are reprinted. Note that in the scope of this appendix, some symbols are used in a different context than in the rest of this thesis.

To facilitate debugging, reference values for all (auxiliary) variables are given in tables D.2 and D.3 for an exemplary structure, whose geometry is given in table D.1.

Let us begin our analysis of the discontinuity by defining the auxiliary variables

$$\begin{aligned}
\alpha &= \tfrac{b-a_1}{b-a_2} \\
\delta &= 1-\alpha \\
b_0 &= b-a_2 \\
d &= b-a_1,
\end{aligned} \tag{D.1}$$

for which reference values can be found in table D.2.

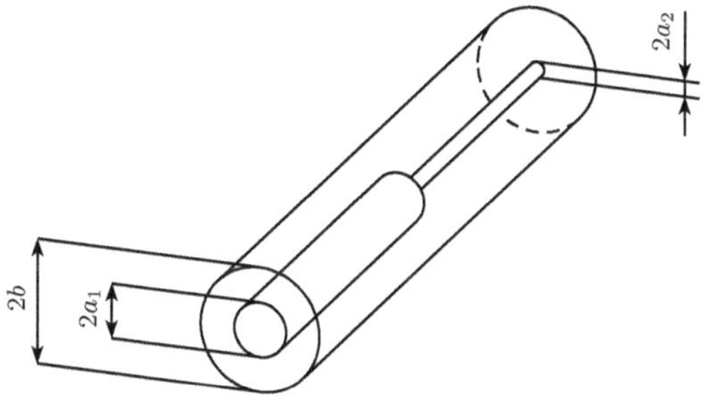

Figure D.1.: Inner conductor step in a coaxial waveguide

We then find the first root χ_1 of the characteristic equation [17, sec. 5.3, p. 230]

$$0 = J_0(\chi)\, N_0(\chi b/a_2) - N_0(\chi)\, J_0(\chi b/a_2), \tag{D.2}$$

which is depicted in figure D.2 for $0 \ll \chi \ll 2$. From χ_1, γ_1 can be calculated as [17, ibid.]

$$\gamma_1 = \chi_1 \frac{b/a_2 - 1}{\pi}. \tag{D.3}$$

Next, two additional auxiliary variables are calculated, namely

$$\begin{aligned}
A_1 &= \frac{a_1}{a_2} \frac{\ln(b/a_2)}{b/a_2 - 1} \left(\frac{b/a_1 - 1}{\ln(b/a_1)} \right)^2 \\
A_2 &= \frac{\pi^2 a_2/a_1}{\gamma_1 \sqrt{1-(2b_0/\gamma_1 \lambda)^2}} \frac{b/a_2 - 1}{\frac{J_0^2(\chi_1)}{J_0^2(\chi b/a_2)} - 1} \left[\frac{J_0(\chi_1)\, N_0(\chi_1 a_1/a_2) - N_0(\chi_1)\, J_0(\chi_1 a_1/a_2)}{b/a_1 - 1} \right]^2 \\
&- \frac{1}{\sqrt{1-\left(\frac{2b_0}{\lambda}\right)^2}} \left(\frac{2}{\pi} \frac{b_0}{d} \sin(\pi d/b_0) \right)^2,
\end{aligned} \tag{D.4}$$

which were introduced in [17, sec. 5.3, p. 229ff] for the treatment of a disk on the inner conductor of a coaxial waveguide.

Reference values for A_1 and A_2 are given in table D.2 for the structure given in table D.1. Also see graphs in [17, p. 231ff].

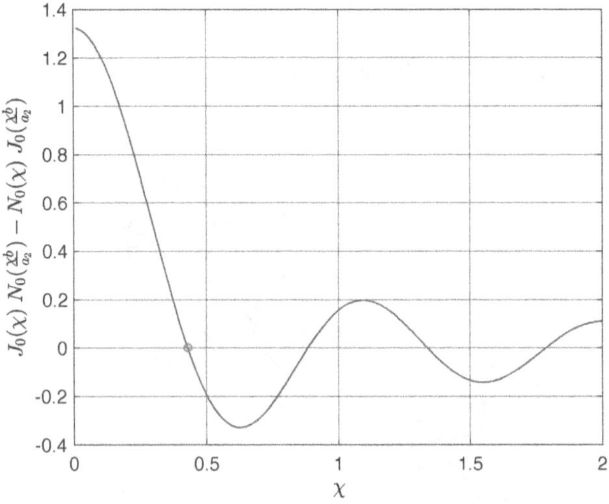

Figure D.2.: Characteristic equation (D.2) plotted for the values given in table D.1.

Finally, we calculate the normalised susceptance B/Y_0 of the step of the inner conductor in a coaxial waveguide using an approximation provided in [17, sec. 5.27, p. 310f]

$$\frac{B}{Y_0} = \frac{2b_0 A_1}{\lambda}\left[2\ln\left(\frac{e}{4\alpha}\right) + \frac{2\alpha^2}{3} + 4\left(\frac{b_0}{\lambda}\right)^2\left(1-\alpha^2\right)^4 + \frac{A_2}{2}\right] \tag{D.5}$$

for [17], (2a) which is valid for $\alpha \ll 1$, that is,

$$\frac{b-a_1}{b-a_2} \ll 1 \iff a_1 \gg a_2. \tag{D.6}$$

Note that the wave admittance

$$Y_0^{-1} = \frac{1}{2\pi}\sqrt{\frac{\mu}{\varepsilon}}\ln\left(b/a_2\right) \tag{D.7}$$

of the coaxial waveguide with the smaller inner conductor was used for normalisation.

a_1	a_2	b	f	λ
3 mm	1 mm	8 mm	3 GHz	0.1m

Table D.1.: Parameters of the structure shown in figure D.1 used to generate reference data

α	δ	b_0	d	χ_1	γ_1	A_1	A_2
0.7143	0.2857	0.0070 m	0.0050 m	0.4300	0.9581	2.5732	0.0985

Table D.2.: Auxiliary variables - reference data for the structure shown in figure D.1 with parameters given in table D.1.

B/Y_0	Y_0	C
0.1048	0.0080 Ω^{-1}	44.5653 fF

Table D.3.: Results - reference data for the structure shown in figure D.1 with parameters given in table D.1.

E. LU Decomposition

In this excursus, LU decomposition is briefly reviewed. Furthermore, we will investigate how the banded structure of the coefficient matrix of an inhomogeneous system of equations (such as the reduced system matrix' $\boldsymbol{\Psi}_{Unknown}$) can be exploited in order to reduce the computational effort to solve the system of equations.

The well-known Gaussian elimination process and LU decomposition are in fact similar processes [157]. Nevertheless, LU decomposition is ideally suited to demonstrate computational benefits of exploiting the banded structure.

Note that the nomenclature in this excursus is mostly independent of the rest of this thesis.

E.1. LU Decomposition of Dense Matrices

Consider the inhomogeneous system of equations

$$\boldsymbol{A} \cdot \vec{x} = \vec{b} \tag{E.1}$$

where $\boldsymbol{A} \in \mathbb{R}^{\mathcal{Q} \times \mathcal{Q}}$ denotes a regular matrix and $\vec{b} \in \mathbb{R}^{\mathcal{Q}}$.

It can be shown that $\boldsymbol{A}$ may be decomposed into a Lower and Upper triangular matrix for which

$$\boldsymbol{A} = \boldsymbol{L} \cdot \boldsymbol{U} \tag{E.2}$$

applies.

If the LU decomposition is known, the system

$$\boldsymbol{A} \cdot \vec{x} = \boldsymbol{L} \cdot \underbrace{\boldsymbol{U} \cdot \vec{x}}_{\vec{y}} = \vec{b} \tag{E.3}$$

may be solved by firstly solving

$$\boldsymbol{L} \cdot \vec{y} = \vec{b} \tag{E.4}$$

and secondly

$$\boldsymbol{U} \cdot \vec{x} = \vec{y}. \tag{E.5}$$

Because both $\boldsymbol{L}$ and $\boldsymbol{U}$ are triangular matrices, (E.4) and (E.5) may easily be solved using forward substitution (E.4) and backward substation (E.5) [86]. Both forward and backward substitution require[1] $\mathcal{Q}^2$ flops[2], that is, they are $\mathcal{O}(\mathcal{Q}^2)$ processes[3] [86, 88].

[1] Note that in order to maintain consistency with the cited textbooks, here we assume real valued quantities as defined at the beginning of this section. However, in Mode Matching, complex quantities are quite common. While real-valued additions and multiplications both require one flop, complex addition requires two flops as can be seen from $(a + ib) + (c + id) = (a + c) + i(b + d)$ and complex multiplication requires 6 flops as is obvious from $(a + ib) \cdot (c + id) = ac + i(ad) + i(bc) - bd$.

[2] Floating Point Operations

[3] The $\mathcal{O}(\mathcal{Q}^x)$ notation implies that the process' operation count is a function of terms of order $\mathcal{Q}^x$ and lower [87].

At this point, a significant advantage of the LU decomposition over "standard" Gaussian elimination becomes obvious: While Gaussian elimination needs to be redone every time the vector $\vec{b}$, that is, the excitation, was changed, LU decomposition only needs to be performed once [88]. The solution for various $\vec{b}$ then can easily be obtained from (E.4) and (E.5).

The obvious question now is how to perform LU decomposition of $\boldsymbol{A}$. Let us firstly assume that we want to put $\boldsymbol{A}$ under upper triagonal form in order to obtain $\boldsymbol{U}$.

To do so, we perform Gaussian elimination on $\boldsymbol{A}$, which can be expressed as matrix multiplication

$$\boldsymbol{L}_{k=Q-1} \dots \boldsymbol{L}_{k=1} \boldsymbol{A} = \boldsymbol{U} \tag{E.6}$$

where L_i denotes so called Frobenius matrices

$$\boldsymbol{L}_k = \begin{bmatrix} 1 & & & & 0 \\ & \ddots & & & \\ & & 1 & & \\ & & -l_{k+1,k} & & \\ & & \vdots & \ddots & \\ 0 & & -l_{Q,k} & & 1 \end{bmatrix} \qquad \boldsymbol{L}_k^{-1} = \begin{bmatrix} 1 & & & & 0 \\ & \ddots & & & \\ & & 1 & & \\ & & l_{k+1,k} & & \\ & & \vdots & \ddots & \\ 0 & & l_{Q,k} & & 1 \end{bmatrix} \tag{E.7}$$

where $l(i,k) = \boldsymbol{A}(i,k)/\boldsymbol{A}(k,k)$ for $k+1 \leq i \leq Q$ [88].

By multiplying a matrix, e.g. $\boldsymbol{A}$ with a matrix under form of $\boldsymbol{L}_k$, all elements below the diagonal element $\boldsymbol{A}(k,k)$ are zeroed. It is thus obvious that calculating (E.6) indeed performs a diagonalisation of $\boldsymbol{A}$ by means of Gaussian elimination.

Let us now rewrite (E.6) as

$$\boldsymbol{A} = \boldsymbol{L}_{k=1}^{-1} \cdots \boldsymbol{L}_{k=Q-1}^{-1} \boldsymbol{U}. \tag{E.8}$$

Interestingly enough, $\boldsymbol{L}_{k=1}^{-1} \cdots \boldsymbol{L}_{k=Q-1}^{-1}$ is indeed a lower triagonal matrix where all diagonal elements are unity[4] [88] so that we have our LU decomposition

$$\boldsymbol{A} = \boldsymbol{L} \cdot \boldsymbol{U}$$

where

$$\begin{aligned} \boldsymbol{L} &= \boldsymbol{L}_{k=1}^{-1} \cdots \boldsymbol{L}_{k=Q-1}^{-1} \\ \boldsymbol{U} &= \boldsymbol{L}_{k=Q-1} \cdots \boldsymbol{L}_{k=1} \boldsymbol{A} \end{aligned} \tag{E.9}$$

readily in place.

A conceptual algorithm to perform LU decomposition (which is not optimised in terms of computational effort and memory requirements) is shown in Algorithm E.1.

[4]The structure of the matrix can be proved using induction as outlined in [88].

Algorithm E.1: Conceptual LU decomposition algorithm - modified from [88]

```
% Sample Matrix, for which LU decomposition
% shall be performed
A=[  2 1 1
     4 3 3
     8 7 9];

% Copy matrix
A_k=A;

% Determine n
Q=size(A,1);

% Predefine B which will contain the result of the multiplication
% of the inverse Frobenius matrices
B_k=eye(Q);

% Perform diagonalisation on all columns
for k=1:Q-1

    % --- Define Frobenius Matrix L_k ---
    % Insert unity elements on the matrix' diagonal
    L_k=eye(Q);
    % Insert non diagonal elements in the kth column
    for i=k+1:Q

        % Calculate element to achieve Gaussian elimination
        L_k(i,k)=-A_k(i,k)/A_k(k,k);

    end
    % --- End Define L_k ---

    % Remove zeros below the A(k,k) element by
    % matrix multiplication
    A_k_1=L_k*A_k;

    % Partial calculation of L=L^-1_1 ... L^-1_N-1
    B_k_1=B_k*L_k^-1;

    % Prepare next iteration
    A_k=A_k_1;
    B_k=B_k_1;

end

% Result for L
L=B_k_1;

% Result for U
U=A_k_1;
```

Before we turn to calculating and optimising the computational effort necessary to perform LU decomposition, let us observe the way algorithm E.1 operates on a simple sample matrix

$$\boldsymbol{A} = \begin{bmatrix} 2 & 1 & 1 \\ 4 & 3 & 3 \\ 8 & 7 & 9 \end{bmatrix} \tag{E.10}$$

The first iteration ($k = 1$) of the algorithm's outer loop (line 18 - 43) is depicted in figure E.1. The diagonal element $\boldsymbol{A}(k, k)$ to be processed is indicated by a green circle.

Regarding the $\boldsymbol{U}$ part of the decomposition, as expected, the matrix multiplication $\boldsymbol{L}_{k=1} \cdot \boldsymbol{A}_{k=1}$ (line 34) zeros all elements below the diagonal element $\boldsymbol{A}(k, k)$ as indicated by the blue box. In addition, it is remarkable that only elements in the red box need to be recalculated while all other elements remain unchanged or must be zero as dictated by Gaussian elimination.

With respect to determining the $\boldsymbol{L}$ part of the decomposition (line 37), it should be noted that non-zero elements only appear below the matrix' diagonal as expected.

The second iteration is shown in figure E.2 and the very same observations regarding the algorithm's matrix manipulations can be made.

We can thus denote the following observations on the algorithm's operation:

- Only matrix elements $\boldsymbol{A}_{k+1}(i, j)$ where $k + 1 \leq i \leq Q$ and $k + 1 \leq j \leq Q$ need to be recalculated in the kth iteration.
- All elements $\boldsymbol{A}_{k+1}(i, k)$ where $k + 1 \leq i \leq Q$ become zero in the kth iteration. The very same elements $\boldsymbol{B}_{k+1}(i, k)$ become populated in the same iteration.
- All other elements remain unchanged.

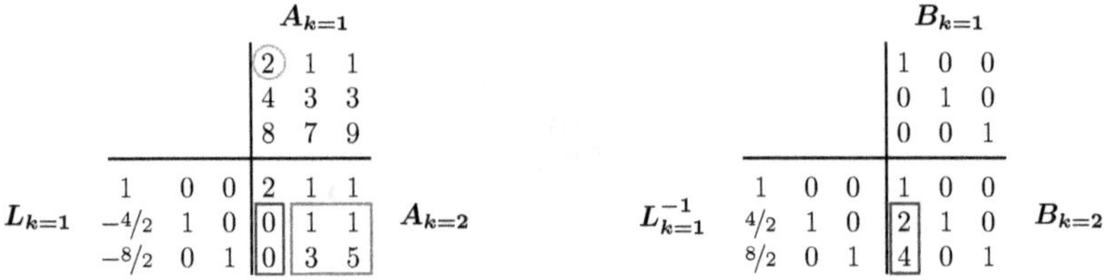

Figure E.1.: First iteration ($k = 1$) of the LU decomposition algorithm E.1

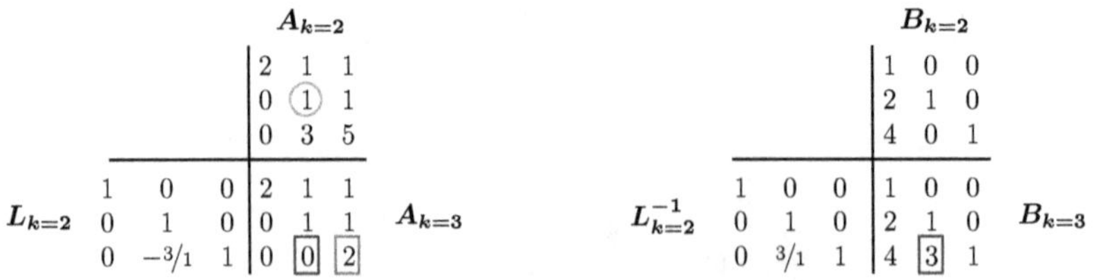

Figure E.2.: Second, final iteration ($k = Q - 1 = 2$) of the LU decomposition algorithm E.1

By exploiting the previously discussed observations on the algorithm's operation, we can design the optimised algorithm E.2 [88], which performs LU decomposition with a minimum number of flops.

In addition, in order to reduced memory requirements, the matrices $\boldsymbol{L}$ and $\boldsymbol{U}$ are stored "in-place" [157], that is, in the memory originally allocated for $\boldsymbol{A}$. Note that here the fact that diagonal elements of $\boldsymbol{L}$ are unity by definition is exploited.

An additional benefit of using "in-place" storage is that elements unchanged during one iteration do not need to be transferred between matrices stored in different memory locations.

Algorithm E.2: In-Place LU Decomposition - modified from [88]

```
% Sample Matrix, for which LU decomposition
% shall be performed
A=[  2 1 1
     4 3 3
     8 7 9];

% Determine n
Q=size(A,1);

% Perform diagonalisation on all columns
for k=1:Q-1

    % --- Optimised matrix multiplication ---

    % Loop over rows
    for i=k+1:Q

        % Calculate l(i,k)
        factor=-A(i,k)/A(k,k);

        % Loop over columns
        for j=k+1:Q

            % Perform matrix multiplications element-wise
            A(i,j)=A(i,j)+factor*A(k,j);

        end

        % Store element of L in-place
        A(i,k)=-factor;

    end

    % --- End optimised matrix multiplication ---

end
```

Let us now assess the computational effort required for algorithms as presented e.g. in [88]: In the innermost loop, assuming real numbers, 2 flops are required, i.e. one multiplication and one addition. The total number of flops required by the algorithm thus may be calculated as

$$\sum_{k=1}^{\mathcal{Q}-1}\sum_{i=k+1}^{\mathcal{Q}}\sum_{j=k+1}^{\mathcal{Q}} 2 = 2\sum_{k=1}^{\mathcal{Q}-1}(\mathcal{Q}-k)^2 = 2\sum_{k=1}^{\mathcal{Q}-1} k^2 = 2\frac{(\mathcal{Q}-1)\mathcal{Q}(2\mathcal{Q}-1)}{6} \approx \frac{2}{3}\mathcal{Q}^3, \quad \text{(E.11)}$$

that is, LU decomposition of dense matrices is an $\mathcal{O}(\mathcal{Q}^3)$ process [88].

For the sake of completeness, it should be noted that the presented algorithm is only applicable[5] if $\boldsymbol{A}(k,k) \neq 0$ [88]. If a matrix violates this criterion, pivoting [86,88], i.e. row and column swapping [86] provides means to overcome this issue. Also, pivoting is crucial for the stability of the LU decomposition algorithm [86].

Pivoting is out of the scope of this excursus. However, as discussed in [88], the effort for LU decomposition with pivoting is about similar to that of algorithm E.1, that is, $2/3 \cdot \mathcal{Q}^3$.

E.2. LU Decomposition of Banded Matrices

Let us now analyse in what way the structure of a banded matrix may be exploited in order to minimise the computational effort required to perform LU decomposition.

It can be shown that when decomposing a banded matrix $\boldsymbol{A}$ with upper bandwidth $\mathcal{B}_u$ and lower bandwidth $\mathcal{B}_l$, $\boldsymbol{L}$ is lower triangular with a lower bandwidth $\mathcal{B}_l$ and $\boldsymbol{U}$ is upper triangular with an upper bandwidth $\mathcal{B}_u$ as illustrated in figure E.3 [88].

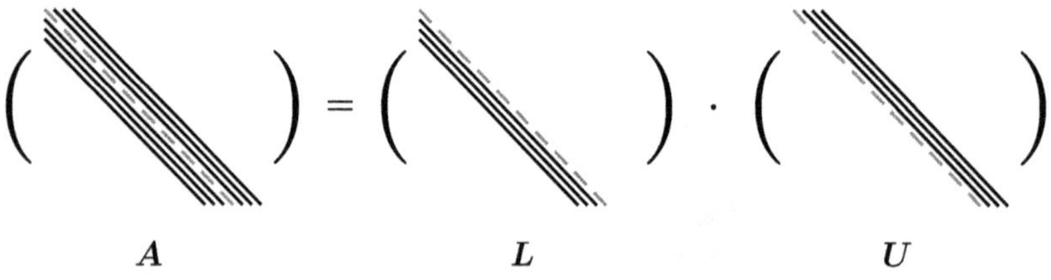

Figure E.3.: LU decomposition of a banded matrix without pivoting. The red line indicates the main diagonal. Reproduced from [88]

[5] Otherwise, in algorithm E.1, line 27 and algorithm E.2, line 19 division by zero would occur.

Assuming that "in-place" LU decomposition shall be performed, algorithm E.2 may be modified as is shown in algorithm E.3 so that only matrix elements within the matrix bandwidth are recalculated. This can easily be achieved by introducing line 15 and 17 and by modifying the two inner loops' upper boundaries in line 29 and 38 [88].

Let us now assess the required computational effort of algorithm E.3: Again in the innermost loop, two flops are required if real numbers are assumed. Thus, the total number of flops may be calculated as [88]

$$\sum_{k=1}^{\mathcal{Q}-1} \sum_{i=k+1}^{\min\{k+\mathcal{B}_l,\mathcal{Q}\}} \sum_{j=k+1}^{\min\{k+\mathcal{B}_u,\mathcal{Q}\}} 2 \approx \sum_{k=1}^{\mathcal{Q}} \sum_{i=k+1}^{k+\mathcal{B}_l} \sum_{j=k+1}^{k+\mathcal{B}_u} 2 = 2\sum_{k=1}^{\mathcal{Q}-1} \mathcal{B}_u\mathcal{B}_l = 2\mathcal{Q}\mathcal{B}_u\mathcal{B}_l. \tag{E.12}$$

Consequently, matrix decomposition of a banded matrix with given upper and lower bandwidths, LU decomposition becomes an $\mathcal{O}(\mathcal{Q})$ process.

If additional pivoting is required in order to ensure feasibility of the LU decomposition, the computational effort slightly rises to

$$2\mathcal{B}_l(\mathcal{B}_l + \mathcal{B}_u)\mathcal{Q} \tag{E.13}$$

when using row pivoting [88]. Nevertheless, the LU decomposition still remains an $\mathcal{O}(\mathcal{Q})$ process.

E.3. Summary on the LU Decomposition's Computational Effort

In this excursus, the computational effort of various LU decomposition algorithms was examined. The corresponding results are shown[6] in table E.1.

	Dense Matrix	Sparse banded Matrix with bandwidths $\mathcal{B}_u$, $\mathcal{B}_l$
LU decomposition w/o Pivoting	$\mathcal{O}(\mathcal{Q}^3)$	$\mathcal{O}(\mathcal{Q})$ and $\mathcal{O}(\mathcal{B}_u\mathcal{B}_l)$
LU decomposition w/ Row Pivoting	$\mathcal{O}(\mathcal{Q}^3)$	$\mathcal{O}(\mathcal{Q})$, $\mathcal{O}(\mathcal{B}_l^2)$ and $\mathcal{O}(\mathcal{B}_u)$
Forward & Backward Substitution	each $\mathcal{O}(\mathcal{Q}^2)$	each $\leq \mathcal{O}(\mathcal{Q}^2)$ (see footnote)

Table E.1.: Computational effort for solving an inhomogeneous system of equations with a coefficient matrix $\boldsymbol{A} \in \mathbb{K}^{\mathcal{Q}\times\mathcal{Q}}$ using LU decomposition

[6] Regarding forward and backward substitution of the LU decomposition of banded matrices, the computational effort depends whether pivoting is used or not. If no pivoting is used, $\boldsymbol{L}$ and $\boldsymbol{U}$ are under form of triangular matrices as shown in figure E.3. Here, additional optimisation could be done for forward and backward substitution. In contrast, if pivoting is used, $\boldsymbol{L}$ looses its band structure and $\boldsymbol{U}$'s bandwidth increases [88].

Algorithm E.3: In-Place LU Decomposition for banded matrices - modified from [88]

```
% Sample Matrix, for which LU decomposition
% shall be performed
A=[ 3 1 0 0 0 0 0
    4 1 5 0 0 0 0
    9 2 6 5 0 0 0
    0 3 5 8 9 0 0
    0 0 7 9 3 2 0
    0 0 0 3 9 4 6
    0 0 0 0 2 4 4];

% Determine n
Q=size(A,1);

% Upper Bandwidth
Bu=1;
% Lower Bandwidth
Bl=2;

% Perform diagonalisation on all columns
for k=1:Q-1

    % --- Optimised matrix multiplication ---

    % Calculate borders for the updating process
    iu=min(k+Bl,Q);
    ju=min(k+Bu,Q);

    % Loop over rows
    for i=k+1:iu

        % Calculate l(i,k)
        factor=-A(i,k)/A(k,k);

        % Loop over columns
        for j=k+1:ju

            % Perform matrix mulitplications element-wise
            A(i,j)=A(i,j)+factor*A(k,j);

        end

        % Store element of L in-place
        A(i,k)=-factor;

    end

    % --- End optimised matrix multiplication ---

end
```

List of References

[1] J. C. Maxwell, "VIII. A dynamical theory of the electromagnetic field," *Philosophical Transactions of the Royal Society of London*, vol. 155, pp. 459–512, 1865.

[2] B. J. Hunt, *The Maxwellians.* Cornell Press University Press, 1991.

[3] T. K. Sarkar, R. J. Mailloux, A. A. Oliner, M. Salazar-Palma, and D. L. Sengupta, *History of Wireless.* Wiley Interscience, 2006.

[4] H. Hertz, "Ueber die Ausbreitungsgeschwindigkeit der electrodynamischen Wirkungen," *Annalen der Physik*, vol. 270, no. 7, pp. 551–569, 1888.

[5] S. Mahmoud, *Electromagnetic Waveguides: Theory and Applications.* Peter Peregrinus Ltd., 1991.

[6] D. M. Pozar, *Microwave Enginnering.* John Wiley & Sons, 2005.

[7] W. J. R. Hoefer, "Fifty Years of Research in Electromagnetics: A Voyage Back in Time," in *Computational Electromagnetics—Retrospective and Outlook: In Honor of Wolfgang J.R. Hoefer*, I. Ahmed and Z. D. Chen, Eds. Springer Singapore, 2015, pp. 1–27.

[8] D. D. Grieg and H. F. Engelmann, "Microstrip-A New Transmission Technique for the Kilomegacycle Range," *Proceedings of the IRE*, vol. 40, no. 12, pp. 1644–1650, Dec 1952.

[9] C. P. Wen, "Coplanar Waveguide, a Surface Strip Transmission Line Suitable for Nonreciprocal Gyromagnetic Device Applications," in *1969 G-MTT International Microwave Symposium*, May 1969, pp. 110–115.

[10] V. E. Boria and B. Gimeno, "Waveguide Filters for Satellites," *IEEE Microwave Magazine*, vol. 8, no. 5, pp. 60–70, Oct 2007.

[11] A. C. Metaxas and R. J. Meredith, *Industrial Microwave Heating.* Institution of Engineering and Technology, 1988.

[12] K. S. Packard, "The Origin of Waveguides: A Case of Multiple Rediscovery," *IEEE Transactions on Microwave Theory and Techniques*, vol. 32, no. 9, pp. 961–969, Sep 1984.

[13] O. Heaviside, *Electromagnetic Theory.* Benn, 1893.

[14] L. Rayleigh, "XVIII: On the Passage of Electric Waves through Tubes, or the Vibrations of Dielectric Cylinders," *The London, Edinburgh, and Dublin Philosophical Magazine and Journal of Science*, vol. 43, no. 261, pp. 125–132, 1897.

[15] G. C. Southworth, "Hyper-Frequency Wave Guides-General Considerations and Experimental Results," *Bell System Technical Journal*, vol. 15, no. 2, pp. 284–309, 1936.

[16] W. L. Barrow, "Transmission of Electromagnetic Waves in Hollow Tubes of Metal," *Proceedings of the Institute of Radio Engineers*, vol. 24, no. 10, pp. 1298–1328, Oct 1936.

[17] N. Marcuvitz, *Waveguide Handbook.* Peter Peregrinus Ltd., 1951.

[18] P.-S. Kildal, *Foundations of Antenna Engineering.* Kildal Antenn AB, 2015.

[19] R. F. Harrington, *Field Computation by Moment Methods.* Robert E. Krieger Publishing Company, Inc., 1982.

[20] M. N. O. Sadiku, *Numerical Techniques in Electromagnetics.* CRC Press LLC, 2000.

[21] D. B. Davidson, *Computational Electromagnetics for RF and Microwave Engineering.* Cambridge Press, 2011.

[22] K. Yee, "Numerical Solution of Initial Boundary Value Problems Involving Maxwell's Equations in Isotropic Media," *IEEE Transactions on Antennas and Propagation*, vol. 14, no. 3, pp. 302–307, 1966.

[23] A. Taflove and S. C. Hagness, *Computational Electrodynamics - The Finite-Difference Time-Domain Method.* Artech House, 2005.

[24] T. Weiland, "A discretization method for the solution of Maxwell's equations for six-component fields," *Electronics and Communications AEÜ*, vol. 31, no. 3, pp. 116–120, 1977.

[25] M. Golio, Ed., *The RF and Microwave Handbook - RF and Microwave Circuits, Measurements and Modelling.* CRC Press LLC, 2008.

[26] A. F. Peterson, S. L. Ray, and R. Mittra, *Computational Methods for Electromagnetics.* IEEE Press, 1998.

[27] Y. Liu, R. Mittra, A. Muto, X. Yang, and W. Yu, *Advanced FDTD Methods: Parallelization, Acceleration, and Engineering Applications.* Artech House, 2011.

[28] T. Weiland, "Finite Integration Method and Discrete Electromagnetism," in *Computational Electromagnetics*, P. Monk, C. Carstensen, S. Funken, W. Hackbusch, and R. H. W. Hoppe, Eds. Springer, 2003, pp. 183–198.

[29] U. van Rienen, *Numerical Methods in Computational Electromagnetics.* Springer, 2001.

[30] A. Prokop, *Interpolation numerisch berechneter elektromagnetischer Felder und Verluste an elektrischen Oberflächen.* Cuvillier Verlag, 2005.

[31] T. Itoh, "Generalized Scattering Matrix Technique," in *Numerical Techniques For Microwave and Millimeter-Wave Passive Structures*, T. Itoh, Ed. John Wiley & Sons, 1989, pp. 622–635.

[32] A. Wexler, "Solution of Waveguide Discontinuities by Modal Analysis," *IEEE Transactions on Microwave Theory and Techniques*, vol. 15, no. 9, pp. 508–517, September 1967.

[33] P. J. B. Clarricoats and K. R. Slinn, "Numerical Method for the Solution of Waveguide-Discontinuity Problems," *Electronics Letters*, vol. 2, no. 6, pp. 226–228, June 1966.

[34] R. Mittra, "Relative Convergence of the Solution of a Doubly Infinite Set of Equations," *Journal of Research of the National Bureau of Standards*, vol. 67D, pp. 245–254, 1963.

[35] S. W. Lee, W. R. Jones, and J. J. Campbell, "Convergence of Numerical Solutions of Iris-Type Discontinuity Problems," *IEEE Transactions on Microwave Theory and Techniques*, vol. 19, no. 6, pp. 528–536, Jun 1971.

[36] M. Leroy, "On the Convergence of Numerical Results in Modal Analysis," *IEEE Transactions on Antennas and Propagation*, vol. 31, no. 4, pp. 655–659, Jul 1983.

[37] T. S. Chu, T. Itoh, and Y.-C. Shih, "Comparative Study of Mode-Matching Formulations for Microstrip Discontinuity Problems," *IEEE Transactions on Microwave Theory and Techniques*, vol. 33, no. 10, pp. 1018–1023, Oct 1985.

[38] A. C. Shih, "The Mode-Matching Method," in *Numerical Techniques For Microwave and Millimeter-Wave Passive Structures*, T. Itoh, Ed. John Wiley & Sons, 1989, pp. 593–621.

[39] Mittra and Lee, *Advanced Techniques in the Theory of Guided Waves.* The MacMillan Company, 1971.

[40] Y. C. Shih and K. G. Gray, "Convergence of Numerical Solutions of Step-Type Waveguide Discontinuity Problems by Modal Analysis," in *1983 IEEE MTT-S International Microwave Symposium Digest*, May 1983, pp. 233–235.

[41] R. Mittra, *Computer Techniques for Electromagnetics: International Series of Monographs in Electrical Engineering.* Pergamon Press Ltd, 1973.

[42] Y.-C. Shih, "Design of Waveguide E-Plane Filters with All-Metal Inserts," *IEEE Transactions on Microwave Theory and Techniques*, vol. 32, no. 7, pp. 695–704, Jul 1984.

[43] H. Patzelt and F. Arndt, "Double-Plane Steps in Rectangular Waveguides and Their Application for Transformers, Irises, and Filters," *IEEE Transactions on Microwave Theory and Techniques*, vol. 30, no. 5, pp. 771–776, May 1982.

[44] J. Bornemann and R. Vahldieck, "Characterization of a Class of Waveguide Discontinuities Using a Modified TE^{x}_{mn} Mode Approach," *IEEE Transactions on Microwave Theory and Techniques*, vol. 38, no. 12, pp. 1816–1822, Dec 1990.

[45] B. Thomas, G. James, and K. Greene, "Design of Wide-Band Corrugated Conical Horns for Cassegrain Antennas," *IEEE Transactions on Antennas and Propagation*, vol. 34, no. 6, pp. 750–757, June 1986.

[46] G. James, "Design of Wide-Band Compact Corrugated Horns," *IEEE Transactions on Antennas and Propagation*, vol. 32, no. 10, pp. 1134–1138, Oct 1984.

[47] T. S. Bird, "Mode Matching Analysis of Arrays of Stepped Rectangular Horns and Application to Satellite Antenna Design," in *1991 Seventh International Conference on Antennas and Propagation, ICAP 91 (IEE)*, Apr 1991, pp. 849–852 vol.2.

[48] R. F. Harrington, *Time-Harmonic Electromagnetic Fields.* IEEE Press, 2001.

[49] I. Wolff, G. Kompa, and R. Mehran, "Calculation Method for Microstrip Discontinuities and T Junctions," *Electronics Letters*, vol. 8, no. 7, pp. 177–179, April 1972.

[50] W. Menzel and I. Wolff, "A Method for Calculating the Frequency-Dependent Properties of Microstrip Discontinuities," *IEEE Transactions on Microwave Theory and Techniques*, vol. 25, no. 2, pp. 107–112, Feb 1977.

[51] R. Mehran, "Computer-Aided Design of Microstrip Filters Considering Dispersion, Loss, and Discontinuity Effects," *IEEE Transactions on Microwave Theory and Techniques*, vol. 27, no. 3, pp. 239–245, Mar 1979.

[52] L. Conti, E. Martini, R. Nesti, G. Pelosi, and S. Selleri, "An Integrated Finite Element-Mode Matching-Plane Wave Expansion Code for Horn Antenna Analysis," in *IEEE Antennas and Propagation Society International Symposium. 2001 Digest.*, vol. 4, July 2001, pp. 298–301 vol.4.

[53] J. R. Montejo-Garai and J. Zapata, "Full-wave Design and Realization of Multicoupled Dual-Mode Circular Waveguide Filters," *IEEE Transactions on Microwave Theory and Techniques*, vol. 43, no. 6, pp. 1290–1297, Jun 1995.

[54] J. M. Reiter and F. Arndt, "Rigorous Analysis of Arbitrarily Shaped H- and E-plane Discontinuities in Rectangular Waveguides by a Full-Wave Boundary Contour Mode-Matching Method," *IEEE Transactions on Microwave Theory and Techniques*, vol. 43, no. 4, pp. 796–801, Apr 1995.

[55] ——, "A Boundary Contour Mode-Matching Method for the Rigorous Analysis of Cascaded Arbitrarily Shaped H-Plane Discontinuities in Rectangular Waveguides," *IEEE Microwave and Guided Wave Letters*, vol. 2, no. 10, pp. 403–405, Oct 1992.

[56] G. Piefke, *Feldtheorie II.* Bilbiographisches Institut Mannheim, Wien, Zürich, 1973.

[57] H.-G. Unger, *Elektromagnetische Theorie für die Hochfrequenztechnik, Bd. 1.* Hüthig Verlag, 1988.

[58] R. E. Collin, *Field Theory of Guided Waves.* IEEE Press, 1991.

[59] K. W. Kark, *Antennen und Strahlungsfelder*, 4th ed. Vieweg+Teubner, 2011.

[60] D. Griffiths, *Elektrodynamik.* Pearson, 2011.

[61] H.-G. Unger, *Elektromagnetische Theorie für die Hochfrequenztechnik, Bd. 2.* Hüthig Verlag, 1989.

[62] Bronstein and Semendjajew, *Taschenbuch der Mathematik.* BSB Teubner, 1989.

[63] C. C. Johnson, *Field and Wave Electrodynamics.* McGraw-Hill, Inc., 1965.

[64] R. Garg, *Analytical and Computational Methods in Electromagnetics.* Artech House, 2008.

[65] C. H. Papas, *Theory of Electromagnetic Wave Propagation.* Dover Publications, Inc., 1988.

[66] H. A. Lorentz, Amsterdammer Akademie van Wetenschappen, vol. 4, pp. 176-187, 1895-1896.

[67] G. V. Eleftheriades, A. S. Omar, L. P. B. Katehi, and G. M. Rebeiz, "Some Important Properties of Waveguide Junction Generalized Scattering Matrices in the Context of the Mode Matching Technique," *IEEE Transactions on Microwave Theory and Techniques*, vol. 42, no. 10, pp. 1896–1903, Oct 1994.

[68] C. Monka and J. Schoebel, "Eigenmodes of partially filled coaxial waveguides," in *2016 German Microwave Conference (GeMiC)*, March 2016, pp. 132–135.

[69] M. Abramowitz and I. A. Stegun, *Handbook of Mathematical Functions.* National Bureau of Standards, 1964, reprinted 2014 by Martino Publishing.

[70] Bronstein and Semendjajew, *Taschenbuch der Mathematik - Ergänzende Kapitel.* BSB Teubner, 1990.

[71] K.-T. Tang, *Mathematical Methods for Engineers and Scientists 3.* Springer, 2007.

[72] U. Scherz, *Quantenmechanik.* Teubner, 1999.

[73] J. Meixner, "Die Kantenbedingung in der Theorie der Beugung elektromagnetischer Wellen an vollkommen leitenden ebenen Schirmen," *Annalen der Physik*, vol. 441, no. 1, pp. 1–9, 1950.

[74] G. Piefke, *Feldtheorie III.* Bilbiographisches Institut Mannheim, Wien, Zürich, 1977.

[75] C. H. Papas, *Randwertprobleme der Mikrowellenphysik.* Springer Verlag, 1955.

[76] S. Bosch, *Lineare Algebra.* Springer, 2014.

[77] K. Kurokawa, "Power Waves and the Scattering Matrix," *IEEE Transactions on Microwave Theory and Techniques*, vol. 13, no. 2, pp. 194–202, Mar 1965.

[78] R. E. Collin, *Foundations for Microwave Engineering.* IEEE Press, 2001.

[79] H. J. Visser, *Approximate Antenna Analysis for CAD.* John Wiley & Sons, 2009.

[80] Anon., *HFSS helpfile entry on Max Delta (Mag S)*, Ansys Inc., 2016.

[81] J. Schoebel, Personal Communication, January 2016.

[82] Anon., *Matlab helpfile on BiCGStab*, The Mathworks, 2017.

[83] W. Dahmen and A. Reusken, *Numerik für Ingenieure und Naturwissenschaftler.* Springer, 2008.

[84] H. R. Schwarz and N. Köckler, *Numerische Mathematik.* Vieweg+Teubner, 2011.

[85] H. Schwandt, *Parallele Numerik.* Teubner, 2003.

[86] Press, Teukolsky, Vetterling, and Flannery, *Numerical Recipes in Fortran.* Cambridge University Press, 1992.

[87] S. C. Chapra and R. P. Canale, *Numerical Methods for Engineers.* McGraw Hill, 2006.

[88] C. Kanzow, *Numerik linearer Gleichungssysteme.* Springer, 2005.

[89] W. Bächtold, *Mikrowellentechnik.* Vieweg, 1999.

[90] R. Rhea, "Exploiting Filter Symmetry," *Microwave Journal*, March 2001, reprint by Agilent EEsof EDA.

[91] I. East, *Computer Architecture and Organization.* Pitman Publishing, 1990.

[92] H. Bethe and J. Schwinger, "Perturbation Theory for Cavities," Massachusetts Institute of Technology, Radiation Laboratory, Tech. Rep., 1943.

[93] V. K. Sharma, *Matrix Methods and Vector Spaces in Physics.* PHI Learning, 2009.

[94] S. Y. Mitsuo Makimoto, *Microwave Resonators and Filters for Wireless Communication.* Springer Verlag, 2001.

[95] C. Kudsia, R. Cameron, and W. C. Tang, "Innovations in Microwave Filters and Multiplexing Networks for Communications Satellite Systems," *IEEE Transactions on Microwave Theory and Techniques*, vol. 40, no. 6, pp. 1133–1149, Jun 1992.

[96] G. L. Matthaei, L. Young, and E. Jones, *Microwave Filters, Impedance Matching Networks, and Coupling Structures.* Norwood, MA: Artech House, 1980, reprinted from the original edition (McGraw-Hill, 1964).

[97] S. B. Cohn, "Direct-Coupled-Resonator Filters," *Proceedings of the IRE*, vol. 45, no. 2, pp. 187–196, Feb 1957.

[98] M. Meyer, *Signalverarbeitung.* Vieweg+Teubner, 2011.

[99] R. W. Rhea, *HF Filter Design and Computer Simulation.* Noble Publishing Corporation, 1994.

[100] J. Schoebel, C. Monka, and J. Fahlbusch, "Direct-coupled resonator filters based on foreshortened coaxial transmission line resonators," in *2018 11th German Microwave Conference (GeMiC)*, March 2018, pp. 127–130.

[101] Y. Zhai, Q. Wang, Z. Wang, and X. x. Gao, "The Design of an Iris Waveguide Filter at 35.75 GHz," in *2008 Global Symposium on Millimeter Waves*, April 2008, pp. 348–350.

[102] S. K. Efstratios Doumanis, George Goussetis, *Filter Design for Satellite Communications: Helical Resonator Technology.* Artech House, 2014.

[103] J. Schoebel, Personal Communication, Summer 2015.

[104] J. Baker-Jarvis, M. Janezic, B. Riddle, R. Johnk, P. Kabos, C. Holloway, R. Geyer, and C. Grosvenor, "Measuring the Permittivity and Permeability of Lossy Materials: Solids, Liquids, Metals, Building Materials and Negative-Index Materials," National Institute of Standards and Technology (NIST), U.S. Dept. of Commerce, NIST Report 1536-R, 2005.

[105] J. Baker-Jarvis, R. G. Geyer, J. H. Grosvenor, M. D. Janezic, C. A. Jones, B. Riddle, C. M. Weil, and J. Krupka, "Dielectric Characterization of Low-loss Materials - A Comparison of Techniques," *IEEE Transactions on Dielectrics and Electrical Insulation*, vol. 5, no. 4, pp. 571–577, Aug 1998.

[106] J. Sheen, "Comparisons of microwave dielectric property measurements by transmission/reflection techniques and resonance techniques," *Measurement Science and Technology*, vol. 20, no. 4, April 2009.

[107] G. Birnbaum and J. Franeau, "Measurement of the Dielectric Constant and Loss of Solids and Liquids by a Cavity Perturbation Method," *Journal of Applied Physics*, vol. 20, no. 8, pp. 817–818, 1949.

[108] A. Kraszewski and S. Nelson, "Observations on Resonant Cavity Perturbation by Dielectric Objects," *IEEE Transactions on Microwave Theory and Techniques*, vol. 40, no. 1, pp. 151–155, Jan 1992.

[109] W. Rueggeberg, "Determination of Complex Permittivity of Arbitrarily Dimensioned Dielectric Modules at Microwave Frequencies," *IEEE Transactions on Microwave Theory and Techniques*, vol. 19, no. 6, pp. 517–521, Jun 1971.

[110] E. Kilic, U. Siart, O. Wiedenmann, U. Faz, R. Ramakrishnan, P. Saal, and T. Eibert, "Cavity Resonator Measurement of Dielectric Materials Accounting for Wall Losses and a Filling Hole," *IEEE Transactions on Instrumentation and Measurement*, vol. 62, no. 2, pp. 401–407, Feb 2013.

[111] W. Courtney, "Analysis and Evaluation of a Method of Measuring the Complex Permittivity and Permeability Microwave Insulators," *IEEE Transactions on Microwave Theory and Techniques*, vol. 18, no. 8, pp. 476–485, Aug 1970.

[112] B. Hakki and P. Coleman, "A Dielectric Resonator Method of Measuring Inductive Capacities in the Millimeter Range," *IRE Transactions on Microwave Theory and Techniques*, vol. 8, no. 4, pp. 402–410, July 1960.

[113] Y. Kobayashi and M. Katoh, "Microwave Measurement of Dielectric Properties of Low-Loss Materials by the Dielectric Rod Resonator Method," *IEEE Transactions on Microwave Theory and Techniques*, vol. 33, no. 7, pp. 586–592, Jul 1985.

[114] W. B. Weir, "Automatic Measurement of Complex Dielectric Constant and Permeability at Microwave Frequencies," *Proceedings of the IEEE*, vol. 62, no. 1, pp. 33–36, 1974.

[115] A. Nicolson and G. F. Ross, "Measurement of the Intrinsic Properties of Materials by Time-Domain Techniques," *IEEE Transactions on Instrumentation and Measurement*, vol. 19, no. 4, pp. 377–382, 1970.

[116] J. Baker-Jarvis, E. Vanzura, and W. Kissick, "Improved Technique for Determining Complex Permittivity with the Transmission/Reflection Method," *IEEE Transactions on Microwave Theory and Techniques*, vol. 38, no. 8, pp. 1096–1103, Aug 1990.

[117] J. Baker-Jarvis, M. Janezic, J. Grosvenor, and R. Geyer, "Transmission/Reflection and Short-Circuit Line Methods for Measuring Permittivity and Permeability," National Institute of Standards and Technology (NIST), U.S. Dept. of Commerce, NIST Report 1355-R, 1993.

[118] C. Monka and J. Schoebel, "Characterisation of low loss dielectrics using a transmission line method," in *Microwave Conference (GeMIC), 2014 German*, March 2014, pp. 1–4.

[119] D. K. Ghodgaonkar, V. V. Varadan, and V. K. Varadan, "Free-Space Measurement of Complex Permittivity and Complex Permeability of Magnetic Materials at Microwave Frequencies," *IEEE Transactions on Instrumentation and Measurement*, vol. 39, no. 2, pp. 387–394, Apr 1990.

[120] C. K. Campbell, "Free-Space Permittivity Measurements on Dielectric Materials at Millimeter Wavelengths," *IEEE Transactions on Instrumentation and Measurement*, vol. 27, no. 1, pp. 54–58, 1978.

[121] A. Kazemipour, M. Hudlicka, S. K. Yee, M. A. Salhi, D. Allal, T. Kleine-Ostmann, and T. Schrader, "Design and Calibration of a Compact Quasi-Optical System for Material Characterization in Millimeter/Submillimeter Wave Domain," *IEEE Transactions on Instrumentation and Measurement*, vol. 64, no. 6, pp. 1438–1445, June 2015.

[122] A. Kazemipour, M. Hudlicka, T. Kleine-Ostmann, and T. Schrader, "A Reliable Simple Method to Extract the Intrinsic Material Properties in Millimeter/Sub-Millimeter Wave Domain," in *29th Conference on Precision Electromagnetic Measurements (CPEM 2014)*, Aug 2014, pp. 576–577.

[123] E. Saenz, L. Rolo, M. Paquay, G. Gerini, and P. de Maagt, "Sub-millimetre Wave Material Characterization," in *Antennas and Propagation (EUCAP), Proceedings of the 5th European Conference on*, April 2011, pp. 3183–3187.

[124] T. P. Marsland and S. Evans, "Dielectric measurements with an open-ended coaxial probe," *IEE Proceedings - Microwaves, Antennas and Propagation*, vol. 134, no. 4, pp. 341–349, August 1987.

[125] A. Kraszewski, M. A. Stuchly, and S. S. Stuchly, "ANA Calibration Method for Measurements of Dielectric Properties," *IEEE Transactions on Instrumentation and Measurement*, vol. 32, no. 2, pp. 385–387, June 1983.

[126] J. M. Anderson, C. L. Sibbald, S. S. Stuchly, and K. Caputa, "Advances in Dielectric Measurements Using an Open-Ended Coaxial Line Sensor," in *Proceedings of Canadian Conference on Electrical and Computer Engineering*, Sep 1993, pp. 916–919 vol.2.

[127] M. D. Janezic and J. Baker-Jarvis, "Full-Wave Analysis of a Split-Cylinder Resonator for Nondestructive Permittivity Measurements," *IEEE Transactions on Microwave Theory and Techniques*, vol. 47, no. 10, pp. 2014–2020, Oct 1999.

[128] T. Nacke, A. Barthel, D. Beckmann, U. Pliquett, J. Friedrich, P. Peyerl, M. Helbig, and J. Sachs, "Messsystem für die impedanzspektroskopische Breitband-Prozessmesstechnik," *Techn. Messen*, vol. 78-1, pp. 3–14, 2011.

[129] C. Monka, S. Brueckner, and J. Schoebel, "M-sequence-based material characterisation," in *GeMiC 2015 - The 9th German Microwave Conference (GeMiC 2015)*, Nuremberg, Germany, Mar. 2015.

[130] D. Kajfez, "Linear Fractional Curve Fitting for Measurement of High Q Factors," *IEEE Transactions on Microwave Theory and Techniques*, vol. 42, no. 7, pp. 1149–1153, Jul 1994.

[131] ——, "Data Processing For Q Factor Measurement," in *ARFTG Conference Digest-Spring, 43rd*, vol. 25, May 1994, pp. 104–111.

[132] S. Shahid, J. Ball, C. Wells, and P. Wen, "Reflection type Q-factor measurement using standard least squares methods," *IEE Proceedings - Microwaves, Antennas and Propagation*, vol. 5, no. 4, pp. 426–432, March 2011.

[133] D. Kajfez, "Random and Systematic Uncertainties of Reflection-Type Q-Factor Measurement with Network Analyzer," *IEEE Transactions on Microwave Theory and Techniques*, vol. 51, no. 2, pp. 512–519, Feb 2003.

[134] ——, *Q Factor*. Vector Fields, 1994.

[135] ——, *Q Factor Measurements Using Matlab*. Artech House, 2011.

[136] U. M. Ascher and C. Greif, *A First Course on Numerical Methods*. Society for Industrial and Applied Mathematics, 2011.

[137] R. Waldron, *The Theory of Waveguides and Cavities*. Maclaran and Sons, 1967.

[138] Anon., "JCGM 100: Evaluation of measurement data - Guide to the expression of uncertainty in measurement," Joint Committee for Guides in Metrology, Tech. Rep., 2010.

[139] ——, "JCGM 100: Evaluation of measurement data - Guide to the expression of uncertainty in measurement - Supplement 1: Propagation of distributions using a Monte Carlo method," Joint Committee for Guides in Metrology, Tech. Rep., 2008.

[140] von Hippel, *Dielectric Materials and Applications.* M.I.T. Press, 1966.

[141] W. J. Ellison, "Permittivity of Pure Water, at Standard Atmospheric Pressure, over the Frequency Range 0-25 THz and the Temperature Range 0-100°C," *Journal of Physical and Chemical Reference Data*, vol. 36, no. 1, pp. 1–18, 2007.

[142] U. Kaatze, "Complex Permittivity of Water as a Function of Frequency and Temperature," *Journal of Chemical & Engineering Data*, vol. 34, no. 4, pp. 371–374, 1989.

[143] T. Meissner and F. J. Wentz, "The Complex Dielectric Constant of Pure and Sea Water From Microwave Satellite Observations," *IEEE Transactions on Geoscience and Remote Sensing*, vol. 42, no. 9, pp. 1836–1849, Sept 2004.

[144] J. Hamelin, J. B. Mehl, and M. R. Moldover, "The Static Dielectric Constant of Liquid Water Between 274 and 418 K Near the Saturated Vapor Pressure," *International Journal of Thermophysics*, vol. 19, no. 5, pp. 1359–1380, Sep 1998.

[145] J. Sheen, "Amendment of Cavity Perturbation Technique for Loss Tangent Measurement at Microwave Frequencies," in *2006 7th International Symposium on Antennas, Propagation EM Theory*, Oct 2006, pp. 1–3.

[146] ——, "Amendment of Cavity Perturbation Technique for Loss Tangent Measurement at Microwave Frequencies," *Journal of Applied Physics*, vol. 102, no. 1, p. 014102, 2007.

[147] S. Groiss, I. Bardi, O. Biro, K. Preis, and K. R. Richter, "Parameters of Lossy Cavity Resonators Calculated by the Finite Element Method," *IEEE Transactions on Magnetics*, vol. 32, no. 3, pp. 894–897, 1996.

[148] G. T. Smith, *Industrial Metrology: Surfaces and Roundness.* Springer Verlag, 2002.

[149] M. J. Madou, *Manufacturing Techniques for Microfabrication and Nanotechnology.* CRC Press, 2011.

[150] H. M. Pickett, J. C. Hardy, and J. Farhoomand, "Characterization of a Dual-Mode Horn for Submillimeter Wavelengths (Short Papers)," *IEEE Transactions on Microwave Theory and Techniques*, vol. 32, no. 8, pp. 936–937, Aug 1984.

[151] J. Pawlan, "Compact and Easy to Manufacture Dual Mode Feed Horn with Ultra-low Backlobes," in *2015 German Microwave Conference*, March 2015, pp. 355–358.

[152] C. Sanderson and R. Curtin, "Armadillo: a template-based C++ library for linear algebra," *Journal of Open Source Software*, vol. 1, p. 26, 2016.

[153] ——, "A User-Friendly Hybrid Sparse Matrix Class in C++," *International Congress on Mathematical Software*, 2018.

[154] J. L. Volakis, A. Chatterjee, and L. C. Kempel, *Finite Element Method for Electromagnetics.* IEEE Press, 1998.

[155] J. B. Davies, "The Finite Element Method," in *Numerical Techniques For Microwave and Millimeter-Wave Passive Structures*, T. Itoh, Ed. John Wiley & Sons, 1989, pp. 33–132.

[156] G. Strassacker and R. Süße, *Rotation, Divergenz und Gradient.* B. G. Teubner, 2006.

[157] M. T. Heath, *Scientific Computing.* McGraw Hill, 2002.

[158] T. Itoh, Ed., *Numerical Techniques For Microwave and Millimeter-Wave Passive Structures.* John Wiley & Sons, 1989.

Publications

- **Carsten Monka** and Joerg Schoebel, "*Eigenmodes of partially filled coaxial waveguides,*" GeMiC 2016 The 10th German Microwave Conference (GeMiC 2016), Bochum, Germany
- **Carsten Monka**, Sebastian Brueckner and Joerg Schoebel, "*M-sequenced-based material characterisation,*" GeMiC 2015 The 9th German Microwave Conference (GeMiC 2015), Nuremberg, Germany
- **Carsten Monka** and Joerg Schoebel, "*Characterisation of low loss dielectrics using a transmission line method,*" GeMiC 2014 The 8th German Microwave Conference (GeMiC 2014), Aachen, Germany
- Fabian Schwartau, **Carsten Monka**, Markus Krueckemeier and Joerg Schoebel, "*Aircraft window attenuation measurements at 60 GHz for wireless in-cabin communication,*" GeMiC 2018 The 11th German Microwave Conference (GeMiC 2018), Freiburg, Germany
- Joerg Schoebel, **Carsten Monka** and Jan Fahlbusch, "*Direct-coupled resonator filters based on foreshortened coaxial transmission line resonators,*" GeMiC 2018 The 11th German Microwave Conference (GeMiC 2018), Freiburg, Germany
- Sebastian Raabe, **Carsten Monka**, Sebastian Franke, Reinhard Caspary and Joerg Schoebel, "*Resonant properties of mismatched ring circuits,*" 2016 46th European Microwave Conference (EuMC), London, 2016, pp. 983-986.
- Sebastian Franke, Matthias Baumkötter, **Carsten Monka**, Sebastian Raabe, Reinhard Caspary, Hans-Hermann Johannes, Wolfgang Kowalsky, Sebastian Beck, Annemarie Pucci, "*Alumina films as gas barrier layers grown by spatial atomic layer deposition with trimethylaluminum and different oxygen sources,*" Journal of Vacuum Science & Technology A 2017 35:1

Papers published after thesis submission

- Markus Krueckemeier, Fabian Schwartau, **Carsten Monka** and Joerg Schoebel, "*Synchronisation of multiple USRP SDRs for coherent receiver applications,*" The Sixth IEEE International Conference on Software Defined Systems (SDS 2019), Rome, Italy

Abstract

Over the past decades computational electromagnetics tools have become an indispensable aid for engineers and scientists. Partial Differential Equation techniques, with the Finite Element method being an example, are recognised as being the standard approach for the electromagnetic analysis of closed structures such as waveguide components and cavities.

A major drawback of these techniques results from the fact that they require a space-segmentation of the domain under consideration. If complex geometries or structures with dimensions exceeding few wavelengths are to be solved, the resulting meshes tend to become very large, which is undesirable in terms of memory requirements and computation time.

In contrast, a striking feature of the Mode Matching technique discussed in this thesis is its ability to efficiently solve the field problem imposed by such structures, provided they can be decomposed into sub-domains whose spectrum of Eigenmodes is known analytically.

Using the Mode Matching technique, orthogonal expansion of the yet unknown tangential fields at the interfaces between these sub-domains is performed, which leads to a system of equations that can be solved for the Eigenmodes' amplitudes.

Several applications such as waveguide filter design strongly benefit from the Mode Matching technique's computational efficiency rather than they suffer from the geometric limitations resulting from the method's quasi-analytical approach.

The present thesis provides a complete treatise of the Mode Matching technique. While literature has focused almost exclusively on the method's underlying electromagnetic concept, this work offers in-depth insights into implementation-related aspects such as efficient matrix population and proper scattering parameter calculation.

To illustrate the advantages of the Mode Matching technique, the solvers developed in the scope of this thesis are used to analyse problems from two fields of application:

The analysis of waveguide filters represents the first application studied in this thesis: Various filter structures have been investigated using the newly developed solvers. Both the simulation methodology as well as the obtained results are discussed with a special focus on the close interplay between filter design and full-wave analysis of waveguide filters.

Moreover, the analysis of partially filled coaxial waveguides, which may be included in tubular filters to increase the power handling capability, is another focus of this work.

The second application of the Mode Matching technique examined in this thesis is the analysis of electromagnetic cavities used for cavity perturbation material measurements. This thesis presents a concise outline of the cavity perturbation material measurement technique and contributes a rigorous analysis of "secondary effects" afflicting this method.

To illustrate the superiority of the Mode Matching technique, a performance comparison between the Mode Matching solvers developed in the scope of this thesis and a commercial Finite Element tool was conducted. The findings from this comparison illustrate for suitable structures that Mode Matching by far outperforms the commercial Finite Element implementation in terms of computation time and accuracy.

Kurzfassung

Im Laufe der letzten Jahrzehnte sind Methoden zur numerischen Berechnung elektromagnetischer Felder zu einem unverzichtbaren Werkzeug für Ingenieure und Naturwissenschaftler geworden. Numerische Methoden, die direkt auf Grundlage partieller Differentialgleichungen arbeiten, stellen die Standardverfahren zur Simulation von geschlossenen Strukturen wie Hohlleiterkomponenten und Hohlraumresonatoren dar.

Ein gewichtiger Nachteil dieser Verfahren folgt aus der Notwendigkeit, den zu betrachtenden Raum mit einem Gitternetz zu überziehen. Sofern komplexe Geometrien oder Strukturen, deren Größe einige Wellenlängen überschreitet, untersucht werden, ergeben sich sehr große Gitternetze, was sich nachteilig auf den Speicherbedarf und die Rechenzeit auswirkt.

Im Gegensatz dazu ist das im Rahmen dieser Arbeit behandelte Mode-Matching-Verfahren in der Lage, das sich für diese Strukturen ergebende Feldproblem effizient zu lösen, sofern die Geometrie in Bereiche zerlegbar ist, deren diskretes Eigenmodenspektrum bekannt ist.

Hierzu werden Orthogonalentwicklungen der noch unbekannten tangentialen Felder an den Übergängen zwischen den einzelnen Bereichen vorgenommen, die auf ein Gleichungssystem führen, dessen Lösung die Amplituden der Eigenmoden der jeweiligen Bereiche liefert.

Verschiedene Anwendungen, wie beispielsweise der Entwurf von Wellenleiterfiltern, für die die sich aus dem quasi-analytischen Ansatz ergebenden Geometrieeinschränkungen oft unproblematisch sind, profitieren erheblich von der effizienten Feldberechnung.

Die vorliegende Dissertation bietet eine vollständige Darstellung der Mode-Matching-Methode. Im Gegensatz zur einschlägigen Literatur, die sich nahezu ausschließlich auf den theoretischen Hintergrund des Verfahrens beschränkt, bietet diese Arbeit umfangreiche Erkenntnisse zu Aspekten der Implementierung dieser Methode. Hierzu zählen beispielsweise die effiziente Aufstellung der Systemmatrix und die Berechnung von Streuparametern.

Zur Verdeutlichung der Vorteile des Mode-Matching-Verfahrens werden im Rahmen dieser Arbeit entwickelte Löser auf Probleme aus zwei Anwendungsbereichen eingesetzt:

Die Untersuchung verschiedener Wellenleiterstrukturen unter Verwendung der neu entwickelten Mode-Matching-Löser stellt einen Schwerpunkt dieser Arbeit dar. Sowohl die Methodik als auch die Ergebnisse dieser Simulationen werden mit einem besonderen Augenmerk auf das Zusammenspiel zwischen Entwurf und Simulation der Filter diskutiert.

Weiterhin werden in dieser Arbeit teilgefüllte koaxiale Wellenleiter betrachtet, die in koaxialzylindrischen Filtern zur Erhöhung der transportierbaren Leistung eingesetzt werden.

Als weitere Anwendung des Mode-Matching-Verfahrens steht die Untersuchung von Hohlraumresonatoren, die zur Materialcharakterisierung mittels eines Störungsverfahrens eingesetzt werden, im Fokus dieser Dissertation. Die vorliegende Arbeit gibt zunächst einen Überblick über dieses Verfahren. Im Anschluss werden mit dieser Methode verbundene „sekundäre Effekte“ unter Zuhilfenahme von Simulationsergebnissen untersucht, welche mit einem neu entwickelten Mode-Matching-Löser ermittelt wurden.

Ein Vergleich der im Rahmen dieser Arbeit entwickelten Mode-Matching-Löser mit einer kommerziellen Finite Elemente-Implementierung zeigt die Überlegenheit des Mode-Matching-Verfahrens sowohl im Hinblick auf Rechenzeit als auch auf Genauigkeit.

Terminology, Nomenclature and List of Symbols

1. Terminology

Ports and Waveports

In this thesis the term *port* is used in a scattering parameter sense, that is, it refers to the terminals of a superordinate structure. The term *waveport* is used in the context of interfaces and waveguides as introduced in chapter 4 and following.

Measurement Error and Measurement Uncertainty

According to the Guide to the Expression of Uncertainty in Measurement (GUM) [138], the term error is defined as "result of a measurement minus a true value of the measurand" [138, sec. B.2.19] while the term uncertainty is defined as "parameter, associated with the result of a measurement, that characterizes the dispersion of the values that could reasonably be attributed to the measurand" [138, sec. B.2.18].

Thus, in this thesis, when discussing the difference between material parameters calculated from simulation results which were carried out for predefined material parameters the term error is used.

In contrast, when discussing actual measurement data or the ability of a given measurement technique to provide trustworthy measurement results, the term uncertainty is used as recommended by the GUM [138].

2. Abbreviations

DOF	Degrees of Freedom
GUM	Guide for the Expression of Uncertainty of Measurement (see [138])
LSE	Longitudinal Section Electric
LSM	Longitudinal Section Magnetic
ODE	Ordinary Differential Equation
PDE	Partial Differential Equation
TE	Transverse Electric
TEM	Transverse Electromagnetic
TM	Transverse Magnetic

3. Nomenclature

The following general remarks should help the reader to better understand the nomenclature of this thesis.

- Matrices are denoted using bold symbols, e.g. $\boldsymbol{A}$.
- Vectors are denoted using vector arrows, e.g. $\vec{x}$.
- Waveport numbering is done using regular Roman numbers $\{I,\ II,\ III,\ ...\}$ while junction and segment numbering is carried out using calligraphic Roman numbers $\{\mathcal{I},\ \mathcal{II},\ \mathcal{III},\ ...,\mathcal{N}\}$.
- General cylinder coordinates denote x_1, x_2, z.
- Cylinder coordinates denote ρ, ϕ, z.
- The set of imaginary numbers is denoted $\mathbb{I}$.
- The complex quantity is usually denoted $j = \sqrt{-1}$.
- The Laplace operator denotes ∇^2.
- ∇_t^2 denotes the transverse Laplacian.
- The differential elements are denoted dA (surface), dV (volume) and dl (line).

4. List of Symbols

In the following, most relevant symbols used in the scope of this thesis are listed. Every symbol is either briefly explained or a reference to the text passage where the symbol was originally defined is provided.

Symbols are grouped into Roman, calligraphic, and Greek letters.

4.1. Roman Letters

$\boldsymbol{A}^{+}$	Pseudoinverse Matrix
$\boldsymbol{A}^{H}$	Hermitian Matrix (conjugate transpose matrix)
${}^{\eta/\nu}A_{m,n}^{TE/TM}$	Amplitude of a forward travelling TE or TM Eigenmode on waveport η or at junction ν respectively. Indices may be omitted where appropriate.
$\vec{A}^{\,TE/TM}$	Vector of forward travelling TE or TM Eigenmodes' amplitudes. Indices may be omitted where appropriate.
$\breve{A}^{\,TE/TM}$	Vector of forward travelling TE or TM Eigenmodes' junction amplitudes. Indices may be omitted where appropriate.
a	Waveguide dimension
a_i	Coefficient of the model function used for linear fractional curve fitting
$\vec{a}$	Total junction amplitudes, that is, amplitudes to all Eigenmodes considered in the field solution, denoted under vector form.
$\vec{a}_{Known}$	Predefined junction amplitudes, i.e. the excitation of the structure, denoted under vector form.
$\vec{a}_{Unknown}$	Unknown junction amplitudes to be determined by Mode Matching.
${}^{\eta/\nu}B_{m,n}^{TE/TM}$	Amplitude of a backward travelling TE or TM Eigenmode on waveport η or at junction ν respectively. Indices may be omitted where appropriate.

$\vec{B}^{\,TE/TM}$	Vector of backward travelling TE or TM Eigenmodes' amplitudes. Indices may be omitted where appropriate.
$\check{B}^{\,TE/TM}$	Vector of backward travelling TE or TM Eigenmodes' junction amplitudes. Indices may be omitted where appropriate.
b	Waveguide dimension
c_0	Speed of light in vacuum, $c_0 = {}^{1}/\sqrt{\varepsilon_0\mu_0}$
c	Speed of light in media, $c = {}^{c_0}/\sqrt{\varepsilon'_r\mu'_r}$
C	Number of columns of a single segment matrix. $C = 8 \cdot \mathcal{M}$ applies.
C	Capacitance
d	Waveguide dimension, dielectric tube's inner diameter See chapter 14
$\vec{D}$	Electric Displacement field
$\vec{E}$	Electric field
$\vec{E}_t$	Transverse electric field, see (2.13)
E_{tan}	Tangential electric field
${}^{\eta/\nu}E^{TE/TM}_{x_i\|m,n}$	Electric field distribution of the x_i component of the $TE_{m,n}$ or $TM_{m,n}$ Eigenmode on waveport η or at junction ν respectively. This field distribution is signed. Indices may be omitted.
f	Frequency
f_c	Cut-off Frequency
$\vec{F}$	Arbitrary field with field components F_{x_i}, see section 3.2
G	Conductance
g_i	Element values of a lowpass prototype filter
$\vec{H}$	Magnetic Field
$\vec{H}_t$	Transverse magnetic field, see (2.15)
${}^{\eta/\nu}H^{TE/TM}_{x_i\|m,n}$	Magnetic field distribution of the x_i component of the $TE_{m,n}$ or $TM_{m,n}$ Eigenmode on waveport η or at junction ν respectively. This field distribution is signed. Indices may be omitted.

i	General index
I_i	Total current on port i of a structure
$I_i^{+/-}$	Current wave. The + or - superscript indicates wave propagation towards or away from the structure, i.e. whether the wave is incident or scattered.
$\vec{I}$	Current vector. Contains the total voltages of all ports of a structure.
I	Functional
$J_n(x_i)$	Bessel function of first kind, nth order
k	Free space wavenumber, $k = \omega\sqrt{\mu\varepsilon}$
k_c	Cut-off wavenumber, $k_c = \frac{2\pi}{c} f_c$
k_x, k_y, k_z, k_ρ	Additional wavenumbers obtained by separation of variables
K	Inverter constant K of an impedance inverter
L	Inductance
M, m	Mode index, see (4.1) to (4.4)
N, n	Mode index, see (4.1) to (4.4)
N	Filter order
$N_n(x_i)$	Bessel function of second kind, nth order. Also known as Neumann function.
P, p	Mode index, see (4.1) to (4.4)
Q, q	Mode index, see (4.1) to (4.4)
R	Residual
R	Number of rows of a single segment matrix. $R = 4 \cdot \mathcal{M}$ applies.
R_a	Arithmetic average roughness
R_s	Surface resistance $R_s = \sqrt{\omega_0\mu/2\sigma}$
R_q	Root mean square roughness
$\vec{S}$	Poynting vector, $\vec{S} = E \times \vec{H}^*$
S, s	Mode index, see (4.1) to (4.4)

$\boldsymbol{S}$	Scattering parameter matrix, $\vec{\mathfrak{b}} = \boldsymbol{S} \cdot \vec{\mathfrak{a}}$ applies.
s_{ij}	Scattering parameter $s_{ij} = \mathfrak{b}_i/\mathfrak{a}_j$. By definition, $\mathfrak{a}_k = 0$ for all $k \neq j$.
Δs_{ij}	Maximum difference between scattering parameters obtained from two Mode Matching solutions with $\mathcal{M}$ and $\mathcal{M}-1$ Eigenmodes retained. $\Delta s_{ij} = \lvert s_{ij}^{\mathcal{M}} - s_{11}^{\mathcal{M}-1} \rvert$
Δs_{max}	Maximum difference between sets of scattering parameters obtained from two Mode Matching solutions with $\mathcal{M}$ and $\mathcal{M}-1$ Eigenmodes retained. $\Delta s_{max} = \max(\Delta s_{11}, \Delta s_{21}, \Delta s_{12}, \Delta s_{22})$.
T, t	Mode index, see (4.1) to (4.4)
U_i	Total voltage on port i of a structure
$U_i^{+/-}$	Voltage wave. The + or - superscript indicates wave propagation towards or away from the structure, i.e. whether the wave is incident or scattered.
$\vec{U}$	Voltage vector. Contains the total voltages of all ports of a structure.
V, v	Mode index, see (4.1) to (4.4)
W, w	Mode index, see (4.1) to (4.4)
$\boldsymbol{Y}$	Admittance matrix, see e.g. [6]
Y	Admittance
Z_0	Characteristic impedance. Ratio of a voltage and a current wave in a waveguide.
Z_{wave}	Wave impedance. Ratio of an Eigenmode's electric and a magnetic field component perpendicular to the direction of propagation.
$\boldsymbol{Z}$	Impedance matrix, see e.g. [6]
$\boldsymbol{Z}_0$	Characteristic Impedance matrix, see (10.5).
Z	Impedance
Z_l	Low characteristic impedance of capacitive transmission lines used in stepped-impedance filter designs
Z_h	High characteristic impedance of inductive transmission lines used in stepped-impedance filter designs

4.2. Calligraphic Roman Letters

$\mathcal{B}$	Total bandwidth of a matrix. The total bandwidth represents the total number of non-zero diagonals contained in the matrix. $\mathcal{B} = \mathcal{B}_u + \mathcal{B}_l + 1$
$\mathcal{B}_l$	Lower bandwidth of a matrix. The lower bandwidth represents the number of non-zero diagonals contained in the matrix below the main diagonal.
$\mathcal{B}_u$	Upper bandwidth of a matrix. The upper bandwidth represents the number of non-zero diagonals contained in the matrix above the main diagonal.
$\mathcal{B}_\Psi$	Lower and upper bandwidth of the reduced system matrix. $\mathcal{B}_\Psi = \mathcal{B}_u = \mathcal{B}_l$ applies.
$\mathcal{C}$	Number of columns of the reduced system matrix
$\mathcal{C}_\Psi$	Number of columns of the system matrix
${}^{\eta/\nu}\mathcal{E}^{TE/TM}_{x_i\|m,n}$	Unsigned electric field distribution of the x_i component of the $TE_{m,n}$ or $TM_{m,n}$ Eigenmode on waveport η or at junction ν respectively. The indices may be omitted.
$\mathcal{F}$	Arbitrary unsigned field distribution, see section 3.2
${}^{\eta/\nu}\mathcal{H}^{TE/TM}_{x_i\|m,n}$	Unsigned magnetic field distribution of the x_i component of the $TE_{m,n}$ or $TM_{m,n}$ Eigenmode on waveport η or at junction ν respectively. The indices may be omitted.
$\mathcal{L}$	Linear operator
${}^{\nu}\mathcal{M}$	Total Number of Eigenmodes considered per junction. May be denoted without index if equal for all junctions.
${}^{\eta}\mathcal{M}$	Total Number of Eigenmodes considered per waveport. May be denoted without index if equal for all ports.
$\mathcal{N}$	Number of segments (i.e. waveguides and waveguide interfaces) contained in the structure.
$\mathcal{O}$	The $\mathcal{O}(\mathcal{Q}^x)$ notation implies that the process' operation count is a function of terms of order $\mathcal{Q}^x$ and lower.
$\mathcal{P}$	Permutation matrix with dimensions $8 \cdot \mathcal{M} \times 8 \cdot \mathcal{M}$ (for full solutions including both TE and TM Eigenmodes) employed for column swapping when interface reusing with flipping is performed.

$\mathcal{P}$	Complex power transported in a waveguide, see appendix C
$\mathcal{P}_{LR}$	Power loss of a filter
$\mathcal{Q}$	Number of unknown Eigenmode amplitudes
$\mathcal{R}$	Number of rows of the reduced system matrix
$\mathcal{S}$	Surface of a structure under consideration
$\mathcal{V}$	Volume of a structure under consideration
$\mathcal{V}_0$	Volume of a cavity
$\mathcal{V}_S$	Sample volume

4.3. Black Letters

$\vec{\mathfrak{A}}$	Magnetic vector potential, $\vec{H} = \nabla \times \vec{\mathfrak{A}}$
$\mathfrak{a}_i$	Incident power waves
$\vec{\mathfrak{a}}$	Vector of incident power waves
$\mathfrak{b}_i$	Scattered, i.e. reflected or transmitted power waves
$\vec{\mathfrak{b}}$	Vector of scattered power waves
$\mathfrak{C}$	Contour of the generalised waveguide, see figure 2.1
$\vec{\mathfrak{F}}$	Electric vector potential, $\vec{E} = -\nabla \times \vec{\mathfrak{F}}$
$\mathfrak{d}$	Diameter of a resonator's circle in the Smith chart
$\mathfrak{f}$	Function
$\mathfrak{G}_0$	Filter admittance
$\mathfrak{G}$	Cavity geometry factor
$\mathfrak{g}$	Function
$\mathfrak{I}$	$\mathfrak{I}$th iteration of a Mode Matching solution
$\mathfrak{K}$	Coupling coefficient
$\mathfrak{N}$	Total number of ports
$\mathfrak{P}_l$	Power dissipated in a cavity
$\mathfrak{P}_c$	Power dissipated in a cavity due to ohmic loss
$\mathfrak{P}_d$	Power dissipated in a cavity due to dielectric loss
$\mathfrak{Q}$	Quality factor
$\mathfrak{Q}_0$	Unloaded quality factor
$\mathfrak{Q}_c$	Quality factor due to ohmic loss
$\mathfrak{Q}_d$	Quality factor of a dielectric sample
$\mathfrak{Q}_{ext}$	External quality factor
$\mathfrak{Q}_L$	Loaded quality factor

$\mathfrak{R}_0$	Filter impedance
$\mathfrak{S}$	Cross-section of a waveguide
$\mathfrak{T}$	Temperature
$\mathfrak{T}_c$	Critical temperature
$\mathfrak{u}$	Unknown function expanded in basis functions $\mathfrak{u}_i$, see appendix A
$\tilde{\mathfrak{u}}$	Approximate solution to $\mathcal{L}\,\mathfrak{u} = \mathfrak{f}$, see appendix A
$\mathfrak{w}$	Relative filter bandwidth $\mathfrak{w} = {}^{BW}/_{\omega_0}$
$\mathfrak{w}_\lambda$	Guided-wavelength fractional bandwidth, see p. 130
$\mathfrak{W}$	Energy stored in a cavity
$\mathfrak{W}_e$	Energy stored in a cavity electrically
$\mathfrak{W}_m$	Energy stored in a cavity magnetically
$\mathfrak{z}$	Normalised input impedance
$\mathfrak{z}_m$	m^{th} zero of the Bessel function J_0 of first kind

4.4. Greek Letters

Alpha	Beta	Gamma	Delta	Epsilon	Zeta	Eta	Theta	Ito	Kappa	Lambda	My
A	B	Γ	Δ	E	Z	H	Θ	I	K	Λ	M
α	β	γ	δ	ϵ, ε	ζ	η	θ, ϑ	ι	$\kappa, \varkappa$	λ	μ
Ny	Xi	Omikron	Pi	Rho	Sigma	Tau	Ypsilon	Phi	Chi	Psi	Omega
N	Ξ	O	Π	P	Σ	T	Y,Υ	Φ	X	Ψ	Ω
ν	ξ	o	π, ϖ	ρ, ϱ	σ, ς	τ	υ	ϕ, φ	χ	ψ	ω

$\vec{\alpha}$	Total amplitudes, that is, amplitudes for all Eigenmodes considered in the field solution, denoted under vector form.
$\vec{\alpha}_{Known}$	Predefined amplitudes, i.e. the excitation of the structure, denoted under vector form.
$\vec{\alpha}_{Unknown}$	Unknown amplitudes to be determined by Mode Matching.
α	Attenuation constant
α_i	Coefficients of the linear model for the geometry factors
$\boldsymbol{\beta}_{PEC}$	PEC matrix, see section 12.2.1
$\boldsymbol{\beta}_{PMC}$	PMC matrix, see section 12.2.1
β	Phase constant
β	Admittance slope
Γ	Reflection coefficient
Γ_S	Detuned reflection coefficient of a resonator
Γ_L	Reflection coefficient of a resonator at resonance
γ	Waveguide matrix describing a non-zero length waveguide of uniform cross-section. The coefficients are obtained as discussed in chapter 5.

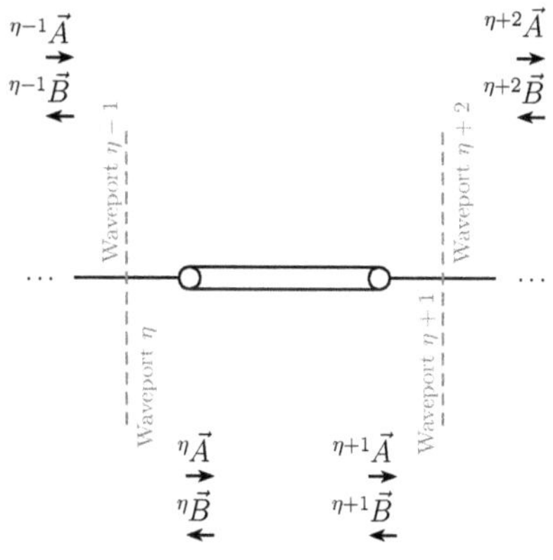

γ	Propagation Constant
Δ	Spectral amplitude of a Debye relaxation process
Δf	Frequency shift of a perturbed cavity
δ	Filter passband ripple
ε	Permittivity
ε_0	Vacuum permittivity, $\varepsilon_0 = 8.854 \cdot 10^{-12}\,\frac{\text{As}}{\text{Vm}}$
ε_r	Relative permittivity. If lossy media is considered, this quantity becomes complex, that is, $\varepsilon_r = \varepsilon_r' - j\varepsilon_r''$.
ε_s	Static Permittivity
ε_∞	Permittivity at infinite frequency, corresponds to the optical refractive index $\sqrt{\varepsilon_\infty}$
η	Index for the waveports of a segment
${}^{\nu}\mathbf{\Theta}$	Segment matrix of the νth segment. A segment may either represent an interface between two waveguides of different geometry or a waveguide segment of uniform geometry. For a segment Eigenmodes are defined as shown below.

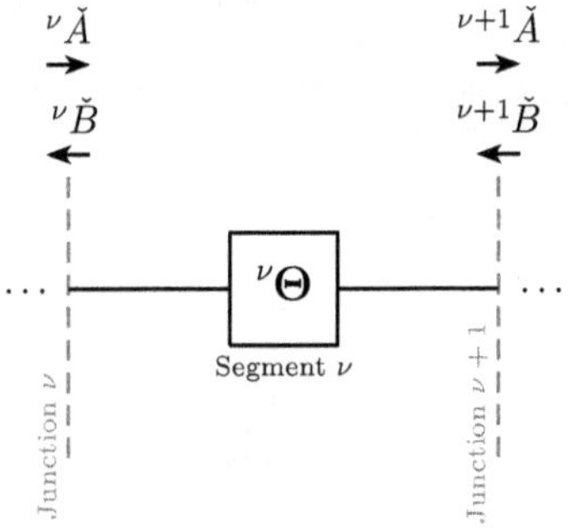

$\boldsymbol{\vartheta}$ Interface matrix describing an interface between two waveguides of different geometry. The interface matrix's coefficients are obtained calculating inner products between appropriate Eigenmodes as described in chapter 4.

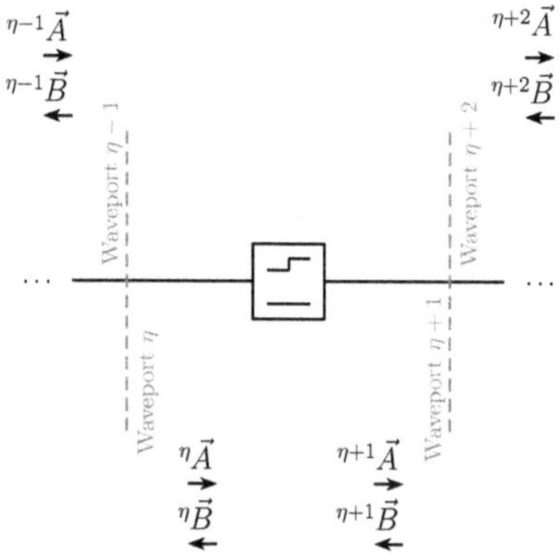

$\boldsymbol{\vartheta}_{Known}$ see section 4.5

$\boldsymbol{\vartheta}_{Unknown}$ see section 4.5

$\vartheta^{\mathfrak{f}}_{\phi}$ Coefficient resulting from the orthogonal expansion of $\mathfrak{f}$ for a set of basis functions containing the basis function ϕ. $\vartheta^{\mathfrak{f}}_{\phi}$ is calculated using inner products, i.e.

$$\vartheta^{\mathfrak{f}}_{\phi} = \frac{\langle \mathfrak{f} | \phi \rangle}{\langle \phi | \phi \rangle}$$

as discussed in section 4.3.2.

For example

$$\vartheta^{{}^{I}\mathcal{H}^{TM}_{y|v,w}}_{{}^{II}\mathcal{H}^{TM}_{y|p,q}}$$

denotes the expansion of the Eigenmode ${}^{I}\mathcal{H}^{TM}_{y|v,w}$ for the mode ${}^{II}\mathcal{H}^{TM}_{y|p,q}$.

$\vartheta_{\eta_1\mathcal{F}}^{\eta_2\mathcal{F}}$	Sub-matrix of an interface matrix containing all coefficients of the expansion of the ${}^{\eta_2}\mathcal{F}$ Eigenmodes for the ${}^{\eta_1}\mathcal{F}$ Eigenmodes. For example $$\vartheta_{{}^{II}\mathcal{H}_y^{TM}}^{{}^{I}\mathcal{H}_y^{TM}}$$ contains the expansion coefficients of the expansion of the contribution of the TM mode-set of waveguide I to the H_y field distribution at the interface for the TM mode-set of waveguide II.
Λ	Eigenvalue of a boundary value problem. See section 2.1.4 and 2.3.1
λ	Wavelength
λ_g	Guided wavelength
μ	Permeability
μ_0	Vacuum permeability, $\mu_0 = 4\pi \cdot 10^{-7}\ \frac{\mathrm{Vs}}{\mathrm{Am}}$
μ_r	Relative permeability. If lossy media is considered, this quantity becomes complex, that is, $\mu_r = \mu_r' - j\mu_r''$.
ν	Index for segment matrices. $1 \leq \nu \leq \mathcal{N}$
ξ	Total number of elements of the reduced system matrix.
$\xi_{Compact}$	Total number of elements included in the reduced system matrix $\mathbf{\Psi}_{unknown}$ under compact form
Π	Scaling factor, see section 14.2 and B.2.1
σ	Weighting function
$\sigma(x)$	Measurement uncertainty of x
σ	Conductivity
τ	Characteristic relaxation time
Υ	Normalised Frequency
$\mathbf{\Phi}$	Coefficient matrix of a partially filled waveguide's homogeneous system of equations, see section 14.2 and B.2.1
$\varphi_{m,n}$	Orthonormal basis function
χ	Impedance slope

$\mathbf{\Psi}$	System matrix corresponding to an homogeneous system of equations completely describing a given structure.
$\mathbf{\Psi}_{Known}$	See figure 6.6
$\mathbf{\Psi}_{Unknown}$	Reduced system matrix with dimensions $\mathcal{R} \times \mathcal{C}$, see figure 6.6
$\mathbf{\Psi}_R$	System matrix resulting from a resonant problem, see p. 102
ψ	Scalar vector potential
ψ_t	Transverse scalar vector potential, see (2.7)
Ω	Detuning, $\Omega = \frac{\omega}{\omega_0} - \frac{\omega_0}{\omega}$
Ω	Complex resonant frequency
ω	Angular frequency
ω_0	Resonant frequency of a cavity or center frequency of a bandpass filter's passband
ω_c	Cut-off frequency of a waveguide or filter
ω_L	Loaded resonant frequency of a cavity

Index

S

T

U

V

W

V

Errata

Carsten Monka-Ewe
Mode Matching Solvers for Filters and Cavities
1. Ed. - Göttingen: Cuvillier, 2019
ISBN: 978-3-7369-7086-1, eISBN: 978-3-7369-6086-2

Despite extensive proofreading, it is rather likely that typos and other minor errors will be spotted in the future. This continuously updated errata will be made available at

http://www.lambda-microwave.de/mmsfc/errata.pdf

in its latest version.

www.ingramcontent.com/pod-product-compliance
Lightning Source LLC
Chambersburg PA
CBHW060351220326
41598CB00023B/2875
* 9 7 8 3 7 3 6 9 7 0 8 6 1 *